Climate-Resilient *Brassicas*

Munir Ozturk • Nafees A. Khan
Sarvajeet Singh Gill • Volkan Altay

Climate-Resilient *Brassicas*

A Pathway To Food Security

Springer

Munir Ozturk
Department of Botany, Faculty of Science
Ege University
Izmir, Izmir, Türkiye

Sarvajeet Singh Gill
Centre for Biotechnology
Maharshi Dayanand University
Rohtak, Haryana, India

Nafees A. Khan
Department of Botany
Aligarh Muslim University
Aligarh, Uttar Pradesh, India

Volkan Altay
Department of Biology, Faculty of Arts and Sciences
Hatay Mustafa Kemal University
Antakya, Hatay, Türkiye

ISBN 978-3-032-12708-2 ISBN 978-3-032-12709-9 (eBook)
https://doi.org/10.1007/978-3-032-12709-9

This Springer imprint is published by the registered company Springer Nature Switzerland AG
The registered company address is: Gewerbestrasse 11, 6330 Cham, Switzerland

農学研究所

This book is dedicated to ***Prof Dr Norindo TAKAHASHI (1923–2014)****, born in Kanagawa Prefecture, Japan. He graduated from the Faculty of Agricultural Science, Tohoku University, in the year 1950, was appointed the same year as an Assistant at the Institute for Agricultural Research, Tohoku University; received his doctoral degree in Agriculture at the same University in 1961; and was appointed as a Lecturer in 1968. In 1971 he became an Assistant Professor and in 1977 full Professor. Prof Takahashi was appointed as Director of Institute of Agricultural Research and Member of the Board of Trustees, Tohoku University, in 1984. He took his retirement from Tohoku University in 1987 and passed away at the age of 90 in 2014. Late Prof Dr Takahashi has contributed a lot in the field of Physiological-Genetics within the concept of Plant Breeding. One of the authors (MO) had an opportunity to collaborate with him at Tohoku University, Institute of Ecological Research, under the JSPS programme for one year on the topic "Eco-Physiological Studies on the Brassica's".*

Preface

"Avicenna," writes that there are humans on our mother earth who have wit but no religion, and those who have religion but no wit, ecologically speaking all things have causes, knowledge of anything is not complete unless you know its causes as such we must know the causes of growth and development in plants as our major benefactors.

The climate is an integral part of our ecosystems and organisms have adapted to their regional climate over time. Almost 20–30% of our species have been assessed to face a risk of extinction from climate change impacts within this century. A recent article published in "NATURE" on the behaviour of living beings points out that almost 80% of the species have shifted towards higher altitudes creating refugee species. The changes in climate have affected our ecosystems at different altitudes. The resilience of the ecosystems is also likely to be affected by the population pollution, land use changes and over-exploitation in near future. Several studies on the effects of regional climate change on our biodiversity at higher altitudes depict that there is a shift in the distributional ranges of our biota, as is clear from the phenological changes, in particular the earlier onset of spring events, migrations and lengthening of the growing seasons.

As per the UN report the world population currently has reached a level of 7.6 billion, in 2030 it is expected to be 8.6 billion, in 2050 9.8 billion and in 2100 11.2 billion. Nearly 83 million humans are added to the global population every year which is directly effecting our biodiversity, a backbone for all living beings and their ecosystems. As a matter of fact, life on mother earth depends on the biodiversity, and living beings depend on it to fulfil their needs. Nearly 2.0 million plant taxa are distributed on our planet, but their use in an unsustainable way has led to a great loss, over 5000 plant species are reported to have gone extinct during the last 200 years, and it is threatening the lives of living beings globally. The reports show that every year nearly 20 plants are lost as such; sustainable use of our resources is needed in order to raise and improve the quality of our lives. We must follow all these in harmony with nature as a valuable system for our future generations. The "NATURE" has been caring for the biota on mother earth by saving them from thousands of abiotic interferences; unfortunately nobody can save these from fools.

Our biota is the earth's endless effort to speak to the listening heavens. Our global green economy needs to make a change in different countries on biodiversity loss, ecosystem management, and environmental sustainability. Simply by planting trees we plant the seeds of hope and peace. As an ecologist I believe that wild plant taxa will continue to disappear at an alarming rate until we leave our human-centred anthropocentric behaviour.

The idea of book on the *Brassica's* took birth in the mind of Munir Ozturk in 1986 following his visit to Tohoku University, Sendai-Japan, as a JSPS Fellow. Following this idea, he started working and collecting data on this economically and ecologically interesting genus. He decided then to approach this idea for a better understanding by pooling up multidisciplinary studies in an integrated manner. The young colleagues from India were approached who were interested on this subject. Their positive approach motivated us to start this book. It covers all important topics related to the genus *Brassica.* Each topic has been dealt with in detail and evaluated from different angles so as to offer a detailed view on this genus and vis-a-vis provide an evidence for its values. The authors have already published some work to fill the gap in connection with the ecological behaviour of this genus. Our biodiversity is an irreplaceable resource, therefore we have tried to associate the detailed findings on genus *Brassica* and correlate the subject with human economy and well-being.

In this book, we have tried to present a global perspective of the genus *Brassica* with a detailed historical background, and attempt has been made to bridge the gap between botany, ecology, molecular aspects, along with the discussions on the recent studies on ecophysiological aspects; some genetic aspects too have been evaluated. The book includes details on the taxonomy of this genus at global level, its industrial use and economical importance. Some information on its ecology, cultivation, genetics and biotechnology has also been included. We hope it will prove of great help to the workers involved in the research aspects of the *Brassicas.* Almost complete or nearly complete whole-genome sequencing of 45 crop plants, 8 model plants, 15 crop progenitors and relatives, and 3 basal plants has been accomplished, majority being in the public domain.

This book titled *Climate-Resilient Brassicas: A Pathway To Food Security* includes nine chapters. Chapter 1 is introduction followed by Chap. 2 discussing the botanical aspects mainly taxonomy and morphology, distribution in the world and fossil records and pollen studies; Chapter 3 deals with ecological and ecophysiological aspects enlightening the facts about the ecological features like; habitat preferences of the *Brassicas*, germination behaviour and role of hormones, eco-physiological aspects, potential abiotic stresses in the *Brassicas*, drought stress, salinity stress, heat stress, pollution stress and heavy metal stress; Chapter 4 presents data on the "*Brassica's*: Mycorrhizal Interactions and Development Related to Other Organisms", covering studies on the mycorrhizal interactions, potential biotic stresses in the *Brassicas*, interactions with pathogen and pest organisms; Chapter 5 deals with the molecular aspects of *Brassica*, including the sections such as introduction, genomic features of *Brassica*, molecular pathways in *Brassica*, stress physiology and molecular mechanisms, genetic resources and biotechnological interventions, and system biology and functional genomics; Chapter 6 presents data on the propagation and production of *Brassica*, including topics like propagation methods: tissue culture and micropropagation, conclusion and future perspectives; Chapter 7 covers studies on the phytochemical constituents, pharmacological activities and human health; Chapter 8 discusses the economic importance of the genus with traditional, industrial and other uses. Chapter 9 is the final chapter dealing with global perspectives and future approaches covering the data available on global perspectives, role in meeting global food and health needs, utilising *Brassica* crops to promote sustainable farming, future approaches for *Brassica*, enhancing the capacity for phytoremediation or resilience to heavy-metal toxicity, techniques for the enhancement of *Brassica* by sequencing, marker-assisted selection (molecular markers) techniques for *Brassica*, enhancement of Indian mustard using genome editing, and *Brassica* and climate resilience.

Attempt has been made to pool up the information on the popularisation of practices for better production, analysis of growth vigour and development, economic evaluation of genotypes for sustainable production, quality vegetable and oil production and development, cost-effective harvesting and conservation practices for varying zones. Attempt has been made to accommodate information on the improved agronomical practices for sustenance of productivity, metabolomics to identify its quality indices and nutritional requirements for improved productivity. The information on the importance of genetic resources in the context of climate change has been evaluated, and the information related to the germplasm collection, need for collecting germplasm and future ways to meet the *Brassica* germplasm requirements related to climate change are discussed. There is a need for germplasm collection from climate stress areas and identification of new traits that can improve the adaptability or increase the tolerance to stress; use of wild relatives for novel gene incorporation has been included as these will contribute towards the adaptability and development of new phenotyping and genomic tools for rapid and accurate screening of bigger populations. Data has been added on the exploration of genetic diversity in order to improve the livelihood of villagers using this genus, integrate genomics and genetic resources for improvement, and increase availability and accessibility of their genetic resources. Readers will find important findings while analysing the data included here.

We believe this book will serve as an important source for the industry, researchers, pharmacy people as well as policymakers to bring up the economic status of local communities together with general economy of a country. We strongly advise students and scientists to make use of this important source in search for innovative and much-needed models to achieve better economic goals. A proper evaluation of the *Brassica* taxa will be a blessing for larger communities in their economic as well as food security. We hope this book will open new perspectives in the genus *Brassica* as it is the genuine proof that the new perspective in the *Brassica*'s will continue to remain very actual in future as well. The book enlightens the fact that plants can play a role in the food consumption, industry as well as the production of potential therapeutics for our health, together with discussions on different approaches towards conservation and evaluation of these plants with commercial feasibility.

The authors hope that information included here will prove useful in upgrading the life of people in several countries because of the economic value of these plants at the global level.

It has been a major task for us all to work on such an economically valuable genus with vast number of varying species. All of us with lifetime experience and expertise have been able to do a good jobs in writing this book. It has provided us an opportunity to read the works of many stalwarts in this field from different parts of the World. This attempt has proved enriching for us to work together, and we all have felt ourselves as students of "SCIENCE" as well as "SCIENCE WORKERS".

We used big time intervals to complete this book besides our other involvements, we could give some hours to our families but they always permitted us to use their hours. Our dedication of this book to them will hopefully do the justice to their sacrifices.

The manuscript has been written as a collective effort by the authors.

Izmir, Izmir, Türkiye — Munir Ozturk
Aligarh, Uttar Pradesh, India — Nafees A. Khan
Rohtak, Haryana, India — Sarvajeet Singh Gill
Antakya, Hatay, Türkiye — Volkan Altay

Acknowledgements

A profound gratitude goes to JSPS for providing fellowship to Munir Ozturk to undertake studies on the *Brassica's* at Tohoku University, Centre for Ecological Studies, and personal thanks are due to **Dr Tadashi SATO** from the same Institute for his help during the stay at Tohoku University and Sendai; he also provided the details about Respected Late **Norindo Takahashi**.

Our special thanks are due to **Prof Dr Azra N. Kamili** and **Dr Rifat John** from Kashmir University, India, for providing photographs as well as **Dr Mustafa Keskin** from Istanbul for his help in upgrading the photographs.

The young colleagues **Prof Dr Hatice Demiray** and **Prof Dr Meliha Gemici** from the Botany Department, Ege University, Turkiye, helped first author in the earlier compilation of data on the *Brassica* species in the beginning upon his return from Tohoku University, Japan, and **Sadaf Saifi** and **Nidhi** from the Botany Department, Aligarh Muslim University, India, for their help extended to **Prof Dr Nafees A. Khan** in chapter compilation on the eco-physiological aspects.

A part of the work cited on plant abiotic stress tolerance has been undertaken at the SSG laboratory and partially supported by University Grants Commission, Science and Engineering Research Board and Council of Scientific & Industrial Research, India; **Dr Gill** acknowledges the support as well as partial support from DBT-BUILDER grant (No. BT/INF/22/SP43043/2021), from MDU Post Seed Grant (No. DRD/25/1076-81).

The authors extend sincere apologies to the colleagues whose work could not be discussed in this book due to page restrictions.

Last and not the least we are thankful to the staff working at the Springer headquarters and publishing section for their full support all through the steps followed during the preparation of this book.

Contents

About the Authors

Munir Ozturk holds Ph.D. and D.Sc. degrees from Ege University, Turkiye, and has served at the same university for nearly 50 years in different positions. He has also been Vice President of the Islamic World Academy of Sciences and has received fellowships from Alexander von Humboldt Foundation, Japanese Society for Promotion of Science and the National Science Foundation of the USA. He has served as Chairman of the Botany Department and Founding Director of the Centre for Environmental Studies, Ege University; Consultant Fellow, Faculty of Forestry, Universitiy Putra Malaysia, Malaysia; and Distinguished Visiting Scientist, ICCBS, Karachi University, Pakistan. His fields of scientific interest are plant ecophysiology, medicinal plant diversity, biosaline agriculture and crops, pollution and biomonitoring. He has published more than 60 books, nearly 100 book chapters, 185 scientific manuscripts in journals, and has served as guest editor for more than 14 Journals. He holds membership of 25 different national and international bodies. Total citations of publications are 22357, h-index: 064, i10-index: 391.

Nafees A. Khan is a Professor and Dean, Faculty of Life Sciences at Aligarh Muslim University, Aligarh, India. His research focuses on identifying strategies for developing abiotic stress-tolerant crop plants through modulation of tolerance mechanisms and strengthening positive source-sink relationships under optimal and limited environmental conditions. He has shown that the use of the signalling molecules enhances the use-efficiency of primary nutrient elements, resulting in the increased efficiency of enzymatic and non-enzymatic antioxidants. The overall effect was seen in the increased potential of photosynthetic machinery and abiotic stress tolerance, with enhancement in crop productivity under stressful environments. He has extensively researched the adaptability of *Brassica* under changing climatic conditions through improving mechanisms using plant hormones/signalling molecules and their synergy with primary nutrient elements. These research findings have been reported in premier journals of *Plant Sciences* and *Plant Physiology*. He has published about 200 research papers and 22 books. He has received recognition as UGC-Career Award, UGC-Mid Career Award, Fellow of the National Academy of Sciences, Fellow of the National Academy of Agricultural Sciences, Fellow of Linnean Society of London, Fellow of Indian Botanical Society and Fellow of Indian Society of Plant Physiology, Highly Cited Researcher 2019–2023, India Research Excellence Citation Award-2023 by Clarivate Analytics (Web of Science), and Plant Science and Agronomy in India Award 2023 and 2024 by Research.com.

Sarvajeet Singh Gill is Professor (Asst.) of Agriculture Biotechnology, at Stress Physiology & Molecular Biology Lab, Centre for Biotechnology, Maharshi Dayanand University, Rohtak, Haryana, India. He has specialised in the fields of Plant Molecular Biology and Genetic Engineering, Plant Biotic & Abiotic Stress Tolerance, Reactive Oxygen Species Signalling and Antioxidant Machinery in Plants, Helicases, Transgenic, Nitrogen & Sulfur Metabolism, Plant Fungal Symbiotic Interactions for Crop Improvement and increasing crop production—now the highest agricultural priority because of increasing world population & changing climatic conditions. He has undertaken research to understand the mechanism of abiotic stress tolerance (heavy metal/salinity/drought) in plants (*Brassica juncea*, etc.) at molecular and physiological levels. Featured consecutively in the "World Google Scholar Citations 22166; h-index: 51; i10index:109; Ranking of Top 2% Scientists" in 2021, 2022 and 2023. Guest Editorship of several journals. He has also published 30 books.

Volkan Altay is a full Professor at the Biology Department, Faculty of Science and Arts, Hatay Mustafa Kemal University, Turkiye. His fields of interest are: Plant Ecology, Vegetation Ecology, Taxonomy, Biodiversity, Biomonitoring and Pollution Ecology, and Sustainable Use of Medicinal Plant Resources. He received his PhD in 2009 and holds a proven track record of nearly 100 publications in highly reputed professional journals. He has also published 8 books, nearly 50 research articles in impact factor journals, more than 65 book chapters (Springer, Elsevier, Cambridge, Wiley, Taylor & Francis, AAP, CRC Press, etc.) and reviews, and more than 70 international and national congress reports. He has been an invited speaker for various conferences. He has been listed in top 2% scientists in the world (Stanford University report published by Elsevier in 2024). Total citations of publications are 5250, h-index: 41, i10-index: 90.

List of Figures

List of Tables

Chapter 1
Introduction

The relationship between plants and people in relation to domestication has changed a lot. Today, domesticated plants can be classified into different groups. Among these is the *Brassica* genus as it includes oil plants, vegetables and spices consumed widely by humans. Their survival and propagation are human-dependent, because we are responsible for increasing crop production but at the same time are strongly affecting the ecological issues. Botanical studies have provided much data on the background of different crops. Sustainable plant cultivation helps us to fulfil the food availability and increase the food security, but we have to be cautious and work hard to reduce our impact on the ecological footprint in order to make arrangements for a sustainable future; otherwise, our ecosystems will crash down. The plants are involved directly as a part of the biodiversity by maintaining ecosystem balance, thereby affecting climate change and mitigating global warming. Our agriculture is facing challenges because crop yields are seriously affected by climate change, and this has been going on for decades now (Vavilov 1949; Ozturk et al. 2019, 2025; Mukherjee et al. 2025).

Looking at the human-agriculture interrelations, the domestication of ancient crops started in four centres which originated 9000–10,000 years ago. A total of 136 species have been listed in the Chinese centre, mainly including central and western China; the Indian centre is shown with two sub-centres with a list of 117 species; the second sub-centre is the Indo-Malayan with 55 crop species listed. The Central Asiatic centre includes the sub-Himalayan region of India and Pakistan, Afghanistan, Tajikistan, and Uzbekistan with 43 listed species including mustard, whereas the near Asian centre includes Asia Minor, Turkmenistan, Iran, and Transcaucasia; several species originate from the Mediterranean centre, the important ones being cabbage and turnip; finally, the Abyssinian centre with 38 species is listed (Gómez-Campo 1999).

During the last five decades, "the roof of our economic development has become too heavy for the walls of ecology" (Ashraf et al. 2017). Although this varies from region to region, landraces have demonstrated resistance to various abiotic and

M. Ozturk et al., *Climate-Resilient* Brassicas,
https://doi.org/10.1007/978-3-032-12709-9_1

biotic stresses depending on their ecological environment. Despite their small grain size, low production and low nutritional value, the fact that they can be stored long term makes them attractive to farmers. A range of new adaptations have been observed which had never existed earlier (Singh and Malik 2025). Conservation strategies for these native varieties can help significantly to improve food security and agriculture at the local levels.

Brassicaceae is one of the largest families of Angiosperms with nearly 338 genera and more or less 4000 species. It embodies important taxa, distributed globally except Antarctica; many taxa are sown as these are well known economically, including edible plants consumed as fresh as well as vegetables, vegetable oils and condiments (Al-Shehbaz et al. 2006). "*Brassica*" was the name for many cabbage-like plants used by Pliny the Elder, and in Latin, the epithet *oleracea* means "vegetable/herbal", derived from *holeraceus* (oleraceus).

Brassica itself has been linked to Celtic names, but it probably comes from "prasikein", vegetable in Greek. Theophrastus (370–285 BC), Roman Cato (234–149 BC) and Plinus (23–79 AD) had already described stemkales and headed cabbages, as such an array of cultivated coles was available to the Romans at the time of Christ. The early name of *B. oleracea* ("krambe") used by the Greeks was probably of Phoenician origin (Gómez-Campo 1999).

The family Brassicaceae appeared in the Irano-Turanian region probably in the Eocene period (Franzke et al. 2011). In the Sanskrit writings, we read about the use of cruciferous crops in India sometime around 3000 BC. At the same time, we also read that *Brassica* crops were first cultivated along the European coastline around 8000 years ago. Today, these vegetables have become highly diverse, particularly along the Mediterranean coast of southern Europe, as well as in Asia and North America (Al-Shehbaz 2012). The valuable genus in the family from an agricultural point is *Brassica*, with many economically important plants such as oilseed crops and vegetables, all with a long history of agricultural use worldwide. These plants were and are globally important for their potential to address major food and vegetable oil production challenges and overcome hunger and malnutrition, following the climatic risks in the agricultural production (Hochman et al. 2022).

The conservation of ancestral species and the preservation of natural environments are a must to prevent the loss of economically valuable species (Maxted et al. 2012). They are facing great challenges due to climate change and pollution, thus directly affecting food security (Rather et al. 2024).

According to Gómez-Campo (1999), as against the worldwide distribution, many *Brassica* species tend to flourish as annuals or biennials, including weedy wild taxa as well as the crops domesticated long ago (Rakow 2004). The origin of the *Brassica* genus probably is found to lie between the Mediterranean and Saharan region, where the climate in early times followed mild winters but hot, dry summers. The drought and salt-adapted taxa within the genus probably resulted from the environment in the Mediterranean helped in their evolution (Tsunoda et al. 1980; Gómez-Campo 1999).

The findings on origin and domestication reveal that earlier records in the ancient literature of different civilizations point out that plant parts excavated, biochemical/

cytogenetical evidence, and distribution in nature have greatly helped in reconstructing the past events of the origin of *Brassica.* Comparative taxonomy and molecular marker studies have proved very helpful in this connection. The origin of a cultivated plant in general has two aspects: the first being the origin of the taxon itself before domestication, following the evolutionary process in the wild, and the second, *Brassica napus* probably appeared in cultivation, thus lacking previous evolutionary history. It probably resulted from natural and artificial selection via human interference. The first domesticated *Brassica* species was probably a diploid turnip rape (*B. rapa*). The uses of *B. nigra* (black mustard) seem to be quite old; the same possibly holds true for amphidiploid *B. juncea*. *B. oleracea* (kale, cabbage) has possibly entered cultivation later, as its natural area was far from the important domestication centres. The last brassica incorporated into agriculture was *B. napus* (rape), an amphidiploid species (Gómez-Campo 1999; Al-Shehbaz 2012).

The *Brassica* crops generally include vegetables and forage forms like *B. oleracea*, *B. napus* and *B. rapa* together with categories like *acephala*, *botrytis*, *capitata*, *capitata rubra*, *costata*, *gemmífera*, *gongylodes*, *italica*, *sabauda* and *viridis*. All these are important cultivars of *B. oleracea* in Europe and are important for the human diet. These are consumed globally and are included among the important food crops in China, Japan, India and Europe. The list includes var. *capitata*—cabbage; var. *acephala*—kale; var. *viridis*—collard green leaves; var. *alboglabra*—Chinese broccoli; var. *gemmifera*—Brussels sprouts, var. *gongylodes*—Kohlrabi stem; var. *botrytis*—cauliflower; var. *italica*—broccoli; var. *botrytis*—Romanesco broccoli; var. *italica x alboglabra*—broccolini, broccoflower; ssp. *rapa*—turnip root; ssp. *pekinensis*—Chinese cabbage, napa cabbage; ssp. *narinosa (or rosularis)*—Asian green leaves; ssp. *chinensis*—bok choy or pak choy leaves; ssp. *pervidis*—komatsuna, Japanese mustard spinach; ssp. *nipposinica* and ssp. *parachinensis* var. *napobrassica*—Mizune leaves, rapini (broccoli rape) and rutabaga (swede) (Al-Shehbaz 2012).

The distribution into sections and lists of *Brassica* species has been proposed at separate times (Salmeen 1979), with different amplitudes; some admit three, others nine sections. In the U-triangle, all species are included in the subgen. *Brassicam* (Tsunoda et al. 1980; Gómez-Campo 1999). The six cultivated species of the classical U-triangle are related to the wild relatives of *Brassica oleracea*, then to the genus *Brassica* as a whole, usually referred to as "crop brassicas". The correct Latin plural for *Brassica* is *Brassicae,* not "*Brassicas*" (Prakash and Hinata 1980). Several investigations have been undertaken lately covering studies on the phylogeny of *Brassica* (Tsunoda et al. 1980; Gómez-Campo 1999; Takahata and Hinata 1986), and five sections have been recognised in the subgen. *Brassica*; however, as per Salmeen (1979), *B. rapa* deserves a sect. *Rapa*, including amphidiploids *B. napus* and *B. juncea*, as all these form tuberous roots. The sects *rapa* and *Brassica* are akin to each other. *B. fruticulosa/spinescens/maurorum* complex, together with *B. procumbens* and *B. cadmea*, clearly share the same characters (Truco and Quiros 1991). *B. barrelieri* and *B. oxyrrhina* also show spherical seeds, seeded beak and lyrate leaves, but due to their tendency to form compact basal rosettes, leaves of *B. barrelieri* and *B. oxyrrhina* are better described as runcinate. These two species resemble each other in the vegetative stage; *B. oxyrrhina* shows small-sized flowers, while *B. barrelieri* has bigger flowers (Gómez-Campo 1999).

The molecular biological studies on the *Brassica's* began with the exploration of the origin of amphidiploid species via cpDNA (Ichikawa and Hirai 1983; Yanagino et al. 1987). The latter authors, for the first time, compared cpDNA of a set of 11 species, reporting clear incongruities with taxonomy. In total, 38 accessions belonging to 15 species of *Brassica* have been studied (Song et al. 1990), and their nuclear RFLPs have been investigated. According to Pradhan et al. (1992) and Warwick and Black (1997), in the phylogenetic tree, two lineages are seen in *Brassica* and its closest allies: these are "nigra" and "rapa/oleracea" lineages (Warwick and Black 1997). The chromosomes of some species possess similar genic sequences (Mizushima 1950). The molecular data support a high coherence degree for Sect. *Micropodium* of *Brassica* with a close relationship (Song et al. 1990; Pradhan et al. 1992; Uchimiya and Wildman 1978; Takahata and Hinata 1983). *B. fruticulosa*, *B. maurorum* and *B. spinescens* are ascribed to the B. *fruticulosa* cytodeme, with *B. maurorum being* an unspecialized weed. *B. fruticulosa* and *B. spinescens* have developed morphological characters as an adaptation to coastal habitats, whereas *B. cossoniana* is an autotetraploid of *B. fruticulosa* (Takahata and Hinata 1983; Harberd 1972), with closer similarities in cp and mt DNA with *B. maurorum* (Pradhan et al. 1992). *B. procumbens* is morphologically similar to other members of this group, shown to have $n = 9$ chromosomes (Baldini 1998), although its lineage remains unknown; the "nigra" lineage includes *B. tournefortii* ($n = 10$), and as per Song et al. (1990), this species evolved from a "rapa/oleracea" ancestor with nuclear introgression from *B. nigra.*

The survey undertaken in Türkiye on the Crucifers has revealed that there is one of the oldest evidence from the delta's agriculture on record in Catalhoyuk, Konya, and it thus could have served as the starting point for the flow of some materials and cultures into Asia (Ozturk et al. 1983). The study on the distribution of *Brassica* species has revealed that *B. campestris* grows as a wild species, being reported for the first time as a weed from the States of Kars and Kastamonu in Türkiye. The *B. cretica* grows all along the coast from Kusadası up to Antalya, *B. nigra* shows a wider distribution and big populations are seen in Adana and its environs, *B. elongata* is found to grow as a common weed in Central and West Anatolia and *B. deflexa* is seen to grow along the road track from Konya to Nevşehir, thus showing better penetration towards the west.

The rapeseed and mustard are shown among the global oilseed crops, providing up to 13% of oil worldwide. The canola seeds contain 40% oil and 30–35% protein. *B. oleracea* is cultivated globally with many wild relatives. The Mediterranean region most probably was concerned with their domestication problems, as the wild *B. rapa* (syn. *B. campestris*) is also found in some parts of the Mediterranean region; cultivated forms were most probably introduced into Asia during ancient times, via the "Silk Road" (Salmeen 1979; Gladis 1989; Sanchez-Yelamo 1990; Eastwood 1996; Lanner 1997).

The old cultivated form *B. alboglabra* from China has puzzled researchers for years because of its phytogeographical problems. According to Bothmer et al.

(1995), it should be kept within *B. oleracea.* It is believed to have reached China from the Mediterranean, and conclusive evidence came later after Hammer et al. (1992) found it locally cultivated in southern Italy, with white flowers, normally, but may be yellow, appearing distinctively during early winter. Infraspecific variability of *B. rapa* is given in accordance with its earlier domestication history (Toxopeus et al. 1984). The domestication in the Far East led to many forms, parallel to those obtained in the Mediterranean with *B. oleracea* (except cauliflower). The importance of *B. rapa* in the West has also been considerable; however, more specialised forms have developed following "competition" by an extensive use of *B. oleracea.*

The herbalist Dodoens believed that cauliflower originated in Cyprus and named it *Brassica cypria.* In fact, economically important plants such as cauliflower and broccoli are thought to have originated in the eastern Mediterranean (Hyams 1971; Snogerup 1980). Although no clear distinction could be made between these cultivated forms, they were known during the early Greek and Roman periods. The first person to distinguish between them was the Spanish-Arab writer Ibn al-Awam (c. 1140), who did so in his book Kitab al-Falaha. In his work, he referred to broccoli as "Syrian or Mosul cabbage" and used the term "quarnabit" for cauliflower, a term which is still in use today (Hyams 1971). Broccoli and cauliflower are popular and economically important members of the *Brassica* genus. The word "broccoli" is Italian in origin, derived from the Latin "brachium". Morphologically, they are spike-like, bearing a single large terminal inflorescence (Boswell 1949). According to the sixteenth-century work *Historia Generalis Plantarum* by Dalechamp, various types of broccoli were introduced to Italy from the Eastern Mediterranean. Cauliflower is thought to be a form of broccoli with more edible inflorescences. Consequently, some researchers hypothesise that cauliflower is descended from broccoli (Schery 1972; Gómez-Campo 1999). Another important *Brassica* crop is the Brussels sprout, which was reportedly served at a fourteenth-century wedding feast and is named after the area near Brussels in Belgium where this variety was developed (Hyams 1971).

According to Burkill (1930), Europe was the place where *B. rapa* was first domesticated as a biennial plant from which annual forms arose through selection. Sun (1946) undertook a comparative morphological study and proposed the existence of two lines: the Western line, which includes oilseed forms and turnip, distributed throughout Europe, Central Asia and the Indian sub-continent, and the Eastern line, comprising East Asian vegetable forms. The isozyme distribution patterns (Denford and Vaughan 1977) and RFLP analysis (Song et al. 1988) support this view.

De Candolle (1886) has proposed its cultivation in Europe around 2500–2000 BC, and it then spread to Asia after 1000 BC. *Shih-ching*, a Chinese book by Confucius (551–479 BC) mentions turnip (Keng 1974). Theophrastus (370–285 BC) has also written about turnip in *Enquiry into Plants*, where he mentions a wild population with long, narrow roots having small hairy leaves; Roman Cato (234–149 BC) has referred to turnip in *On Agriculture* as a vegetable; Columella (42 AD) too has recorded its uses. Plinius (23–79 AD) wrote extensively about turnip in *Natural History* referring to it as *rapa* and *napus* and states that the Greeks distinguish three kinds of turnips: a flat one, a round one and a wild form with a very long root. Columella (c. 60 AD) has mentioned long "Roman", "Round" from Spain, the "Syrian", the "White" and the "Egyptian" (Gómez-Campo 1999).

The differentiation of leafy forms is given as China from oilseed forms of *B. rapa* after its introduction through western Asia or Mongolia in the first century AD. The first to evolve in central China was ssp. *chinensis*, with narrow or wide green-white petioles (Li 1982). It was also the most primitive form of the East-Asian group from which subspecies *parachinensis* developed in Central China. The history and origin of ssp. *pekinensis* (Chinese cabbage) is well documented (Li 1982). The loose-leaved primitive form appeared as a hybrid between ssp. *chinensis* and ssp. *rapa* in the tenth century in the city of Young-Chow, as per *The Classics of Illustrated Medical Herbs*. This primitive form is still grown in southern China. In the twelfth-century literature, heading form with thick petioles appeared for the first time in northern China and is well recorded. The better agronomical practices, irrigation and nutrient supply have helped in the appearance of semi-heading forms. The *Miscellanea of Gardening—Shua-Pu-Tsa-Su* (fourteenth-century book) mentions such forms, which probably were further improved by gardeners and solid heads with fluffy tops or fully solid heads developed. These types have been mentioned in the *Shuin-Tian-Fu-Tse* book too, from the seventeenth century (Li 1982; Gómez-Campo 1999).

Various researchers have suggested that canola (*Brassica napus*), also known as rapeseed, originated in the Mediterranean region of southwestern Europe (Sinskaia 1928; Schiemann 1932). Today, it is an important non-traditional oilseed crop cultivated in a variety of climates and on varying soil types. Furthermore, it is one of the five most widely cultivated oilseed crops worldwide (Karp and Richter 2011; Liersch et al. 2020). However, from an agronomic perspective, if this variety experiences post-ripening, then seed shattering and delayed harvesting will be seen under adverse environmental conditions. This can result in yield losses of up to 50 % (Karp and Richter 2011; Liersch et al. 2020).

Brassica crops are rich in phytochemicals, including polyphenols, triterpenes and vitamins, as well as other health-promoting non-mineral compounds (Jahangir et al. 2009; Šamec et al. 2019). Indeed, various extracts from these plants are already commercially available (Li et al. 2023). Consequently, these vegetables are widely accepted as functional foods and dietary supplements (Faust et al. 2017; Li et al. 2023). Broccoli, in particular, is renowned for its anti-cancer properties (Faust et al. 2017; Li et al. 2023). The scientific evidence suggests that consuming *Brassica* crops can reduce the risk of developing various types of cancer (Wu and Lou 2007; Liu et al. 2013). These significant health benefits have been found to be related to indoles, which are sulphur-containing compounds that are generally glucosinolates and their hydrolysis products (Murillo and Mehta 2001). Thanks to these bioactive components, *Brassica* crops exhibit significant pharmacological effects, including antioxidant, anti-inflammatory, anticancer and cardioprotective features (Murillo and Mehta 2001). However, many *Brassica* crops are still widely consumed through traditional gastronomic practices, such as microwave heating, ice cream production, fermentation, steaming, boiling and pan-frying (Murillo and Mehta 2001).

Many researchers have conducted experimental pharmacological studies on the plant specimens of *Brassica rupestris*, *B. napus*, *B. rapa* and *B. oleracea*, proving that they exhibit anti-cancer and antioxidant properties (Amri 2014; Prasad 2014; Chauhan et al. 2016; Rahman et al. 2018). Cultivated forms of *B. napus* exhibit analgesic, diuretic, anti-catarrhal, anti-scurvy and anti-inflammatory properties and have beneficial effects on the bladder; those of *B. oleracea* show antibacterial, anti-inflammatory and anti-gastric disease characteristics, as well as anti-diabetic, hypolipidaemic, and antihyperglycaemic, and show cardioprotective effects. Important bioactive effects have also been identified in *B. rapa*, including diuretic, analgesic, anti-gout, aphrodisiac, anti-inflammatory and anthelmintic features (Prasad 2014; Ravikumar 2015; Chauhan et al. 2016).

Brassica taxa have served as important sources of food. They have been the subject of great scientific interest, particularly the *B. oleracea* species, which has assumed an important role in human nutrition, as the predominant dietary source of glucosinolates but also with high contents of phenolics and other antioxidant compounds (Tsunoda et al. 1980; Ozturk et al. 1983; Gómez-Campo 1999). *Brassica* offer a promising option for food and nutritional security due to their high nutritional profiles, exceptional resilience to harsh environmental conditions, richness in macro and micronutrients, dietary fibres, essential amino acids, minerals, antioxidants, trace elements, proteins, lipids and carbohydrates (Gómez-Campo 1999). Their ability to thrive under challenging conditions makes them an invaluable resource for sustainable agriculture (Choudhary et al. 2024).

Due to their climate-resilient characteristics and superior nutrient profile, the *Brassica* plants are mentioned as "miracle vegetables" (Ahmad et al. 2015). They prove beneficial for the community health and provide farmers with a source of revenue. Current nutrient deficits of protein, calcium and iron in poor nations can be addressed by the inclusion of *Brassica-based* foods. A superior nutritional profile position makes these plants crucial components of sustainable food systems, and rapeseed is the second most used source for producing vegetable oil for food and biofuel. Its crude extract and phytoconstituents show many pharmacological effects; yellowish seeds produce one of the most popular condiments on Earth, mustard, which has been used by a range of cultures for centuries (Šamec and Salopek-Sondi 2019). Most of the cultivated forms and some of their wild relatives have been used in medicine due to their constituents; some are very effective in the control of soil nematodes, others have been positively tested for the heavy metal removal from contaminated soils (Memon and Zahirovic 2014; Mourato et al. 2015; Dalyan et al. 2017; Meraklı and Memon 2023). The opportuneness of these plants following application of the most recent advances in their biology is expected (Ahmad et al. 2015). However, many significant risks to food safety and human health following bioaccumulation in edible plant parts pose a great threat (Uddin et al. 2021). Some species are an actual or potential source of useful genes to transfer to the cultivated forms in order to develop highly productive lines for seed oil, and unpleasant or unhealthy components are left; all these have increased the economic importance as well as the acreage of some *Brassica* crops (Akhalkatsi et al. 2012).

Modern cultivar forms that are economically valuable, highly productive and in demand have become increasingly popular with farmers (van de Wouw et al. 2010; Mercer and Perales 2010). However, alongside the destruction and degradation of habitats resulting from traditional land use, this has led to the over-recruitment of modern crop plants and an elimination of local native breeds, causing rapid genetic erosion (Mercer and Perales 2010; van de Wouw et al. 2010).

The scientific interest in *Brassica* has always been matched by its economic importance to agriculture, and we humans have serious, positive, negative and confusing feelings about it. Although much has been published on the subject, cytogenetic and molecular methods are shedding new light on *Brassica*, but the relationship is still unclear.

Today, alongside global population growth, there is an increasing demand for high-value agricultural products. *Brassica* plants are a valuable source of genetic material as they can be used to produce vegetables, oil, animal feed and seasonings. Given their vital global importance, there is an urgent need for incorporating efforts to transfer, develop and support *Brassica* genetic materials for breeding projects. We therefore recommend that future studies should examine the most effective processes and approaches for different *Brassica* cultivars.

References

Ahmad P, Hashem A, Abd-Allah EF et al (2015) Role of *Trichoderma harzianum* in mitigating NaCl stress in Indian mustard (*Brassica juncea* L) through antioxidative defense system. Front Plant Sci 6:868

Akhalkatsi M, Ekhvaia J, Asanidze Z (2012) Diversity and genetic erosion of ancient crops and wild relatives of agricultural cultivars for food: implications for nature conservation in Georgia (Caucasus). Perspect Nat Conserv Pattern Press Prospect 3:51–92

Al-Shehbaz IA (2012) A generic and tribal synopsis of the Brassicaceae (Cruciferae). Taxon 61:931–954

Al-Shehbaz I, Beilstein MA, Kellogg E (2006) Systematics and phylogeny of the Brassicaceae (Cruciferae): an overview. Plant Syst Evol 259:89–120

Amri E (2014) The role of selected plant families with dietary Ethnomedicinal species used as anticancer. J Med Plant 2:28–39

Ashraf M, Ejaz K, Arshad MD (2017) Water use efficiency and economic feasibility of laser land leveling in the fields in the irrigated areas of Pakistan. Sci Technol Dev 36(2):115–127

Baldini RM (1998) Rediscovery of *Brassica procumbens* (Poir.) O.E.Schulz (Cruciferae) in Italy with some systematic and distributional observations. Webbia 53:57–68

Boswell VR (1949) Our vegetable travellers. Nat Geogr Magaz 96:134–217

Bothmer R, Gustafsson M, Snogerup S (1995) Brassica sect. Brassica (Brassicaceae). II. Inter- and intraspecific crosses with cultivars of *B. oleracea*. Genet Resour Crop Evol 42:165–178

Burkill IH (1930) The Chinese mustards in the Malay Peninsula. Gad's Bull 1930:5

Chauhan ES, Tiwari A, Singh A (2016) Phytochemical screening of red cabbage (*Brassica oleracea*) powder and juice – a comparative study. J Med Plants Stud 4:196–199

Choudhary RL, Jat RS, Singh HV et al (2024) Low budget natural farming practices for cultivation of rapeseed and mustard. Natural farming: basics and application. Venture Graphics 2024:122–133

Dalyan E, Yuzbasıoglu E, Keskın BC et al (2017) The identification of genes associated with Pb and Cd response mechanism in *Brassica juncea* L. by using *Arabidopsis* expression array. Environ Exp Bot 139:105–115

De Candolle A (1886) Origin of cultivated plants, 2nd edn. Hafner, New York

Denford KE, Vaughan JG (1977) A comparative study of certain seed isoenzymes in the ten chromosome complex of *Brassica campestris* and its allies. Ann Bot 41:411–418

Eastwood A (1996) The conservation and utilization of *Brassica incana* Ten. in Croatia, with particular reference to leaf trichomes as a potential source of insect resistance. MSc thesis. Faculty of Science Univ., Birmingham

Faust D, Nikolova T, Wätjen W et al (2017) The Brassica-derived phytochemical indolo [3, 2-b] carbazole protects against oxidative DNA damage by aryl hydrocarbon receptor activation. Arch Toxicol 91(2):967–982

Franzke A, Lysak MA, Al-Shehbaz IA et al (2011) Cabbage family affairs: the evolutionary history of Brassicaceae. Trends Plant Sci 16:108–116

Gladis T (1989) Die Gattung *Brassica* L. und die Reproduktion entomophiler Pjanzensippen in Genbanken. Dissertation zur Erlangung des Akademischen Grades. Doktor der Naturwissenschaften, Gatersleben

Gómez-Campo C (1999) Biology of Brassica coenospecies. Elsevier

Hammer K, Knüpfer H, Laghetti G, Perrino P (1992) Seeds from the past. A catalogue of crop germplasm in South Italy and Sicily. Germplasm Institute. Consiglio Nazionale delle Ricerche

Harberd DJ (1972) A contribution to cyto-taxonomy of *Brassica* (Cruciferae) and its allies. Bot J Linn Soc 65:1–23

Hochman G, Zhang H, Xia L et al (2022) Economic incentives modify agricultural impacts of nuclear war. Environ Res Lett 17(5):054003

Hyams E (1971) Cabbages and kings. In: Plants in the service of man, London, pp 33–61

Ichikawa H, Hirai A (1983) Search for the female parent in the genesis of *Brassica napus* by chloroplast DNA restriction patterns. Japan J Genet 58:419–424

Jahangir M, Kim HK, Choi YH, Verpoorte R (2009) Health-affecting compounds in Brassicaceae. Compr Rev Food Sci F 8:31–43

Karp A, Richter GM (2011) Meeting the challenge of food and energy security. J Exp Bot 62(10):3263–3271

Keng H (1974) Economic plants of ancient north China as mentioned in Shih Ching (Book of Poetry). Econ Bot 28:391–410

Lanner C (1997) Genetic relationships within the *Brassica* cytodeme. Comparison of molecular marker systems. Acta Universitatis Agriculturae Sueciae Agraria 39:1–57

Li CW (1982) The origin, evolution, taxonomy and hybridization of Chinese cabbage. In: Chinese cabbage, Procedings of 1st international AVRDC symposium 1981, Taiwan, pp 1–10

Li Y, Zhou B, Teng Z et al (2023) Improved metabolomic approach for evaluation of phytochemicals in mustard, kale, and broccoli microgreens under different controlled environment agriculture conditions. J Agric Food Res 14:100719

Liersch A, Bocianowski J, Nowosad K et al (2020) Effect of genotype× environment interaction for seed traits in winter oilseed rape (*Brassica napus* L.). Agriculture 10(12):607

Liu X, Song Q, Tang Y et al (2013) Human health risk assessment of heavy metals in soil-vegetable system: a multi-medium analysis. Sci Total Environ 463:530–540

Maxted N, Kell S, Ford-Lloyd B et al (2012) Toward the systematic conservation of global crop wild relative diversity. Crop Sci 52:774–785

Memon AR, Zahirovic E (2014) Genomics and transcriptomics analysis of Cu accumulator plant *Brassica nigra* L. J Appl Biol Sci 8(2):1–8

Meraklı N, Memon AR (2023) Expression of heavy metal ATPases (HMA1 and HMA3) in *Brassica nigra* and *B. juncea* grown at different Cu levels. Botanica Serbica 47(2):309–316

Mercer KL, Perales HR (2010) Evolutionary response of landraces to climate change in centers of crop diversity. Evol Appl 3(5-6):480–493

Mizushima U (1950) Karyogenetic studies of species and genus hybrids in the tribe Brassiceae of Cruciferae. Tohoku J Agric Res 1:1–14

Mourato PM, Moreira IN, Leitao I et al (2015) Effect of heavy metals in plants of the genus *Brassica*. Int J Mol Sci 16:17975–17998

Mukherjee B, Jha RK, Sattar A et al (2025) Harnessing the potential of millets for climate-resilient and sustainable agriculture. Front Plant Sci 16:1574699

Murillo G, Mehta RG (2001) Cruciferous vegetables and cancer prevention. Nutr Cancer 41(1–2):17–28

Ozturk M, Hinata K, Tsunoda S, Gomez C (1983) A general account of the distribution of the cruciferous plants in Turkey. Ege Univ Sci Faculty J 4:87–98

Ozturk M, Hakeem KR, Ashraf M, Ahmad MSA (eds) (2019) Crop production technologies for sustainable use and conservation – physiological and molecular advances. Apple Academic Press

Ozturk M, Gul A, Rafique K et al (2025) Nature conservation, air pollution and plant diversity across various ecosystems and the impact of climate change. In: Forests and carbon sequestration. Cambridge Scholars Publishing, pp 330–422

Pradhan AK, Prakash S, Mukhopadhyay A, Pental D (1992) Phylogeny of *Brassica* and allied genera based on variation in chloroplast and mitochondrial DNA patterns: molecular and taxonomic classifications are incongruous. Theor Appl Genet 85:331–340

Prakash S, Hinata K (1980) Taxonomy, cytogenetics and origin of crop brassicas, a review. Opera Bot 55:1–57

Prasad M (2014) Antimicrobial potential of Brassicaceae family against clinical isolates. Int J Pure Appl Biosci 2:158–162

Rahman M, Khatun A, Liu L, Barkla B (2018) Brassicaceae mustards: traditional and agronomic uses in Australia and New Zealand. Molecules 23:231

Rakow G (2004) Species origin and economic importance of Brassica. In: Brassica. Springer, pp 3–11

Rather MA, Badroo IA, Taj S (2024) For global food security. Food security, nutrition and sustainability through aquaculture technologies, pp 1–457

Ravikumar C (2015) Therapeutic potential of *Brassica oleracea* (broccoli) – a review. Int J Drug Dev Res 7:9–10

Salmeen O (1979) A systematic revision of the genus *Brassica* L. in the Mediterranean region. PhD thesis, Univ. of Reading

Šamec D, Salopek-Sondi B (2019) Cruciferous (*Brassicaceae*) vegetables. In: Nonvitamin and nonmineral nutritional supplements. Elsevier, pp 195–202

Šamec D, Kruk V, Ivanišević P (2019) Influence of seed origin on morphological characteristics and phytochemicals levels in *Brassica oleracea* var. *acephala*. Agronomy 9(9):502

Sanchez-Yelamo MD (1990) Relaciones quimiotaxon6micas y evolutivas en el complejo Diplotaxis-Erucastrum-Brassica (Cruciferae-Brassiceae) mediante anddisis cromatogr(~ficos y electrofor~ticos. Tesis doctoral, Universidad Politdcnica de Madrid

Schery RW (1972) Plants for man, New Jersey, pp 508–509

Schiemann E (1932) Entstehung der Kulturpflazen. Handb. Verebwis. Lfg., 15

Singh SP, Malik JK (2025) Migratory small ruminant farming system in the himalayas-peculiarities, problems, prospects, climate effects, nutrition and health. Springer

Sinskaia EN (1928) The oleiferous plants and root crops of the family Cruciferae. Bull Appl Bot Genet Plant Breed 9:1–648

Snogerup S (1980) The wild forms of the *Brassica oleracea* group (2n = 18) and their possible relations to the culvated ones. In: Brassica crops and wild allies, biology and breeding. Scientific Society Press, Tokyo, pp 121–132

Song KM, Osborn TC, Williams PH (1988) *Brassica* taxonomy based on nuclear restriction fragment length polymorphisms (RFLPs). 2. Preliminary analysis of subspecies within *B. rapa* (syn. campestris) and *B. oleracea*. Theor Appl Genet 76:593–600

Song KM, Osborn TC, Williams PH (1990) *Brassica* taxonomy based on nuclear restriction fragment length polymorphisms (RFLPs). Theor Appl Genet 79:497–506

Sun VG (1946) The evaluation of taxonomic characters of cultivated *Brassica* with a key to species and varieties. I. The characters. Bull Torrey Bot Cl 73:244–281

Takahata Y, Hinata K (1983) Studies on cytodemes in the subtribe Brassicineae. Tohoku. J Agric Res 33:111–124

Takahata Y, Hinata K (1986) A consideration of the species relationships in subtribe Brassicinae (Cruciferae) in view of cluster analysis of morphological characters. Pl Sp Biol 1:79–88

Toxopeus H, Oost E, Reuling R (1984) Current aspects of the taxonomy of cultivated *Brassica* species. Cruciferae Newsl 9:55–58

Truco MJ, Quiros CF (1991) Evolutionary study of *Brassica nigra* and related species. In: Proceedings of 8th international rapeseed congress, vol 2, Saskatoon, pp 318–323

Tsunoda S, Hinata K, Gómez-Campo C (1980) *Brassica* Crops and wild allies: biology and breeding. Japan Scientific Press, p 354

Uchimiya H, Wildman SG (1978) Evolution of fraction I protein in relation to origin of amphidiploid *Brassica* species and other member of Cruciferae. J Heredity 69:299–303

Uddin MM, Zakeel MCM, Zavahir JS et al (2021) Heavy metal accumulation in rice and aquatic plants used as human food: a general review. Toxics 9(12):360

van de Wouw M, Kik C, van Hintum T et al (2010) Genetic erosion in crops: concept, research results and challenges. Plant Genet Resour 8(1):1–15

Vavilov NI (1949) The origin, variation, immunity and breeding of cultivated plants. Chron Bot 13:1–364

Warwick SI, Black LD (1997) Phylogenetic implications of chloroplast DNA restriction site variation in subtribes Raphaninae and Cakilinae (Brassicaceae, tribe Brassiceae). Can J Bot 75:960–973

Wu YD, Lou YJ (2007) Brassinolide, a plant sterol from pollen of *Brassica napus* L., induces apoptosis in human prostate cancer PC-3 cells. Die Pharmazie Int J Pharm Sci 62(5):392–395

Yanagino T, Takahata Y, Hinata K (1987) Chloroplast DNA variation among diploid species in *Brassica* and allied genera. Japan J Genet 62:119–125

Chapter 2
Botany

2.1 Taxonomy and Morphology

The Mustard family (Brassicaceae or Cruciferae) is one of the large families of flowering plants in the plant kingdom, belonging to the order Brassicales, with a total of 338 genera and about 4000 species distributed globally on all continents except Antarctica (Anjum et al. 2012). Although different species of the Brassicaceae family have limited distribution areas, especially in mountainous regions in the tropics (subalpine and alpine habitats in some East African countries, especially in Kenya, Tanzania and Ethiopia), this family is generally widely distributed in different habitats of the temperate regions of the Northern and Southern Hemispheres (Koch and Kiefer 2006).

The family comprises herbs and shrubs with a sharp, strong taste, due to the presence of glucosinolates. The family is known for its prominent cross-shaped flowers, and the species specific to this family typically have simple, alternately arranged leaves; tetradyn stamens; and a carpel that matures into a capsular fruit known as a silique (Jabeen 2020).

Many plant species in this family, especially the genus *Brassica*, have made significant contributions to almost every field of modern plant biology, especially in the development of model species and model generic systems of economic and agricultural importance (Koch and Mummenhoff 2006; Anjum et al. 2012); many of its species are widely grown cultivated forms, especially cabbage, broccoli, Brussels sprouts and cauliflower, plants of great importance in human health and food safety on a global scale (Bailey et al. 2006).

The phylogenetic analyses have been conducted to determine the relationships between genera and species within this family. The reports show that most species in this family belong to three main lineages and 52 monophyletic tribes (Beilstein et al. 2006, 2008; Al-Shehbaz 2012; Jabeen 2020). Unfortunately, this analysis has

M. Ozturk et al., *Climate-Resilient* Brassicas,
https://doi.org/10.1007/978-3-032-12709-9_2

not clearly revealed the relationships among the species (Jabeen 2020). Another phylogenetic analysis study reports that 32 species in this family, representing 29 of the 51 tribes, form six clades following their analysis using nuclear markers derived from their transcriptomes (Huang et al. 2015). More recently, a study of 79 species belonging to 50 of the 52 Brassicaceae tribes has identified meaningful phylogenetic relationships within six main lineages (Nikolov et al. 2019).

Several studies on the taxonomy of the *Brassica* genus have attempted to compile a list of its natural species (Salmeen 1979; Takahata and Hinata 1980; Tsunoda et al. 1980; Gomez-Campo 1980; Takahata and Hinata 1986; Gladis 1989; Sanchez-Yelamo 1990; Snogerup et al. 1990; Eastwood 1996; Lanner 1997; Lazaro 1997; Gomez-Campo 1999b). The taxonomy of this genus has been revised and updated in a book edited by Gomez-Campo (1999a).

The genus *Brassica* is a monophyletic group within the Brassicaceae but is reported to have diverged 14.5–20.4 million years ago (Yang et al. 1999; Bowers et al. 2003). The taxonomic status of the *Brassica* genus has always been a subject of debate, thought to be due to its constant flux. The terms "genus", "species", "variety", "subspecies", "botanical group" and "cultivar", along with their descriptive characteristics, are often used interchangeably in the literature (Fahey 2016).

Some species of *Brassica* are globally important due to their environmental, social and economic uses, but these species have become taxonomically synonymous and have been transferred to other genera. The examples include *B. nigra*, *B. tournefortii*, *B.* × *juncea* and *B. carinata. B. nigra* is now a synonym of *Mutarda nigra*. The native range of this species is Europe to West and North China, Mediterranean to Ethiopia. It is an annual and grows primarily in the temperate biome. *B. tournefortii* is now a synonym of *Coincya tournefortii*. The native range of this species is Mediterranean to South Pakistan and the Arabian Peninsula. It is an annual and grows primarily in the subtropical biome. *Brassica* × *juncea* is a synonym of × *Brassarda juncea*. The native range of this hybrid is the Caucasus. It is an annual or subshrub and grows primarily in the temperate biome. The hybrid formula is *Brassica rapa* × *Mutarda nigra*. *B. carinata* is a synonym of *Mutarda carinata*. The native range of this species is a cultigen from North-Eastern Tropical Africa. It is an annual or subshrub, growing primarily in the seasonally dry tropical biome. The hybrid formula is a stabilised hybrid derived from *Mutarda nigra* × *B. oleracea* (www.powo.science.kew.org).

Brassica is an old-world genus that includes about 42 species of mostly annual herbs, as well as some perennial herbs and small shrubs (Anjum et al. 2012). In light of the latest data, considering the current online database accepted globally, the total number of accepted species in the genus *Brassica* in the world is reported as 42 (www.powo.science.kew.org). It is possible to come across different species names in different country floras. The oleiferous genus *Brassica* is taxonomically placed within the family Brassicaceae, being the third most important source of vegetable oil in the world after palm and soybean oil (Anjum et al. 2012).

A perusal of the latest taxonomic studies reveals that the distribution of wild Brassica **species in the world is as follows**:

1. *Brassica assyriaca* Mouterde
2. *Brassica aucheri* Boiss.
3. *Brassica baldensis* (Prosser & Bertolli) Prosser & Bertolli
4. *Brassica balearica* Pers.
5. *Brassica barrelieri* (L.) Janka
6. *Brassica beytepeensis* Yıld.
7. *Brassica bourgeaui* (Webb ex Christ) Kuntze
8. *Brassica cadmea* Heldr. ex O.E.Schulz
9. *Brassica cretica* Lam.
10. *Brassica deflexa* Boiss.
11. *Brassica deserti* Danin & Hedge
12. *Brassica desnottesii* Emb. & Maire
13. *Brassica dimorpha* Coss. & Durieu
14. *Brassica elongata* Ehrh.
15. *Brassica fruticulosa* Cirillo
16. *Brassica gravinae* Ten.
17. *Brassica hilarionis* Post
18. *Brassica huseyinduralii* Başköse & Savran
19. *Brassica incana* Ten.
20. *Brassica insularis* Moris
21. *Brassica loncholoma* Pomel
22. *Brassica macrocarpa* Guss.
23. *Brassica maurorum* Durieu
24. *Brassica montana* Pourr.
25. *Brassica napus* L.
26. *Brassica nivalis* Boiss. & Heldr.
27. *Brassica oleracea* L.
28. *Brassica oxyrrhina* Coss.
29. *Brassica procumbens* (Poir.) O.E.Schulz
30. *Brassica rapa* L.
31. *Brassica repanda* (Willd.) DC.
32. *Brassica rupestris* Raf.
33. *Brassica setulosa* (Boiss. & Reut.) Coss.
34. *Brassica somalensis* Hedge & A.G.Mill.
35. *Brassica souliei* (Batt.) Batt.
36. *Brassica spinescens* Pomel
37. *Brassica taiwanensis* S.S.Ying
38. *Brassica taurica* (Tzvelev) Tzvelev
39. *Brassica trichocarpa* C.Brullo, Brullo, Giusso & Ilardi
40. *Brassica tyrrhena* Giotta, Piccitto & Arrigoni
41. *Brassica villosa* Biv.
42. *Brassica* × *harmsiana* O.E.Schulz

2.1.1 *Brassica assyriaca Mouterde*

Synonyms None reported.

Description The native range of this species is Syria, and it grows primarily in the subtropical biome (www.powo.science.kew.org).

2.1.2 *Brassica aucheri Boiss.*

Synonyms *Raphanus aucheri* (Boiss.) Boiss.; *Sinapis aucheri* (Boiss.) O.E.Schulz; *Enarthrocarpus tragicerus Boiss.* & Hausskn.; *Raphanus boissieri* Al-Shehbaz (www.powo.science.kew.org).

Description Hairs retrorsis hispida, stems low, subsimple, oblong crenate leaves briefly petiolate, petals with longer hirsute calyx nails, fruit peduncles contorted, reflexed, curved, multi-angled pods, strangulated between seeds, thin conical snout shortly terminated, densely retrorsum hispid (Boissier 1867).

2.1.3 *Brassica baldensis (Prosser & Bertolli) Prosser & Bertolli*

Synonyms *Guenthera repanda* subsp. baldensis Prosser & Bertolli (www.powo.science.kew.org).

Description Hemicryptophyton (sometimes chamaephyton), a tufted plant with a thick woody root and a branched stem (branches up to 1–1.5 cm thick); the stem is covered with the remains of dead leaves of previous years, upper parts of all branches bear a rosette of leaves; almost all branches have a single aphyllum scape at the apex. The flowering scapes are erect 20–40 cm long, fruiting procumbent, fruits are 30–60 (−80) cm long, smooth (c. 2 mm thick), glabrous, green with a woolly texture, reddish when exposed to the sun, fragile. All leaves arranged in rosettes, usually c. 20 or more, fleshy, dark green, faintly scented, 5–23 cm long, 0.3–3.5 cm wide; lamina gradually attenuated into petiole at base, obovate-lanceolate, very variegated, from sinuous to pinnatifida, very rarely entire, often in the same individual with dissimilar incisions, in this case the lower laminae less incised than the upper ones, margin with c. 4–6 rounded apex sometimes with lobes on either side towards the apex, sparsely ciliate (hairs c. 0.4–0.7 mm long), glabrous side; petiole lamina 1–2 times longer. The raceme bears c. 15–30 very fragrant flowers; anthesis progresses from the base to the apex, together with the extension of the axis; therefore, the inflorescence initially takes the form of a flowering cymbal, then the form of a raceme, at the bottom distant and at the top dense with flowers. The peduncle subanthese is c. 1 cm, under fructification 1.3–3 cm long, glabrous. The flowers have 2 outer sepals cupped, 6–9 mm long, 2 inner sepals saccate, slightly shorter, all bearing 2–8 hairs (c. 0.5 mm long) near the apex, green with pale margin; petals lemon coloured, 1.2–1.6 cm long, 0.5–0.9 cm wide, lamina obovate-orbiculata, abruptly restricted to 3–4 mm long nail, extra-patent, joined to nail at an angle of almost 90°, adapex ± truncate, entire or slightly emarginate; stamens 6, 2 shorter (c. 6–7 mm long) and 4 longer (10–12 mm long), anther c. 2–2.5 mm long; pistil c. 10–12 mm long, stigma 1 mm wide, style c. 3 mm long; 2 green nectaries, c. 0.5 mm long, inserted at the base of the outer sepals, persistent in fruit. The pods are glabrous, ± pendulous or horizontal, at maturity 30–70 mm long and 3–5(−6) mm wide, with maximum width in the upper third, ± clavate, beak 3–5 mm long, slightly attenuated at the apex, c. 1 mm at the base and c. 0.6 mm thick at the apex, at dehiscence, the stramineous valves are clearly prominent in the middle of the vein and reticulate side. The seeds are arranged irregularly in 1–2 series, brown, ± ovate, slightly compressed laterally, c. 2.2 × 1.6 × 1 mm (Prosser and Bertolli 2007).

2.1.4 Brassica balearica Pers.

Synonyms *Brassica oleracea* var. *frutescens* Spach; *Erucastrum balearicum* Bluff, Nees & Schauer (www.powo.science.kew.org).

Description Characterised by its shiny, fleshy leaves with wavy margins which are concentrated at the base of the plant. There is no other plant in the rocky crags with similar leaves; it is unmistakable; subshrub up to 30 cm or more in height; leaves basal or rosetted at the stem tips, up to 10 cm long, long-stalked, pinnately lobed to lyrate, rather like those of common oak, flowers about 1.5 cm wide, bright yellow and sweetly scented, observed during spring; reasonably hardy outside but best regularly replaced by young plants (Snogerup and Persson 1983).

2.1.5 Brassica barrelieri (L.) Janka

Synonyms *B.barrelieri* subsp. *sabularia* Maire; *B. sabularia* Brot.; *Diplotaxis barrelieri* (L.) DC.; *Sisymbrium barrelieri* L.; *S. murale* var. barrelieri (L.) Mérat; *B. barrelieri* var. *laevigata* (Lag.) Maire & Weiller; *B. barrelieri* var. *major* (Gand.) O.E.Schulz; *B. barrelieri* var. *papillaris* (Boiss.) O.E.Schulz; *B. barrelieri* var. *psammophila* (Pomel) Batt.; *B. bunias* DC.; *B. grandiflora* Balansa ex Coss.; *B. laevigata* Lag.; *B. laevigata* var. *major* Gand.; *B. papillaris* Boiss.; *B. parra* (L.) T.Durand & Schinz; *B. psammophila* Pomel; *B. sabularia* var. *laevigata* (Lag.) Samp.; *B. sabularia* var. *papillaris* (Boiss.) Boiss.; *B. sabularia* var. *psammophila*

(Pomel) Batt.; *B. subularia* Pers.; *Erucastrum oleraceum* C.Presl; *E. sabularia* Link; *Rapistrum halleri* Crantz; *Sisymbrium parra* L. (www.powo.science.kew.org).

Description An annual, weakly hispid plant with a non-napiform taproot. The plant can exceed 80 cm in height, but its growth varies greatly depending on soil depth and the presence of nutrients; small forms are common. Leaves: Lower leaves up to 20 cm, numerous, arranged in a dense rosette, very shortly petiolate, runcinate-pinnatifid, with a thick, white rachis and 7–10 pairs of lateral segments, triangular-acute, more or less auriculate, the terminal one very slightly larger, acuminate. Stem leaves are few, sessile, more or less entire and glabrous. Flowers: Clusters of 16–40 flowers. Pedicels 5–18 mm at anthesis, 15–25 mm at fruiting. Sepals 4–5.5 mm, erect, almost glabrous. Petals 9–12 mm, yellow. Median nectaries oblong. Fruits 20–50 × 1.8–3 mm, with a very pronounced midrib, containing 8–15 seeds in each cell; beak 5–12 mm, conical, containing 0–1 seed; carpophore 0.5–1 mm. Seeds 0.8–1.2 mm in diameter, spherical, reddish-brown (Bruno and Christophe 2002).

2.1.6 *Brassica beytepeensis* Yıld.

Synonyms None reported.

Description Annual to biennial herb; stem 30–40 cm long, divaricately branched from base, antrorse adpressed setosa hairy, terete, striate. Basal leaves 3–10 × 0.5–1.5 cm including 1–3.5 cm petioles, oblanceolate, apex obtuse, base attenuate, entire to repandly or regularly dentate, adpressed setose hairy; stem leaves like basal leaves but smaller and petiolate. Infrutescence corymbose, glabrous. Fruiting pedicels 10–15 × 0.5 mm, patent, glabrous. Siliquae 20–50 × 1 mm, linear, patent, glabrous, tetragonal, constricted between the seeds (lomentoid), beaked with a lower segment 2.5 mm, upper seedless stylar part 1 mm, valves with one prominent middle nerve. Stigma capitate, shallowly 2-lobed. Seeds 1.2–1.5 × 0.4 mm, oblong, glabrous, green (Yıldırımlı 2010).

2.1.7 *Brassica bourgeaui* (Webb ex Christ) Kuntze

Synonyms *Sinapidendron bourgeaui* Webb ex Christ (www.powo.science.kew.org).

Description Suffrutescent perennial, 40–100 cm, cultivated specimens up to 150 cm; young plant first develops a basal rosette of leaves, gradually prolongs into a thick and up to 1-m-tall main stem with densely placed leaves; flowering first moderate in size or in dwarfed plants simple, a pical inflorescence, often followed by some later developed branches; usually perennating for several flowering seasons and then becoming a moderately branched shrub, later inflorescences usually smaller than the first, life span unknown for wild specimens. Stem terete, as old lignified withy ear-rings; leaf scars prominent, often visible even on older stem parts. Leaves glabrous, comparatively thick, with a conspicuous greyish surface similar to that of *B. oleracea*'s, lightly bluish green; basal and cauline leaves including petiole usually 15–20 × 12–17 cm, petiole 1–5 cm, lamina orbicular to broadly obovate, 1–2 pairs of small lobes on petiole and a semi-amplexicaul foot, margin undulate with very small pointed teeth, midrib broad, side veins thin, branching and anastomosing, some of the larger veinlets reaching the margin in the teeth; upper leaves and those of the branches smaller, often with a longer petiole sometimes almost equalling the lamina, with very small lobes, foot often lacking wing, bracts of inflorescence sessile, semi-amplexicaul, undivided, with up to 3-mm teeth. First inflorescence usually of only a few partial racemes, usually 20–50 (−100 flowered with several buds overtopping the flowers, flowers opening 2–10 cm below the inflorescence top. Pedicels 15–20 mm, at anthesis erecto-patent. Flowers usually conspicuously protandrous, anthers ripening 1–2 days before the stigma. Sepals 9–11(−12) mm, at anthesis erect, slightly distant from the petal claws, the outer 2.5–4 mm broad, saccate at base, shallowly boat-shaped, the inner 2.5–3 mm broad, flatter, cucullate especially in bud. Petals at anthesis usually patent, whitish yellow to white; claw 8.5–11 mm, much widened upwards, limb 11–13.5 × 7.5–9 mm, comparatively narrowly elliptic. Stamens exserted c. 5 mm above the petal disc; outer filaments 8.5–10.5 mm, inner 11–12.5 mm, usually longer than the outer; anthers 3.5–4.1 mm, outer sometimes slightly longer, connective tip up to 0.15 mm, pale. Pistil exserted above anthers at maturity, ripe ovary 10–11 mm, style 4–5 mm, stigma c. 1.5 mm broad. Fruiting pedicels 18–25 mm, erecto-patent, often weak and more or less reflexed. Siliqua except rostrum (40–)65–80(−85) mm long, (3–)4–4.5 mm broad with 12–20 seeds, valves with a strong midvein and usually conspicuous side veins, but veins often inconspicuous on unripe siliquas. Rostrum (7–)11–14 mm long, c. 3 mm broad, seedless but often thick and spongy. Seeds 2.3–2.5 mm, with a network of c. 0.05 mm wide mashes (Snogerup et al. 1990).

2.1.8 Brassica cadmea Heldr. ex O.E.Schulz

Synonyms None reported.

Description The root seems to be annual. Galls from the root numerous, 0.30–0.60 m long, ascending, branched, teretiuscule, at the base with short hairs 0.25–0.75 mm long, glabrous above, often violet; branches long, erect-open. Lower leaves subrosulate, shortly petiolate, pinnately divided, 5–3-jointed: terminal lobe shortly ovate, obtuse, unevenly coarsely callose-crenate, lateral lobes ovate, gradually smaller, lowest entire and descending into a broad winged petiole; upper leaves much smaller, 4–1.5 cm long, sub- or sessile, oblong or linear, few-jointed or simple, remotely denticulate; all fleshy, on both sides covered with hairs arising from tubercles, scattered bristle-like and ciliate. Raceme corymbose during anthesis, later very elongate, 30–50-flowered. Pedicels long, 6–8 mm long, filiform, apex hairy. Sepal 3.5 mm long, erect-open, oblong, shortly villous. Petals 6.5 mm long, yellow; lamina oblong-obovate, Stamens inner 6 mm, outer ones clearly shorter, 3–4 mm long; anthers 1.5 mm long, oblong. Ovary 18–24-ovulate; style very short; stigma barely wider than style. Pods on pedicels 2.5–1 cm long, thin, erect, 1.9–2.9 cm long, 1.5–2 mm wide, torulose, in short rostrum 3–5 mm long, wider, ancipes monospermum barely attenuated; valves very membraneous, apex obtuse, grey-green or pale violet; stigma with a beak equilateral. Seeds subglobose, ca. 1 mm diam., slightly brown, scarcely alveolate, pendulous, erect on the beak (Schulz 1919).

2.1.9 Brassica cretica Lam.

Synonyms *Brassica oleracea* var. *cretica* (Lam.) Coss. (www.powo.science.kew.org).

Description Suffrutescent perennial, 50–150 cm high. During the first year, it develops a rosette or a series of leaves usually considerably larger than those formed later and often different in shape. Stem becoming hard, woody, up to 10(−30) mm thick with conspicuous year rings. Leaves glabrous, semi-coriaceous, keeping their texture on drying, veining pinnate with a strong midrib and anastomosing veinlets not running directly into the margin; basal and cauline leaves usually pinnatifid to pinnatipartite in the basal part, petiolate, very variable in shape, margin usually entirely or with blunt teeth only. First inflorescence variable in size, usually with several branches, later inflorescences smaller, usually simple, partial racemes usually many-flowered with several buds overtopping the flowers; flowers usually conspicuously protandrous, white or different shades of yellow, anthers ripening 1–4 days before the stigma. Style elongated at anthesis, stigma elevated above the stamens at maturity. Siliqua 35–105 mm long, rostrum of variable size, seedless or with 1–2 seeds. Seeds 1.6–2.8 mm, globose, with a more or less conspicuous reticulum (Snogerup et al. 1990).

B. cretica also includes three subspecies (www.powo.science.kew.org).

(a) Brassica cretica **subsp.** aegaea **(Heldr. & Halácsy) Snogerup, M.A.Gust. & Bothmer**

Synonyms *B. cretica* var. *aegaea* Heldr. & Halácsy; Brassica cretica var. attica Onno (www.powo.science.kew.org).

(b) Brassica cretica **subsp.** creti**ca**

Synonyms: *B. botrytis* (L.) Mill.; *B. cretica* convar. *botrytis* (L.) G.H. Loos; *B. cretica* subsp. *nivea* (Boiss. & Spruner) M.A.Gust. & Snogerup; *B. cretica* var. *nivea* (Boiss. & Spruner) O.E.Schulz; *B. leucantha* Boiss. ex Nyman; *B. nivea*

Boiss. & Spruner; *B. oleracea* subsp. *botrytis* (L.) Metzg.; *B. oleracea* stirps *botrytis* (L.) DC.; *B. oleracea* proles *botrytis* (L.) Duchesne; *B. oleracea* convar. *botrytis* (L.) Alef.; *B. oleracea* var. *botrytis* L.; *B. oleracea* subvar. *nivea* (Boiss. & Spruner) Coss.; *B. oleracea* var. *nivea* (Boiss. & Spruner) Fiori;

B. rupicola Boul. (www.powo.science.kew.org).

(c) Brassica cretica **subsp.** laconica **M.A.Gust. & Snogerup**

Synonyms: None reported.

2.1.10 Brassica deflexa Boiss.

Synonyms *Erucastrum deflexum* (Boiss.) V.I. Dorof.; *Guenthera deflexa* (Boiss.) V.I.Dorof.; *Brassica deflexa* var. *auranthiaca* Bornm.; *B. deflexa* var. *glabrescens* O.E.Schulz; *B. deflexa* var. *lasiocalycina* (Boiss. & Hausskn.) O.E.Schulz; *B.deflexa* subsp. *leptocarpa* (Boiss.) Hedge; *B. deflexa* var. *leptocarpa* (Boiss.) O.E.Schulz; *B. deflexa* var. *mollis* O.E.Schulz; *B.deflexa* var. *tigridis* (Boiss.) Boiss.; *B. iranica* Rech.f., Aellen & Esfand.; *B. lasiocalycina* (Boiss. & Hausskn.) Boiss.; *B. leptocarpa* Boiss.; *B. tigridis* Boiss.; *Erucastrum lasiocalycinum* Boiss. & Hausskn.; *Guenthera deflexa* var. *auranthiaca* (Bornm.) V.I.Dorof. (www.powo.science.kew.org).

Description Annual. Stem erect, 0.20–0.60 m, branched from the base, terete, with unequal open or subdeflexed hairs below up to 1 mm long, densely hairy above, glabrous; branches ascending, in the axils white-spotted. Lower leaves with a long hairy petiole 3–3.5 cm, oblong-ovate, rounded at the apex, unevenly repando-denticulate margin, acute at the base; middle leaves more shortly petiolate, ovate, acute at the apex, sinuate-dentate, sublobed below, often unequally cordate at the base; upper leaves minute, oblong, denticulate, narrowed into a short petiole cuneately or subsessile; all membraneous, both sides sparsely hairy. The raceme initially corymbose, then very elongate, 25–40-flowered, pedicels 4–7 mm long, sparsely hairy, sepals 5–6 mm long, erect-spreading, outer narrowly oblong, rounded, inner oblong-ovate, acute, all densely hairy or villous, petals 1–1.2 cm long, lemon or intensely yellow; lamina broadly obovate, apex rounded, at the base suddenly narrowed into a claw, multiveined. Stamens inner 7–8.5 mm, outer 6–7 mm long; anthers 2–2.5 mm long, oblong, acute and recurved at the apex. The ovary is ovate 25–45 long; style ca. 1.5 mm long; stigma style equilateral; pods on pedicels 2–0.5 cm long, rectangular, open or ascending or recurved, ascending or often deflexed, 3.5–7.5 cm long, ca. 1.5 mm wide, very torulose, glabrous, in the proportionally short rostrum 3–6 mm long, a- or monospermum thick, barely attenuated; valves greenish-yellow, membraneous; stigma rostrum narrower. The seeds are small, globose, 0.75 mm diam., slightly brown, blackish at the tip, sublaevous, pendulous, erect in the rostrum (Boissier 1867).

2.1.11 *Brassica deserti* Danin & Hedge

Synonyms *Erucastrum deserti* (Danin & Hedge) V.I.Dorof. (www.powo.science.kew.org).

Description Annual or perennial. The stems erect, 15–60 cm tall, branched at the base, terete, grey-green, hairy, densely covered with weak, flat hairs 0.5–2 mm long; leaves petiolate, often lyrate, flat-hairy; lower leaves with two lateral lobes, 0.5–1 cm long, arranged suboppositely; terminal lobe largest, ovate, 5 × 3–55 cm, margin unevenly crenate or bi-crenate; middle leaves smaller, lyrate, 2–4-lobed; upper leaves lyrate or entire, 7–10 × 3–6 mm, with obscurely pinnate nerve leaves. Raceme subanthese corymbose then elongate, 20–40-lobed. Pedicels villous, filiform, erect-open, 5–10 mm long, slightly elongate in fruit; sepals villous, covered with a few glandular hairs; outer sepals broad, 5 × 1 mm, inner ones 5 × 2 mm, base slightly saccate; petals yellow, glabrous, 14 mm long; lamina obovate, 8 × 4 mm, attenuated to a nail 6 × 0.7 mm; stamens inner 7 mm long, outer ones only 5 mm long; anthers obtuse; median glands 0.5 mm long, much smaller on the lateral ones; ovary 20–26-ovulate, not or shortly stipitate; style 1.5 mm long, 0–5 mm in diameter; stigma capitate, shortly bilobate, mm in diameter. The pods are 2–2.5 cm long, 2 mm wide, base up to 1 mm stipitate, biconvex, attenuated at the rostrum 3–5 mm long, rough (rarely monospermous); sub-membraneous valves constricted between the seeds, with a prominent median nerve; seeds globose, 1 mm in diameter, brown, sub-smooth, uniseriate (Danin and Hedge 1973).

2.1.12 *Brassica desnottesii* Emb. & Maire

Synonyms *Guenthera desnottesii* (Emb. & Maire) Gómez-Campo (www.powo.science.kew.org).

Description Elongated stems covered with leaf vestiges. Submembranous leaves, all basally rosulate pinnatifida, gradually tapering to a petiole subequal to the limb, loosely hairy, green, with acute lacinis at the apex, terminally larger; flowering stems emerge from the apex of the rosette, erect, glabrous. The flowers are arranged in a corymbiform raceme after anthesis, very elongate; peduncles sub anthesis calyces c. equal, fruiting 10–17 mm long, flat, glabrous; sepals oblong, loosely hairy above, rounded at the apex, laterals slightly gibbous at the base; petals c. 10 mm long, yellow, obovate, apex rounded, base subequal limb slightly attenuated; lateral stamens c. 5 mm long; median c. 7 mm long; anthers c. 1.5 mm long, obovate, glabrous, excluding the rostrum 7–8 × 4–5 mm; basal valves more or less attenuate, apex rounded, median nerve prominent up to the apex, otherwise slightly reticulate-nervous; rostrum compressed 5 mm long, rough; stigma capitate-bilobum; biseriate subglobose seeds, moist mucosa, smooth, c. 1.5 mm in diameter (Emberger and Maire 1930).

2.1.13 *Brassica dimorpha* Coss. & Durieu

Synonyms *Guenthera dimorpha* (Coss. & Durieu) Gómez-Campo (www.powo.science.kew.org).

Description The plant subshrubby at base, often multi-stemmed; stems erect, simple or branched, glaucous, at least with stiff, open, hispid hairs below; leaves hispid below, oblong, attenuate below, coarsely and remotely toothed or subpinnatifid; upper leaves glaucous, ovate or ovate-oblong, undivided, base broadly cordate-amplecaulid; flowers small, long-pediculated, arranged in subcorymbiform racemes; sepals erect; petals yellow, calyces twice as long; fruit racemes elongated, pedicels open, pods slightly shorter; pods ascending, glaucous, oblong-linear, subtetragonous, subtorulose, locules 4–7 sperm; valves endowed with a single strong nerve, lateral veins obsolete, beak subtriple longer; beak suberete, rough; seeds yellow-brown, globose, smooth; cotyledons conduplicate (Cosson and Durieu 1955).

2.1.14 *Brassica elongata* Ehrh.

Synonyms *Brassica elongata* f. *genuina* Thell.; *Brassicastrum elongatum* (Ehrh.) Link: *Erussica elongata* (Ehrh.) G.H. Loos; *Guenthera elongata* (Ehrh.) Besser (www.powo.science.kew.org).

Description Biennial to perennial (Fig. 2.1), often with a thick woody base, up to 1 m; stem usually glabrous (rarely hispid), pale green, often violet-coloured at the

Fig. 2.1 Flowering branches of *Brassica elongata* Ehrh

base; lower leaves petiolate, oblong-ovate, often bright green, entire or coarsely toothed, with tubercled curved bristles on both sides; upper leaves oblong or lanceolate, shortly stalked. The petals are yellow, 6–10 × 3–4 mm; ovary c. 15-ovulate; fruiting pedicels 6–14 mm, spreading, glabrous; siliquae linear, 1.5–4.5 cm × 1–2.5 mm, torulose; stipe 1–4 mm; beak seedless, 1–2.5 mm; seeds mucilaginous (Davis 1965).

Brassica elongata includes seven subspecies (www.powo.science.kew.org).

(a) Brassica elongata **subsp.** duralii **Yıld.**
Synonyms: None reported.

(b) Brassica elongata **subsp. e**longata
Synonyms: *Brassica elongata* var. *astyla* Rupr.; *B. elongata* var. *brevisiliqua* N.Busch; *B. elongata* var. *macrocarpa* Thell.; *B. elongata* f. *mesocarpa* Thell.; *B. elongata* var. *scabra* O.E.Schulz; *B. elongata* f. *subintegrifolia* Thell.; *B. elongata* var. *turcomanica* O.E.Schulz; *Crucifera elongata* E.H.L. Krause; *Eruca elongata* Baumg.; *Erucastrum elongatum* Rchb.; *Macropodium elongatum* Schur ex Nyman; *Sinapis elongata* Spach ex Griseb. & Schenk.; *Sinapis laevigata* Pall.; *Sisymbrium elongatum* (Rchb.) Prantl (www.powo.science.kew.org).

(c) Brassica elongata **subsp.** imdrahsiana **Quézel**
Synonyms: *Guenthera elongata* subsp. *imdrahsiana* (Quézel) Gómez-Campo; *Guenthera subscaposa* subsp. *imdrahsiana* (Quézel) D.A.German (www.powo.science.kew.org).

(d) Brassica elongata **subsp.** integrifolia **(Boiss.) Breistr.**
Synonyms: *Brassica elongata* var. *integrifolia* Boiss.; *Guenthera elongata* subsp. *integrifolia* (Boiss.) Gómez-Campo; *Brassica armoracioides* Czern. ex Turcz.; *Brassica brevirostrata* Z.X.An; *Brassica elongata* subsp. *armoracioides* (Czern. ex Turcz.) Asch. & Graebn.; *Brassica elongata* var. *armoracioides* (Czern. ex Turcz.) Schmalh.; *Brassica elongata* var. *persica* (Boiss. & Hohen.) Paol.; *Brassica obtusangula* subsp. *integrifolia* Arcang.; *Brassica persica* Boiss. & Hohen.; *Erucastrum armoracioides* (Czern. ex Turcz.) Cruchet; *Erussica armoracioides* (Czern. ex Turcz.) G.H. Loos; *Guenthera persica* (Boiss. & Hohen.) D.A.German (www.powo.science.kew.org).

(e) Brassica elongata **subsp.** pinnatifida **(Schmalh.) Greuter & Burdet**
Synonyms: *Brassica elongata* var. *pinnatifida* Schmalh.; *Guenthera elongata* subsp. *pinnatifida* (Schmalh.) Gómez-Campo; *Erucastrum cretaceum* Kotov; *Guenthera cretacea* (Kotov) D.A. (www.powo.science.kew.org).

(f) Brassica elongata **subsp.** subscaposa **(Maire & Weiller) Maire**
Synonyms: *Brassica subscaposa* Maire & Weiller; *Guenthera elongata* subsp. *subscaposa* (Maire & Weiller) Gómez-Campo; *Guenthera subscaposa* (Maire & Weiller) D.A.German (www.powo.science.kew.org).

(g) Brassica elongata **subsp.** takhtajanii **(V.I. Dorof.) D.A.German**
Synonyms *Erucastrum takhtajanii* V.I. Dorof.; *Guenthera persica* subsp. *takhtajanii* (V.I. Dorof.) D.A.German (www.powo.science.kew.org).

2.1.15 *Brassica fruticulosa Cirillo*

Synonyms *Erucastrum fruticulosum* (Cirillo) C.Presl (www.powo.science.kew.org).

Description Annuals, biennials or perennials; glabrous or nearly so. Stems branched distally, 3–9 dm. The basal leaves early deciduous; petiole 1.5–6 cm; blade lyrate-pinnatifid, 3–10(−15) cm × 10–65 mm, lobes 1–3 (or 4) each side; cauline leaves shortly petiolate; blade (often lanceolate, reduced in size distally), base tapered or cuneate, not auriculate, apex acute. The racemes paniculately branched; fruiting pedicels spreading to divaricately ascending, (slender), (5–)10–15(−25) mm; flowers possess 3–8 × 1–1.7 mm sepals; petals pale yellow, narrowly obovate, 7–15 × 3–4 mm, claw 2–3 mm, apex rounded; filaments 3–6 mm; anthers 1.5–2 mm; gynophore 1–1.5 mm in fruit. The fruits stipitate, spreading to divaricately ascending, strongly torulose, linear, subcylindric, 1.5–3 cm × 1.5–2 mm; valvular segment with 5–13 seeds per locule, 1.2–2.5 cm, terminal segment seedless or 1-seeded, (conic), 3–6 mm; seeds brown or yellow, 0.6–1.2 mm diam.; seed coat finely reticulate-alveolate, not mucilaginous when wetted (www.efloras.org).

Brassica fruticulosa includes five subspecies (www.powo.science.kew.org).

(a) Brassica fruticulosa **subsp.** cossoniana **(Boiss. & Reut.) Maire**
Synonyms: *B. fruticulosa* var. *cossoniana* (Boiss. & Reut.) Coss.; *B. fruticulosa* subvar. *christobalii* Maire & Sennen; *B. fruticulosa* f. *christobalii* (Maire & Sennen) Maire; *B. fruticulosa* subvar. *christobalis* Maire & Sennen; *B. fruticulosa* var. *hispida* Faure & Maire; *B. fruticulosa* var. *leucantha* Coss.; *B. fruticulosa* subsp. *mauritanica* (Coss.) Maire; *B. fruticulosa* var. *mauritanica* Coss.; *B. fruticulosa* f. *platycarpa* (Maire & Sennen) Maire; *B. fruticulosa* subvar. *platycarpa* Maire & Sennen; *B. fruticulosa* var. *rubecula* Costa; *B. fruticulosa* var. *subacaulis* Maire; *B. fruticulosa* var. *tenuisiliqua* Faure & Maire; *B. cossoniana* Boiss. & Reut.; *B. christobalis* Sennen; *B. cossoniana* var. *subacaulis* Maire; *Diplotaxis heterophylla* Porta & Rigo (www.powo.science.kew.org).

(b) Brassica fruticulosa **subsp.** fruticulosa
Synonyms: *B. fruticulosa* var. *vera* Boiss.; *B. parra* var. *psammophila* T. Durand & Schinz
(www.powo.science.kew.org).

(c) Brassica fruticulosa **subsp.** numidica **(Coss.) Maire**
Synonyms: *B. fruticulosa* var. *numidica* Coss. (www.powo.science.kew.org).

(d) Brassica fruticulosa **subsp.** pomeliana **Maire ex Greuter**
Synonyms: None reported.

(e) Brassica fruticulosa **subsp.** radicata **(Desf.) Batt.**
Synonyms: *B. fruticulosa* var. *radicata* (Desf.) Coss.; *B. fruticulosa* subsp. *glaberrima* (Pomel) Batt.; *B. glaberrima* Pomel; *B. radicata* Batt. *Sinapis radicata* Desf (www.powo.science.kew.org).

2.1.16 *Brassica gravinae* Ten.

Synonyms *B. gravinae* var. *italica* Maire & Weiller; *B. gravinae* var. *brachyloma* (Boiss. & Reut.) Batt.; *B. gravinae* var. *caespitosa* (Pomel) Batt.; *B. gravinae* f. *djurdjurae* Batt.; *B. gravinae* var. *djurdjurae* (Batt.) Maire; *B. gravinae* var. *rupicola* (Pomel) Batt.; *B. boissieri* Munby; *B. brachyloma* Boiss. & Reut.; *B. caespitosa* Pomel; *B. humilis* var. *caespitosa* Batt.; *B. rupicola* Pomel; *Erucastrum gravinae* (Ten.) Calest.; *Guenthera gravinae* (Ten.) Gómez-Campo; *Sinapis recurvata* Desf. (www.powo.science.kew.org).

Description Pubescent, hairy, stem slightly suffruticosa, leaves lower oblong lyrato-runcinate, lobules subdiscrete, often obtuse, subintegerrimus terminally dentate, obtuse or acute, calyx subpatulo, sliiquis glabrous teretebus, style filiform, stigma capitate terminating (www.meteofunghi.it).

2.1.17 *Brassica hilarionis* Post

Synonyms *B. cretica* subsp. *hilarionis* (Post) H. Lindb.; *B. cretica* var. *hilarionis* (Post) O.E.Schulz (www.powo.science.kew.org).

Description Suffrutescent perennial plants, usually 50–80 cm, rarely taller, old specimens densely branched; young plants first form a more or less dense rosette of large leaves, soon prolonged, main stem ending in an apical inflorescence, growth then continued by branches, sometimes as thick as the stem; flowering several years. Stem strongly lignified, becoming at least c. 10 mm thick, with prominent year rings; leaf scars conspicuous, becoming obscure on older stems. The leaves are glabrous, semi-coriaceous, keeping their texture on drying, young leaves sometimes with a slightly glaucous greyish surface, lamina undivided, or at least in the plants from Ag. Hilarion, some lower cauline leaves sparingly pinnatifid to pinnatipartite, with shallow but usually acute teeth; basal and lower cauline leaves including petiole 5–20(−30) × 3.5 15(−20) cm, petiole 1–12 cm, usually broadly winged, lamina obovate-orbicular, base often semi-amplexicaul-auriculate; basal part of midrib up to 1 cm broad, veining pinnate, side veins branching, forming a network in outer part but sometimes reaching the margin in the largest teeth; middle cauline leaves more or less narrowly obovate, similar in size, lacking a distinct petiole, upper leaves gradually smaller and more lanceolate to narrowly ovate, bracts of inflorescence branches linear to ovate. First inflorescence with few branches, later inflorescences usually smaller, partial racemes often with c. 50 flowers, buds overtopping flowers; the pedicels (7–)15–30 mm, at anthesis erecto-patent to almost patent; sepals 8–12 × 2–3 mm, erect or slightly diverging, whitish green with membraneous margin, often tinged with pink or purple, as old turning purplish white; petals white or light pink, claw (8–)10 12(−15) mm, limb 17–20 × (5–)7–10 mm, narrow, margin often somewhat undulate; stamens exserted c. 5 mm above the petal disc; filaments 10–12 mm, the outer shorter; anthers (3–)5–6 mm, connective tip small, light; pistil exserted above anthers at maturity, ripe ovary 8–10 mm, style 5–6 mm, stigma 1–1.3 (−1.5) mm broad; fruiting pedicels 15–55 mm, patent to erecto-patent; siliqua except rostrum (35–)40–50(−60) mm long, 6–8(−10) mm broad, with 20–40 seeds, seeds in a more or less conspicuous 2-row arrangement in each loculus, septum flat, midrib of valves a thin line, no side veins present, surface of siliqua almost smooth, inconspicuously rugose and reticulate, the rostrum (7–)10–15 mm long, tapering from a 4–5-mm-broad base, normally seedless and seeds 1.5–2.0 mm, finely and obscurely reticulate (Schulz 1919).

2.1.18 Brassica huseyinduralii Başköse & Savran

Synonyms None reported.

Description The plants are caespitose perennial herb; caudex woody; stem 20–45 cm, often unbranched or rarely the lower parts of the stem are branched, cylindrical, lower parts hairy, middle and upper parts glabrous, green or greenish yellow but light brown at maturity; leaves both at the base and on the stem (1–3 pieces); rosette leaves, 5.0–9.0 × 0.5–1.2 cm, petiolate, 1–3 cm, succulent, blade oblong, oblong-elliptic, oblanceolate or obovate, rarely spatulate, both surfaces densely covered strigose hairs, hairs spreading or retrorse; margin entire or rarely inconspicuously toothed, apex acute or obtuse, base attenuate; stem leaves 1–3 and only present near the base, 2.0–4.0 × 0.15–0.40 cm, blade linear or narrow oblong, hairy, margin entire, apex acute or obtuse, base attenuate. The inflorescence racemose or rarely panicle, 10–30 flowered, bract absent, all flowers pedicellate, pedicel (5.0–) 6.0–12.5 (−15.0) mm, glabrous or hairy; sepals 4, free, 4.0–6.75 × 1.0–2.5 mm, obovate or oblong, yellow, yellowish-green or rarely maroon, apex obtuse, margin entire, near the tip hispid loosely hispid; petals 4, free, bright yellow, 7.5–11.5 × 2.0–3.5 mm, spatulate, blade obovate, ovate or partly orbicular, glabrous; stamens 6, 5.0–12 mm; anthers pale yellow, 2.0–3.0 × 0.7–1.2 mm, ovate-oblong; filaments 4.0–9.0 × 0.30–0.80 mm, cylindrical, pale yellow, glabrous; pistil 1, 2-carpellate, 5.5–9.5 mm long, yellowish green, glabrous; stigma capitate; style very short, about 1–1.5 mm long; carpophore cylindrical and 2.0–5.75 mm, glabrous. The fruit is siliqua, 20–45 × 2.0–3.75 mm, 1-veined, beak absent, sometimes torulose, spreading in the early period but recurved at maturity, glabrous; seeds 1.3–2.0 × 1.0–1.5 mm, ovoid or rarely globose, uniseriate, pale brown and sometimes slightly black the parts near the base (Başköse et al. 2023).

2.1.19 *Brassica incana* Ten.

Synonyms *B. oleracea* subvar. *incana* (Ten.) Coss.; *B. oleracea* var. *incana* (Ten.) Paol. (www.powo.science.kew.org).

Description Suffrutescent perennials, usually 80–150 cm high, in cultivation often reaching 2–3 m, usually with a conspicuous main stem branching only apically to form the inflorescence; young plant first developing a more or less dense rosette of large leaves, later prolonged and after 1–5 years forming a large inflorescence, often perennating for several flowering periods, in cultivation usually withering in fruit. Stem strongly lignified with conspicuous year rings, often 10–25 mm thick in the basal part, in younger parts with conspicuous leaf scars, sometimes remaining visible on older parts but usually disappearing. Indumentum dense or very dense and usually soft, densest and with the longest hairs on veins and rachis, hairs simple or rarely a few branched, long hairs straight, short hairs straight or apically curved. The leaves comparatively thin, herbaceous, losing their texture during drying, lobes rounded, lower and middle cauline leaves including petiole 18–30(−40) × 8–15(−26) cm or when cultivated under favourable conditions up to 50 × 25 cm, petiole usually 60–150 mm, lamina usually pinnatifid or sometimes pinnatisect with several pairs of lobes on the petiole and a usually broad semi-amplexicaul wing on the leaf foot, sometimes entire petiole winged, wings less developed in the upper cauline leaves; margin usually with minute teeth, side veins conspicuous, usually reaching the margin in the largest teeth, branching and anastomosing. Upper cauline leaves gradually smaller and less lobed, usually some very broad-winged, of 10 simple but conspicuously serrate, semi-amplexicaul, bracts of inflorescence branches entire, usually oblong. First inflorescence usually very large, repeatedly branched, under favourable conditions sometimes 1 m long and wide with thousands of flowers, if growth continues for more seasons later inflorescence of 10 also rather large; partial racemes often 50–100-flowered, usually with densely placed buds, flowers thus open close to the top, in some populations buds are on a more elongated axis. The pedicels (12–)14–25(−35) mm, at anthesis erecto-patent to patent; flowers usually protandrous, anthers ripening 1–2 days before the stigma, in at least one Greek population stigma ripening almost simultaneously with the anthers; sepals 10–16 mm, at anthesis erect or diverging at an angle of up to 50–60°, the outer 3–5 mm broad, boat-shaped, saccate, green to yellowish green, the inner 2–3.5 mm broad, flatter, narrower, cucullate especially in bud; the petals at anthesis usually patent, rarely deflexed in late male phase, bright yellow or in some Italian and at least one Greek population whitish yellow, in one E Sicilian population white from an ochroleucous start; claw 7.5–13 mm, limb 12–14(−21) × 8–12(−14) mm. The stamens exserted (2–)3–5 mm above the petal disc; outer filaments 8–11 mm, inner 10–13 mm, ratio inner/outer filament (1.05–)1.1–1.2; anthers (3–)4–4.5(−5) mm, yellow, connective tip inconspicuous or usually 0.1–0.15 mm, pale or violet; pistil exserted above anthers at maturity, ripe ovary 6–9(−13) mm, style 2.5–4.5(−5.5) mm, stigma 1–1.5 mm broad; fruiting pedicels 12–25(−35) mm, patent or up to c. 60° erecto-patent.

Siliqua except rostrum (35–)60–90(−100) mm long, 3–5 mm broad, 4–7 mm thick, well-filled with (15–)20–30(−40) seeds. The rostrum (7–)10–15(−20) mm, if seed-bearing up to 5 mm broad, if empty usually 1.5–2 mm broad, usually seedless, sometimes (e.g. in some Greek material) with 1 seed; seeds generally 1.9–2.6(−2.8) mm, with usually inconspicuous reticulum (Snogerup et al. 1990).

Brassica incana also includes three subspecies (www.powo.science.kew.org).

(a) *B. incana* subsp. *hirta* Raimondo & Spadaro: **Synonyms:** None reported
(b) *B. incana* subsp. *incana*: **Synonyms:** *B. botterii* Vis.; *B. cazzae* Ginzb. & Teyber; *B. mollis* Vis.; *B. oleracea* var. *pilosa* Caruel; *Erucaria persica* W.Baxter; *Hirschfeldia inflexa* C.Presl; *H. integrifolia* C.Presl; *Sinapis inflexa* (C.Presl) Walp.; *S. integrifolia* Vahl ex Walp.; *S. panormitana* Spreng. (www.powo.science.kew.org)
(c) *B. incana* subsp. *raimondoi* (Sciandr., C.Brullo, Brullo, Giusso, Miniss. & Salmeri) Raimondo & Spadaro
Synonyms: *B. raimondoi* Sciandr., C.Brullo, Brullo, Giusso, Miniss. & Salmeri (www.powo.science.kew.org)

2.1.20 Brassica insularis Moris

Synonyms *B. oleracea* subsp. *insularis* (Moris) Rouy & Foucaud; *B. oleracea* subvar. *insularis* (Moris) Coss.; *B. amblyphylla* Jord.; *B. atlantica* (Coss.) O.E.Schulz; *B. atlantica* Coss. & Durieu ex Munby; *B. calcarea* Jord.; *B. conferta* Jord.; *B. corsica* Jord.; *B. cretica* subsp. *atlantica* (Coss.) Onno; *B. erigens* Jord.; *B. flexicaulis* Jord.; *B. hololeuca* Jord.; *B. insularis* var. *angustiloba* (Briq.) O.E.Schulz; *B. insularis* var. *aquellae* Widler & Boquet; B. *insularis* var. *atlantica* (Coss.) Batt.; *B. insularis* var. *conferta* (Jord.) O.E.Schulz; *B. insularis* var. *latiloba* (Briq.) O.E.Schulz; *B. luteola* Jord.; *B. oleracea* subvar. *angustiloba* Briq.; *B. oleracea* subvar. *atlantica* Coss.; *B. oleracea* var. *atlantica* (Coss.) Bonnet & Barratte; *B. oleracea* proles *atlantica* (Coss.) Rouy & Foucaud; *B. oleracea* var. *corsica* Briq.; *B. oleracea* subvar. *latiloba* Briq.; *B. praeruptorum* Jord.; *B. recurva* Jord.; *B. revelieri* Jord. (www.powo.science.kew.org).

Description Suffrutescent perennials, height usually 50–100 cm, in cultivation reaching at least 150 cm, old specimens richly branched; young plants forming a more or less dense rosette of large leaves, stem ending in an apical inflorescence after one to several years, growth continued by branches, often as thick as the main stem; usually flowering several times and becoming at least 5 years old in nature. Stem strongly lignified with easily visible year rings, often 10–20 mm thick in the basal part; leaf scars are conspicuous on young stem parts, disappearing on older. The leaves are usually glabrous some marginal hairs observed in one Corsican population, semi-coriaceous, keeping their texture after drying, as young sometimes slightly greyish, leaf margin with acute teeth; basal and cauline leaves including petiole 15–30(−35) × 7–15(−20) cm, petiole (3–)6–12(−20) cm, lamina orbicular to obovate, ovate or lanceolate, in main part undivided, only basally pinnatisect to pinnatifid, petiole usually with 1–2(−4) pairs of small free lobes, foot usually without, or in some Tunisian material with, small semi-amplexicaul auricles; midrib strong, side veins prominent, reaching the margin in the largest teeth, veins of higher order anastomosing, forming a network; upper leaves gradually smaller and simpler, uppermost usually oblong, often with a broad amplexicaul base. First inflorescence varying in size among populations, later inflorescences smaller, partial racemes usually many-flowered and dense with 50–100 flowers, buds usually densely placed and flowers thus opening close to the top. The flowers are conspicuously fragrant, usually protandrous with anthers ripening 1–2 days before the stigma, pedicels (8–)12–24(−30) mm, at anthesis erecto-patent to paten; the sepals 9–13 mm long, sometimes protruding above the petals, erect or diverging at an angle of up to 450, often conspicuously whitish (or in one Corsican population yellowish), the outer (2.5–)3.5–4 mm broad, boat-shaped, saccate at base, often slightly cucullate, the inner 2–3 mm broad, flatter, cucullate especially in bud. The petals at anthesis usually patent, rarely deflexed up to 450 in male phase, in one population slightly erecto-patent, white, though slightly greenish in bud, in the population at Punta Corbaghiola, Corsica light yellow; claw 7–9(−11) mm, limb (11–)14–18(−20) × (9–)10–13(−15) mm. The stamens are exserted (2–)3–6 mm above the petal disc, outer filaments (8–)9–10(−11) mm, inner (9–)9.5–11 mm, usually longer, ratio inner/outer filament 1.0–1.2; anthers (3–)3.5–4.5(−5) mm, outer often slightly longer, yellow, connective tip usually 0.1–0.2 mm, pale white; pistil exserted above anthers at maturity, ripe ovary 6–10 mm, style length very variable, usually 2–7 mm, stigma 0.8–1.0 mm broad. The fruiting pedicels (12–)20–30(−40) mm, patent or erecto-patent to c. 700; siliqua except rostrum (35–)45–65(−75) mm long, 4–6 mm broad, usually slightly thicker, if well-filled usually 15–35-seeded, seeds in one row in each loculus, septum zig-zag bent with the seeds, midrib and side veins of valves usually conspicuous in dry fruits, valves often reticulate. Rostrum (5–)10–20(−23) mm long, in Sardinian material usually seedless or 1-seeded, in Tunisian material usually with 1–2(−3) seeds, if seed bearing 4–6 mm broad, if empty only 3–3.5 mm. The seed size is 2.5–3.5 mm with a conspicuous reticulum (Snogerup et al. 1990).

2.1.21 Brassica loncholoma Pomel

Synonyms *Eruca loncholoma* (Pomel) O.E.Schulz; *Guenthera loncholoma* (Pomel) Gómez-Campo; *Brassica aurasiaca* Coss. & Kralik (www.powo.science.kew.org).

Description Short, pauciflorous raceme; a stout fruiting peduncle 4 mm long; silique with a thickened robust podogyne projecting from the base of the valves, slightly erect, linear, sublanceolate, compressed, with a small, slightly attenuated style, as long as the width of the valves. The valves are keeled by the prominent dorsal vein, with rugulose veins on the sides, slightly bumpy, attenuated at the base, 15 mm long by 4 mm wide; seeds subuniseriate, globose ovoid, matt leaves are fasciculate at the apex of the stump divisions, bristling with sparse hairs, attenuated into a long, serrate petiole, linear lanceolate, pinnatidentate, with acute, porrate teeth. Stems scapiform, naked, erect, 40 cm long; stalk with short, ceptile divisions; covered with old petiolar fibres (Pomel 1874).

2.1.22 Brassica macrocarpa Guss.

Synonyms *Eruca macrocapa* (Guss.) Caruel; *Brassica macrocarpa* f. *glabra* Lojac.; *B. macrocarpa* var. *villosissima* Damanti ex O.E.Schulz; *B. macrocarpa* f. *villosoincana* Lojac. (www.powo.science.kew.org).

Description Suffrutescent perennials, usually 1–1.5 m high, in cultivation often reaching 2–3 m or in flower up to 4 m, usually with a conspicuous main stem branching only apically to form a large inflorescence; young plant first forming a rosette of large leaves, soon prolonged and after 1–5 years flowering, in nature often perennating for more flowering seasons, in cultivation usually withering in fruit. Stem strongly lignified with conspicuously ear rings, often 15–25 mm thick in the basal part; leaf scars conspicuous, often remaining on older stem parts. Plant usually glabrous, rarely sparsely hairy, especially as young and on veins. The leaves are comparatively thin, herbaceous, losing their texture during drying; basal and cauline leaves including petiole (12–)20–35(−40) × (6–)10–13(−18) cm, petiole (6–)10–15(−20) cm, lamina ovate to broadly lanceolate, undivided or usually basally pinnatifid to pinnatisect, often 1–3 pairs of more or less free lobes on upper part of petiole; petiole not winged and without any auricles on leaf foot; margin with acute teeth, midrib prominent, side veins also conspicuous, reaching the margin in the largest teeth; upper leaves gradually smaller and more undivided, variable in shape, some of the upper most usually with semi-amplexicaul base, bracts of inflorescence branches often linear or almost so. First inflorescence usually very large, repeatedly branched, under good conditions about 1 m long and wide with thousands of flowers, if growth continues for more seasons, later inflorescences sometimes almost as large but often simpler, partial racemes many-flowered, bud usually densely placed, flowers thus open close to the top. Pedicels (8–)12–20(−35) mm, at anthesis usually erecto-patent. Flowers are usually conspicuously protandrous, anthers ripening 1–2 days before the stigma; sepals (8–)10–12(−14) mm, slightly distant from the petal claws, the outer 3–4 mm broad, boat-shaped, saccate at base, often slightly cucullate, the inner 2–3(−3.5) mm broad, flatter, distinctly cucullate especially in bud; petals light yellow; claw 7–9(−1 1) mm, limb (10–)13–15(−17) × (8–)9–11 mm; the stamens are exserted 1–4.5 mm above the petal disc; outer filaments 7–9(−11) mm, inner 8–11 mm, usually longer; anthers (2–)3–4(−4.5) mm, yellow, connective tip inconspicuous or c. 0.1 mm, white; pistil is exserted above anthers at maturity, ripe ovary 4–6(−8.5) mm, style 2–4.5(−5.5) mm, stigma c. 1 mm broad; fruiting pedicels are (12–)15–25(−40) mm, the lowest in large racemes sometimes almost erect; the siliqua except rostrum 25–35(−45) mm long, 8–10(−12) mm broad with 12–20(−25) seeds, seeds in a more or less conspicuous 2-row arrangement in each loculus, septum flat, midrib of valves a thin line, no side veins visible on ripe siliqua, surface almost smooth, reticulately rugose; the rostrum is (6)10–15(−20) mm long, if seed-bearing 5–8 mm broad, if empty 2–3 mm, usually 1-seeded, rarely empty or with 2 seeds; seeds 2.1–2.8 mm, with a network of c. 0.1 mm wide meshes formed by low ridges (Snogerup et al. 1990).

2.1.23 Brassica maurorum Durieu

Synonyms *Diplotaxis maurorum* (Durieu) M.B.Crespo & M.Fabregat (www.powo.science.kew.org).

Description Annual, erect, branched, glaucous, stem fistulous or glabrous; leaves radical and lower stem partly setulose, lyrate, lobes few, approximate, subinteger, ovate-lanceolate, terminal at the very top, subrotund or ovate, upper leaves glabrous, lanceolate-oblong, last few linear; racemes elongate, pedicels very patent; sepals inwardly bisetose, exteriorly muticous; petals calyces twice as long, limb late obovate; each ascending, compressed, pedicel scarcely longer, beak ancipitous, acute, three-nerved, valves half long, monospermous; seeds subbiseriate, small, globose, punctate, red; flowers yellow (Schulz 1919).

2.1.24 Brassica montana Pourr.

Synonyms *B. allionii* Moretti; *B. balearica* Loisel.; *B. montana* var. *apenninica* (Cavara) O.E.Schulz; *B. montana* f. *ligustica* Burnat; *B. oleracea* var. *leronensis* (Burnat) Rouy & Foucaud; *B. oleracea* subvar. *ligustica* Burnat; *B. oleracea* var. *ligustica* (Burnat) Rouy & Foucaud; *B. oleracea* subsp. pourretii Rouy & Foucaud; *B. oleracea* var. *pourretii* (Rouy & Foucaud) O.Bolòs & Vigo; *B. oleracea* subsp. *robertiana* (J.Gay) Rouy & Foucaud; *B. oleracea* subvar. *robertiana* (J.Gay) Coss.; *B. pourretii* (Rouy & Foucaud) Lév.; *B. robertiana* J.Gay; *B. robertiana* var. *apenninica* Cavara; *B. robertiana* var. *leronensis* Burnat (www.powo.science.kew.org).

Description Suffrutescent perennials, usually 80–150 cm high, usually branched from near the base, rarely with a tall and dominating stem; young plant first developing a rosette or dense sequence of large leaves, after 12 to several years starting flowering with an apical inflorescence followed by lower branches; rarely, especially in disturbed spots and cultivation, rapidly developing an up to 100(−120) cm tall main stem before flowering; usually perennating for several flowering seasons and reaching an age of at least 6 years in nature. The stem is strongly lignified, up to 20 mm thick with easily visible year-rings; leaf scars first prominent, later disappearing; leaves are glabrous, rather thick, usually keeping their texture during drying; lower and middle cauline leaves including petiole 5–15(−20) cm, petiole usually much shorter than lamina, lamina obovate to broadly lanceolate, in basal part pinnatisect with 1(−3) pairs of almost free lobes, petiole usually with some pairs of free large lobes and often at least one pair of very small ones; leaf foot usually without any wings or auricles, rarely with a narrow semi-amplexicaul wing; margin usually with blunt or mucronulate, rarely acute teeth, often somewhat undulate; midrib thick, side veins conspicuous, a nastomosing, usually not reaching the margin, and if so often not in the largest teeth, only some Alpi Apuanae populations with a venation similar to that of *B. rupestris*; upper leaves gradually smaller and simpler, variable, often more sharply dentate, at least some of the upper most with a semi-amplexicaul base, bracts of inflorescence branches oblong-linear. First inflorescence is apical, large but with few branches, followed by lower secondary branches; partial racemes usually large, 50–100-flowered, with several buds overtopping the flowers; pedicels 10–20(−30) mm, at anthesis usually erecto-patent to 45–60°, rarely patent. The flowers are often slightly scented, usually conspicuously protandrous, anthers ripening 1–2 days before the stigma. Sepals (8.5–)9–12(−16.5) mm, usually erect though slightly distant from the petals, green or yellowish green, the outer (2.5–)3–4(−4.5) mm broad, boat-shaped, more or less saccate at base, the inner 2–3(−3.5) mm broad, flatter, cucullate especially in bud; petals whitish yellow to light yellow; claw (6.5–)8–10(−12) mm, limb 11–15(−19) × (7–)9–11(−12) mm; stamens exserted (3–)4–6 mm above the petal disc, outer filaments (7–)8–11(−11.5) mm, inner (8–)9–12(−13) mm, conspicuously longer, or filaments rarely equal and inner stamens then higher because of higher insertion; anthers (2.5–)3–5(−5.5) mm, the outer often slightly longer, yellow, connective tip in conspicuous or up to 0.15 mm, whitish. Pistil exserted above anthers at maturity, ripe ovary 9–11(−15) mm, style 3–6 mm, stigma (1–)1.2–1.5 mm broad. The fruiting pedicels (5–)10–30 mm, patent or slightly erecto-patent; siliqua except rostrum (55–)60–90(−150) mm long, (2–)3–5 mm broad, with (14–)18–25(−30) seeds, often conspicuously constricted between the seeds, seeds in one row in each loculus, septum zig-zag bent with the seeds, midrib of valves usually inconspicuous. The rostrum is 9–15(−18) mm, if seed bearing 4 mm broad, if empty 2–4 mm, seedless or one-seeded, often swollen even as seedless; seeds are 1.6–2.5(−3.1) mm, spherical, with a usually rather inconspicuous network of 0.05–0.1 mm broad meshes (Snogerup et al. 1990).

2.1.25 *Brassica napus* L.

Synonyms *B. macrorhiza* Gray; *B. oleifera* Moench; *B. rapa* subsp. *napus* (L.) Briq.; *B. sativa* subsp. *napus* (L.) Clavaud; *Crucifera napus* (L.) E.H.L.Krause; *Napus oleifera* K.F.Schimp. & Spenn.; *Rapa napus* (L.) Mill.; *Raphanus napus* (L.) Crantz; *Sinapis napus* (L.) Brot.; *B. campestris* var. *napus* (L.) Bab.; *B. campestris* subsp. *napus* (L.) Hook.f.; *B. campestris* var. *oleifera* (K.F.Schimp. & Spenn.) Prain; *B. campestris* var. *akana* Makino; *B. campestris* subvar. *alba* DC.; *B. campestris* subsp. *annua* Schübl. & G.Martens; *B. campestris* subsp. *biennis* Schübl. & G.Martens; *B. campestris* var. *biennis* (Schübl. & G.Martens) Rchb.; *B. campestris* var. *bullatopetsai* Makino; *B. campestris* var. *communis* DC.; *B. campestris* var. *dentatopetsai* Makino; *B. campestris* var. *dichotoma* Duthie & Fuller; *B. campestris* f. *glauca* (Roxb.) Prain; *B. campestris* var. *glauca* (Roxb.) Duthie & Fuller; *B. campestris* f. *luteoalba* Makino; *B. campestris* stirps *napobrassica* (L.) DC.; *B. campestris* var. *napobrassica* (L.) Prain; *B. campestris* subsp. *napobrassica* (L.) Schübl. & G.Martens; *B. campestris* var. *nippo-oleifera* Makino; *B. campestris* stirps *pabularia* DC.; *B. campestris* subvar. *purpurascens* DC.; *B. campestris* f. *quadrivalvis* (Hook. f. & Thomson) Prain; *B. campestris* var. *quadrivalvis* (Hook.f. & Thomson) Duthie & Fuller; *B. campestris* subsp. *rapifera* (Metzg.) Sinskaya; *B. campestris* var. *rutabaga* DC.; *B. campestris* f. *rutabaga* DC.; *B. campestris* f. *simplex* Prain; *B. campestris* f. *spontanea* Makino; *B. campestris* var. *toria* Duthie & Fuller; *B. campestris* subvar. *ulti* Prain; *B. gongylodes* Mill.; *B. macrorhiza* var. *sativa* Gray; *B. napobrassica* Mill.; *B. napobrassica* convar. *sibirica* Scheb.; *B. napus* var. *oleifera* Delile; *B. napus* f. *typica* Posp.; *B. napus* var. *aestiva* Endl.; *B. napus* var. *alba* DC.; *B. napus* var. *annua* W.D.J.Koch; *B. napus* f. *annua* (Schübl. & G. Martens) Thell.; *B. napus* var. *arabica* O.E.Schulz; *B. napus* var. *arvensis* (Duchesne) Thell.; *B. napus* var. *biennis* (Schübl. & G.Martens) Rchb.; *B. napus* f. *biennis* (Schübl. & G.Martens) Thell.; *B. napus* var. *brevirostris* Borbás; *B. napus* f. *dissecta* Peterm.; *B. napus* f. *dolichocarpa* O.E.Schulz; *B. napus* var. *edulis* Delile; *B. napus* var. *esculenta* DC.; *B. napus* stirps *esculenta* DC.; *B. napus* var. *flava* DC.; *B. napus* var. *glauca* (Roxb.) O.E.Schulz; *B. napus* var. *hiemalis* Döll; *B. napus* var. *italica* Alef.; *B. napus* var. *leptorhiza* Spach; *B. napus* f. *micrantha* O.E.Schulz; *B. napus* subsp. *napobrassica* (L.) O.Schwarz; *B. napus* var. *nigricans* DC.; *B. napus* convar. *oleifera* (DC.) Alef.; *B. napus* stirps *oleifera* DC.; *B. napus* subsp. *oleifera* (DC.) Metzg.; *B. napus* convar. *pabularia* (DC.) Holub; *B. napus* subsp. *pabularia* (DC.) Janch.; *B. napus* var. *pabularia* (DC.) Rchb.; *B. napus* var. *quadrivalvis* (Hook.f. & Thomson) O.E.Schulz; *B. napus* subsp. *rapifera* Metzg.; *B. napus* convar. *rapifera* (Metzg.) Alef.; *B. napus* var. *rapifera* Metzg.; *B. napus* var. *rossica* Alef.; *B. napus* f. *rubescens* Peterm.; *B. napus* var. *rutila* Alef.; *B. napus* var. *sahariensis* A.Chev.; *B. napus* var. *sarcorhiza* Spach; *B. napus* var. *schebalinae* Bondartseva & Pivov.; *B. napus* convar. *sibirica* (Scheb.) Bondartseva & Pivov.; *B. napus* var. *syntomocarpa* O.E.Schulz; *B. napus* var. *trimestris* Boenn.; *B. napus* var. *ulti* (Prain) O.E.Schulz; *B. oleracea* proles *arvensis* Duchesne; *B. oleracea* var. *hongnoensis* H.Lév.; *B. oleracea* var. *napobrassica* L.; *B. oleracea* proles *napobras-*

sica (L.) Duchesne; *B. oleracea* f. *praecox* Spreng.; *B. oleracea* var. *suecica* Gray; *B. praecox* Schult.; *B. praecox* Kit. ex Hornem.; *B. quadrivalvis* Hook. f. & Thomson; *B. rapa* subsp. *rapifera* Metzg.; *B. rapa* convar. *rapifera* (Metzg.) Alef.; *B. rutabaga* (DC.) A.Vilm.; *B. stricta* Nestl. ex DC.; *B. napus sylvestris* A.Vilm.; *S. glauca* Roxb. (www.powo.science.kew.org).

Description Annuals or biennials; (taproot slender or swollen); (glaucous), glabrous, glabrescent, or pubescent, (trichomes coarse); stems branched distally, 3–13 dm; basal leaves (rosulate when biennial); petiole (often winged), to 15 cm; blade lyrate-pinnatifid, ± pinnately lobed, 5–25(−40) cm × 20–70(−100) mm, lobes 0–6 each side, (smaller than terminal), surfaces (glaucous), glabrous or sparsely hairy when immature, glabrescent, or, rarely, pubescent; cauline leaves (middle and distal) sessile; blade base auriculate or amplexicaul, (margins entire). The racemes are not paniculately branched, (buds overtopping or equal to open flowers); fruiting pedicels spreading to ascending (slender), 1–3 cm. Flowers: sepals (5–)6–10 × 1.5–2.5 mm; petals golden or creamy to pale yellow, broadly obovate, 10–16 × (5–)6–9(−10) mm, claw 5–9 mm, apex rounded; filaments (5–)7–10 mm; anthers 1.5–2.5 mm; fruits spreading to ascending, smooth or slightly torulose, terete, (3.5–)5–10(−11) cm × (2.5–)3.5–5 mm; valvular segment with 12–20(−30) seeds per locule, (3–)4–8.5(−9.5) cm, terminal segment usually seedless, rarely 1 or 2-seeded (attenuate-conic, thin), (5–)9–16 mm. The seeds are dark brown to black, light brown, or reddish in colour, 1.8–2.7(−3) mm diam.; seed coat finely reticulate-alveolate, not mucilaginous when wetted (Schulz 1919).

2.1.26 Brassica nivalis Boiss. & Heldr.

Synonyms *Brassicella nivalis* (Boiss. & Heldr.) O.E.Schulz; *Coincya nivalis* (Boiss. & Heldr.) Greuter & Burdet; *Guenthera nivalis* (Boiss. & Heldr.) Gómez-Campo; *Hutera nivalis* (Boiss. & Heldr.) Gómez-Campo; *Rhynchosinapis nivalis* (Boiss. & Heldr.) Heywood (www.powo.science.kew.org).

Description Stems pale green, leaves with setose margins, sublyrate-pinnate, very obtuse segments, thin stems, subsimple, subphyllous, few-flowered racemes, medium-sized yellow flowers, (junior) linear, torulose pods, descending into a thin stipule (Boissier 1867).

B. nivalis also includes two subspecies (www.powo.science.kew.org).

(a) B. nivalis **subsp.** jordanoffii **(O.E.Schulz) Akeroyd & Leadlay**
Synonyms *B. jordanoffii* O.E.Schulz; *Guenthera nivalis* subsp. *jordanoffii* (O.E.Schulz) Gómez-Campo (www.powo.science.kew.org).

(b) B. nivalis **subsp.** nivalis
Synonyms None reported.

2.1.27 *Brassica oleracea* L.

Synonyms *Crucifera brassica* E.H.L.Krause, *Napus oleracea* (L.) K.F.Schimp. & Spenn., *Raphanus brassica* Crantz, *Brassica alba* Boiss., *B. alboglabra* L.H.Bailey, *B. arborea* Steud., *B. bullata* Pasq., *B. campestris* subsp. *sylvestris* (DC.) Janch., *B. capitala* DC. ex H.Lév., *B. cauliflora* Garsault, *B. caulorapa* (DC.) Pasq., *B. cephala* DC. ex H.Lév., *B. cretica* convar. *italica* (Plenck) G.H.Loos, *B. fimbriata* Steud., *B. gemmifera* H.Lév., *B. laciniata* Steud., *B. maritima* DC. ex Tardent, *B. millecapitata* H.Lév., *B. muscovita* Steud., *B. odorata* Schrank ex Steud., *B. oleracea* var. *aberdeenea* Alef., *B. oleracea* var. *acephala* DC., *B. oleracea* convar. *acephala* (DC.) Alef., *B. oleracea* subsp. *acephala* (DC.) Metzg., *B. oleracea* stirps *acephala* DC., *B. oleracea* var. *asparagoides* DC., *B. oleracea* var. *atrorubens* Alef., *B. oleracea* var. *aurata* (DC.) Alef., *B. oleracea* subvar. *aurata* DC., *B. oleracea* f. *azurea* Uspenski, *B. oleracea* var. *bardovicensis* Alef., *B. oleracea* subvar. *brumalis* Duchesne, *B. oleracea* var. *bullata* (Pasq.) DC., *B. oleracea* stirps *bullata* DC., *B. oleracea* subsp. *bullata* (DC.) Metzg., *B. oleracea* convar. *bullata* (DC.) Alef., *B. oleracea* convar. *capitata* (L.) Alef., *B. oleracea* proles *capitata* (L.) Duchesne, *B. oleracea* var. *capitata* L., *B. oleracea* subsp. *capitata* (L.) Schübl. & G.Martens, *B. oleracea* stirps *capitata* (L.) DC., *B. oleracea* var. *cauliflora* DC., *B. oleracea* convar. *caulorapa* (DC.) Alef., *B. oleracea* subsp. *caulorapa* (DC.) Metzg., *B. oleracea* stirps *caulorapa* DC., *B. oleracea* var. *chamaegongyla* Alef., *B. oleracea* var. *chamjodusa* Alef., *B. oleracea* var. *chlorusa* Alef., *B. oleracea* subvar. *communis* DC., *B. oleracea* var. *communis* DC., *B. oleracea* subvar. *compressa* Duchesne, *B. oleracea* var. *conica* (Duchesne) DC., *B. oleracea* subvar. *conica* Duchesne, *B. oleracea* convar. *costata* (DC.) Gladis ex Diederichsen, *B. oleracea* var. *costata* DC., *B. oleracea* subvar. *crassa* Duchesne, *B. oleracea* var. *crispa* Vill., *B. oleracea* subvar. *crispa* Duchesne, *B. oleracea* subvar. *crispa* Duchesne, *B. oleracea* var. *culinarum* Alef., *B. oleracea* subvar. *cymosa* Duchesne, *B. oleracea* subvar. *cymosa* Duchesne, *B. oleracea* subvar. *cymosa-albida* Duchesne, *B. oleracea* subvar. *cymosa-violacea* Duchesne, *B. oleracea* var. *depressa* DC., *B. oleracea* var. *dinojodusa* Alef., *B. oleracea* f. *dissecta* Peterm., *B. oleracea* var. *elliptica* DC., *B. oleracea* var. *erythrobotrys* Alef., *B. oleracea* f. *exaltata* (Rchb.) Thell., *B. oleracea* var. *fimbriata* Mill., *B. oleracea* var. *flandrica* Alef., *B. oleracea* var. *frutescens* Vis., *B. oleracea* subsp. *fruticosa* Metzg., *B. oleracea* convar. *fruticosa* (Metzg.) Alef., *B. oleracea* convar. *gemmifera* (DC.) Gladis ex Diederichsen, *B. oleracea* var. *gemmifera* DC., *B. oleracea* subsp. *gemmifera* (DC.) O.E.Schulz, *B. oleracea* var. *germanorum* Alef., *B. oleracea* var. *glomerosa* Gray, *B. oleracea* var. *gongylodes* L., *B. oleracea* subsp. *gongylodes* (L.) Schübl. & G.Martens, *B. oleracea* proles *gongylodes* (L.) Duchesne, *B. oleracea* var. *grangei* Alef., *B. oleracea* var. *herbacea* Spach, *B. oleracea* subvar. *humilis* DC., *B. oleracea* var. *italica* Plenck, *B. oleracea* var. *jodocephala* Alef., *B. oleracea* var. *jodogongyla* Alef., *B. oleracea* var. *kashmiriana* Naqshi & Javeid, *B. oleracea* subvar. *laciniata* (L.) O.E.Schulz, *B. oleracea* var. *laciniata* L., *B. oleracea* subsp. *laevis* Metzg., *B. oleracea* var. *longa* Alef., *B. oleracea* var. *loudonii* Alef., *B. oleracea* var. *luteola* Alef., *B. oleracea* var. *mac-*

rocephala Alef., *B. oleracea* var. *microcephala* Alef., *B. oleracea* var. *macrogongyla* Alef., *B. oleracea* var. *magniportana* Alef., *B. oleracea* var. *major* (Duchesne) DC., *B. oleracea* subvar. *major* Duchesne, *B. oleracea* var. *maritima* Coss., *B. oleracea* subvar. *maxima* Duchesne, *B. oleracea* var. *media* Alef., *B. oleracea* var. *medullosa* Thell., *B. oleracea* var. *micropodium* Kuntze, *B. oleracea* subvar. *minor* Duchesne, *B. oleracea* subvar. *nepenthiformis* DC., *B. oleracea* var. *oblonga* DC., *B. oleracea* var. *obovata* DC., *B. oleracea* var. *opsibotrys* Alef., *B. oleracea* var. *opsigongyla* Alef., *B. oleracea* var. *opsjodusa* Alef., *B. oleracea* var. *palmifolia* DC., *B. oleracea* subvar. *parva-praecox* Duchesne, *B. oleracea* var. *pentovillea* Alef., *B. oleracea* var. *percrispa* Alef., *B. oleracea* var. *phaeusa* Alef., *B. oleracea* subvar. *pinnata* DC., *B. oleracea* var. *plumacea* Alef., *B. oleracea* subvar. *praecox* DC., *B. oleracea* var. *praecox* Alef., *B. oleracea* var. *procera* Alef., *B. oleracea* subvar. *procerior* Duchesne, *B. oleracea* var. *protobotrys* Alef., *B. oleracea* var. *protojodusa* Alef., *B. oleracea* subvar. *purpurascens* DC., *B. oleracea* subvar. *purpurascens* DC., *B. oleracea* f. *pyramidalis* Mill., *B. oleracea* var. *pyramidula* Alef., *B. oleracea* var. *quercifolia* DC., *B. oleracea* var. *ramosa* DC., *B. oleracea* subvar. *rubra* (L.) Duchesne, *B. oleracea* f. *rubra* (L.) Peterm., *B. oleracea* var. *rubra* L., *B. oleracea* subvar. *rubra-minor* Duchesne, *B. oleracea* var. *sabauda* L., *B. oleracea* subvar. *sabauda* (L.) O.E.Schulz, *B. oleracea* var. *sabellica* L., *B. oleracea* subvar. *sabellica* (L.) O.E.Schulz, *B. oleracea* f. *salinaria* Kitt., *B. oleracea* var. *schizogongyla* Alef., *B. oleracea* var. *scotica* Alef., *B. oleracea* var. *selenisia* L., *B. oleracea* f. *selenisia* (L.) O.E.Schulz, *B. oleracea* var. *selinoides* Gray, *B. oleracea* var. *semiclausa* Alef., *B. oleracea* var. *serotissima* Alef., *B. oleracea* var. *sibirica* Alef., *B. oleracea* var. *sphaerica* DC., *B. oleracea* subvar. *sphaerica* (DC.) G. Martens, *B. oleracea* subvar. *subacuta* Duchesne, *B. oleracea* var. *sublaciniata* Alef., *B. oleracea* var. *subrubens* Gray, *B. oleracea* var. *sylvestris* L., *B. oleracea* stirps *sylvestris* DC., *B. oleracea* var. *theiusa* Alef., *B. oleracea* subvar. *turionensis* DC., *B. oleracea* var. *ulmensis* Alef., *B. oleracea* var. *veris* Alef., *B. oleracea* subvar. *versicolor* DC., *B. oleracea* subvar. *violacea* Duchesne, *B. oleracea* f. *violascens* G.Martens, *B. oleracea* subvar. *viridis* Duchesne, *B. oleracea* f. *viridis* Duchesne, *B. oleracea* var. viridis L., *Brassica oleracea* proles *viridis* (L.) Duchesne, *B. oleracea* subvar. *viridis* (L.) DC., *B. oleracea* var. *vulgaris* DC., *B. oleracea* subvar. *vulgaris* Duchesne, *B. oleracea* var. *vulgaris* DC., *B. peregrina* Steud., *B. quercifolia* DC. ex H.Lév., *B. rubra* Steud., *B. rupestris* var. *gongylodes* (L.) Janch., *B. rupestris* subsp. *gongylodes* (L.) Janch., *B. sabellica* Pers., *B. suttoniana* H.Lév., *B. sylvestris* (L.) Mill., *Rapa rotunda* Mill., *Raphanus brassica-officinalis* Crantz, *R. erucoides* L.f., *Sinapis brassicata* L. (www.powo.science.kew.org).

Description Suffrutescent perennial, usually 50–150 cm tall, usually developing a tall main stem before flowering; several populations dependent on rather strong vernalisation for flowering, length of vegetative phase therefore often dependent on weather; young plant first forming a basal rosette of large leaves, later prolonged, but leaves sometimes again aggregate apically in a head-like manner; usually starting flowering on the main stem, but some individuals with side inflorescences only, habit thus often much variable in the same population; usually perennating for several flowering seasons, in nature sometimes reaching an age of at least 10 years. The stem is lignified with easily visible year-rings, up to 30(−50) mm thick; leaf scars prominent, often remaining on older stems; the leaves are usually glabrous, only in a few populations young plants slightly hairy, rather thick, with a conspicuously greyish surface; basal and cauline leaves including petiole (15–)20–35(−45) × 7–15(−25) cm, petiole 5–20 cm, lamina circular to obovate or ovate or rarely lanceolate, undivided or pinnatifid, petiole often with 1–3(−5) more or less distinct leaflets, usually winged at the base; margin serrate with low, blunt teeth, midrib broad, side veins also conspicuous, soon anastomosing, forming a network, often reaching the margin in the teeth; upper leaves gradually smaller, usually undivided and with a broad amplexicaul base. First inflorescence often 1/2–1 m with 100–200 flowers, usually with only 2–5 branches, lateral side inflorescences sometimes almost as large; partial racemes often with 25–50 flowers, buds usually densely placed, flowers thus opening close to the top, in some populations, buds overtopping the flowers. The pedicels are 10–22 mm, at anthesis erecto-patent up to 45° to quite patent; flowers usually conspicuously protandrous, anthers ripening 1–2 days before the stigma; sepals 9–11(−12.5) mm, slightly deflexed, up to 300, or erect and then parallel-distant, the outer 3–4(−5) mm broad, boat-shaped, slightly saccate at base, inconspicuously cucullate, the inner (2–)2.5–3(−4) mm broad, flatter, cucullate especially in bud, petals at anthesis patent to slightly deflexed or in some French populations quite deflexed during male phase, whitish yellow or rarely light yellow; claw (7–)8–9(−10) mm, limb 11–15(−16.5) × 8–10.5(−12.5) mm. Stamens are exerted 3–5 mm above the petal disc; outer filaments (7–8.5–11.5(−13) mm, inner 9–12(−14) mm and usually conspicuously longer; anthers (3–)3.5–4.5(−5.5) mm, yellow, connective tip usually 0.1–0.2 mm, white to dark violet. The pistil is exserted above the anthers at maturity, ripe ovary 7–10 mm, style (1–)2–3(−5) mm, stigma 1–1.2 mm broad. Fruiting pedicels (9)12–25(−35) mm, usually patent or almost so. The siliqua except rostrum (40–)50–90(−120) mm long, 3.5–5(−6) mm broad, often slightly thicker than broad, when well-filled with (15–)25–40(−50) seeds. Rostrum (3–)6–11(−15) mm, if seed-bearing 4–5 mm broad, if empty 1.5–3 mm, usually seedless, rarely some with one seed; seeds 1.7–2.5(−2.8) mm, spherical, with a rather obscure network (Snogerup et al. 1990).

2.1.28 *Brassica oxyrrhina* (Coss.) Gomes Mach.

Synonyms *B. barrelieri* subsp. *oxyrrhina* (Coss.) Regel; *B. barrelieri* var. *oxyrrhina* (Coss.) O.E.Schulz; *B. repanda* subsp. *oxyrrhina* (Coss.) P.W.Ball & Heywood; *B. sabularia* subsp. *oxyrrhina* (Coss.) Maire; *Coincya oxyrrhina* (Coss.) Rivas Mart.; *Sinapis oxyrrhina* Coss.; *B. bursipastorifolia* (Pourr. ex Willk. & Lange) Samp.; *B. nostalgica Samp.*; *B. oxyrrhina* proles *nostalgica* Samp.; *B. oxyrrhina* var. *nostalgica* (Samp.) Samp.; *B. oxyrrhyncha* Coss. ex Regel; *B. sabularia* Willk.; *Heliophila bursipastorisfolia* Pourr. ex Willk. & Lange (www.powo.science.kew.org).

Description Slender, with a thin, whitish root perpendicular to the ground, leaves are subadpressed, runcinate- and pectinate-pinnate-partite, the margin and below the costa densely setose, the lobes are irregularly sinuate-dentate, the terminal ones are often larger; the cauline leaves are absent or very few remote smaller ones, the tops are linear or squamae-shaped; the stem is erect, subnude, the base is setose at the back, the rest is glabrous, simple or slightly branched above. The flowers are large, long-pedicellate. racemose, pods with pedicel subequal to the rostrum, patent, narrowly linear, compressed-subtetragonous, glabrous, rostrum valves 3–5-nerved, median nerve carinated, rostrum teretiusculo conico-subulate often curved, base seminiferous, equaling or exceeding the length of the valves; basal leaves shortly petiolate (Willkomni and Lange 1880).

2.1.29 Brassica procumbens (Poir.) O.E.Schulz

Synonyms *Erucastrum procumbens* (Poir.) Calest.; *Sinapis procumbens* Poir.; *S. choulettiana* Coss. & Durieu (www.powo.science.kew.org).

Description Annual; galls 0.20–0.50 m long, numerous at base, spreading or ascending, branched, terete, the shorter ones barely 0.5 mm long, curving upwards, loosely obtuse, pale green, violet in colour; rarely ascending. Lower leaves sessile, iridescent-pinnatifida, terminal lobe obovate, apex obtuse, unequally 4–6-toothed, lateral lobes slightly smaller, upper acuminate, partly denticulate, lower triangular, entire, rhizome broadly winged; leaves medium broadly oblong, manifestly dentate, often recurved at the base of the lobes, sessile; upper leaves linear, acute, entire; all scattered briefly hairy or glabrous on both sides, bright green. The raceme is corymbose under inflorescence, then elongate, very lax, 20–30-flowered, with lower flowers (1–15) often in the axils of the upper bracteoles, only 4 mm long; pedicels 4 mm long, hairy; sepals 4 mm long, erect, narrowly oblong, rounded at the apex, scattered hairy; petals 8 mm long, dilutely yellow; blade widely obovate, at basin briefly cuneate an-apice taste, rounded; stamens as inner and outer 5.75 only mm, 3.75 mm long; anthers minute, 1 mm long, subovate; the pistil especially with ovarium hairs curved hairy; ovary upwards 12-ovulate; style 2–2.5 mm long; stigma stylus wide. The nectariferous glands are large; the pods on pedicels 1.4–0.4 cm long, erect-patent, spreading or recurved, 1.6–2.3 cm long, 1.8 mm wide, tortuose, in a trinervated beak 5–8.5 mm long or monosperm attenuated, with sparsely acumbulating hairs, scattered hairy or conspicuously hairy; valves single-nerved, lateral nerves thin, yellow; stigma equal, seeds subglobose, 1 mm diam., light brown, black at the tip, minutely alveolate (Schulz 1919).

2.1.30 *Brassica rapa* L.

Synonyms *B. campestris* proles *rapa* (L.) Samp.; *B. campestris* var. *rapa* (L.) C.Hartm.; *B. rapa* var. *communis* Schübl. & G.Martens; *B. rapa* var. *typica* Posp.; *B. sativa* Clavaud; *B. sativa* subsp. *rapa* (L.) Bonnier & Layens; *B. sativa* var. *rapa* (L.) Clavaud; *B. sphaerorhiza* Gray; *B. amplexicaulis* Hochst. ex A.Rich.; *B. antiquorum* H.Lév.; *B. arvensis* Hablitz; *B. asperifolia* Lam.; *B. asperifolia* var. *esculenta* Gren. & Godr.; *B. asperifolia* var. *oleifera* Godr.; *B. asperifolia* var. *sylvestris* Lam.; *B. brassicata* A.Chev.; *B. briggsii* Varenne; *B. campestris* L.; *B. campestris* var. *afghanica* Sinskaya; *B. campestris* var. *agrestis* Prain; *B. campestris* var. *amplexicaulis* Makino; *B. campestris* var. *annua* (W.D.J.Koch) Rchb.; *B. campestris* f. *auriculata* (DC.) O.E.Schulz; *B. campestris* subvar. *autumnalis* DC.; *B. campestris* f. *briggsii* (Briggs) O.E.Schulz; *B. campestris* var. *chinensis* (L.) N.Busch; *B. campestris* subsp. *chinensis* (L.) Makino; *B. campestris* subvar. *depressa* (DC.) O.E.Schulz; *B. campestris* subsp. *dichotoma* (Roxb.) G.Olsson; *B. campestris* var. *dissecta* O.E.Schulz; *B. campestris* var. *genuina* Čelak.; *B. campestris* var. *glabra* Sinskaya; *B. campestris* var. *mana* Makino; *B. campestris* var. *musifolia* Makino; *B. campestris* subsp. *narinosa* (L.H.Bailey) G.Olsson; *B. campestris* subsp. *nipposinica* (L.H.Bailey) G.Olsson; *B. campestris* f. *oblonga* (Mill.) Prain; *B. campestris* subvar. *oblonga* (Mill.) O.E.Schulz; *B. campestris* var. *ohsakana* Makino; *B. campestris* stirps *oleifera* DC.; *B. campestris* subsp. *oleifera* (DC.) Sinskaya; *B. campestris* var. *orientalis* Sinskaya; *B. campestris* var. *osakana* Makino; *B. campestris* subsp. *pekinensis* (Lour.) G.Olsson; *B. campestris* f. *pinnata* Makino; *B. campestris* subvar. *praecox* DC.; *B. campestris* var. *purpuraria* L.H.Bailey; *B. campestris* subvar. *pygmaea* Alef.; *B. campestris* var. *rubra* Sinskaya; *B. campestris* var. *sarson* Prain; *B. campestris* var. *setulosa* Caruel; *B. campestris* var. *shigatsuna* Makino; *B. campestris* var. *sotetsuna* Makino; *B. campestris* f. *tenuis* O.E.Schulz; *B. campestris* f. *trilocularis* (Roxb.) Prain; *B. campestris* var. *trilocularis* (Roxb.) Duthie & Fuller; *B. campestris* subsp. *trilocularis* (Roxb.) G.Olsson; *B. campestris* subsp. *vulgaris* Schübl. & G.Martens; *B. campestris* var. *wakana* Makino; *B. celerifolia* (Tsen & S.H.Lee) Y.Z.Lan & T.Y.Cheo; *B. chinensis* L.; *B. chinensis* f. *aichi* M. Hiroe; *B. chinensis* f. *chifu* M.Hiroe; *B. chinensis* var. *communis* Tsen & S. H. Lee; *B. chinensis* var. *crispa* André; *B. chinensis* var. *dentata* (Matsum. & Nakai) M.Hiroe; *B. chinensis* f. *hiberna* M.Hiroe; *B. chinensis* f. *hiroshimana* (Makino) M.Hiroe; *B. chinensis* f. *honsaitai* M.Hiroe; *B. chinensis* f. *indivisa* (Makino) M.Hiroe; *B. chinensis* f. *komatsuna* (Matsum. & Nakai) M.Hiroe; *B. chinensis* var. *komatsuna* Matsum. & Nakai; *B. chinensis* f. *mana* (Makino) M.Hiroe; *B. chinensis* var. *musifolia* (Makino) M.Hiroe; *B. chinensis* var. *narinosa* (L.H.Bailey) M.Hiroe; *B. chinensis* f. *nigra* M.Hiroe; *B. chinensis* f. *nozawensis* M.Hiroe; *B. chinensis* var. *oleifera* Makino & Nemoto; *B. chinensis* f. *osakana* (Makino) M.Hiroe; *B. chinensis* subf. *peking* M.Hiroe; *B. chinensis* f. *piaoh-tsai* (Kitam.) M.Hiroe; *B. chinensis* var. *rosularis* Tsen & S.H.Lee; *B. chinensis* subf. *shigatsuna* (Makino) M.Hiroe; *B. chinensis* f. *taito* M.Hiroe; *B. chinensis* f. *youngtungtsai* (Kitam.) M.Hiroe; *B. colza* H.Lév.; *B. cyrenaica* Spreng.; *B. dubiosa*

L.H.Bailey; *B. japonica* Makino; *B. japonica* var. *indivisa* Makino; *B. japonica* var. *isena* Makino; *B. japonica* var. *suigikuna* Makino; *Brassica* × *juncea* var. *celerifolia* Tsen & S.H.Lee; *B. lutea* Gilib.; *B. mizuna* Siebold; *B. musifolia* (Makino) Makino; *B. napella* Chaix; *B. napus* var. *chinensis* (L.) O.E.Schulz; *B. napus* var. *dichotoma* (Roxb.) Prain; *B. napus* var. *longirostris* Borbás; *B. napus* var. *trilocularis* (Roxb.) O.E.Schulz; *B. narinosa* L.H.Bailey; *B. nipposinica* L.H.Bailey; *B. nipposinica* var. *laciniifolia* L.H.Bailey; *B. oleracea* var. *chinensis* (L.) Prain; *B. oleronensis* A.Sav. ex Foucaud; *B. parachinensis* L.H.Bailey; *B. pe-tsai* L.H.Bailey; *B. pe-tsai* var. *dentata* Matsum. & Nakai; *B. pekinensis* (Lour.) Rupr.; *B. pekinensis* var. *cephalata* Tsen & S.H.Lee; *B. pekinensis* var. *cylindrica* Tsen & S.H.Lee; *B. pekinensis* var. *hiroshimana* Makino; *B. pekinensis* var. *laxa* Tsen & S.H.Lee; *B. perfoliata* Crantz; *B. perviridis* (L.H.Bailey) L.H.Bailey; *B. petsai* L.H.Bailey; *B. polymorpha* Syme; *B. pseudocolza* H.Lév.; *B. purpuraria* (L.H.Bailey) L.H.Bailey; *B. rapa* subsp. *afghanica* (Sinskaya) Shebalina; *B. rapa* subvar. *alba* DC.; *B. rapa* var. *alborosea* Shebalina; *B. rapa* var. *amplexicaulis* Yo.Tanaka & Ono; *B. rapa* var. *annua* W.D.J.Koch; *B. rapa* var. *arenaria* Alef.; *B. rapa* var. *baubryana* Alef.; *B. rapa* var. *biennis* Alef.; *B. rapa* var. *briggsii* Briggs; *B. rapa* subsp. *campestris* (L.) A.R. Clapham; *B. rapa* var. *campestris* (L.) Peterm.; *B. rapa* convar. *cephalata* (Tsen & S. H. Lee) G.H.Loos; *B. rapa* var. *cephalata* (Tsen & S.H.Lee) Hanelt; *B. rapa* var. *chinensis* (L.) Kitam.; *B. rapa* subsp. *chinensis* (L.) Hanelt; *B. rapa* var. *chloroplax* Alef.; *B. rapa* var. *cinerascens* Alef.; *B. rapa* var. *communis* (Tsen & S. H. Lee) Hanelt; *B. rapa* convar. *cylindrica* (Tsen & S.H.Lee) G.H.Loos; *B. rapa* var. *cylindrica* (Tsen & S.H.Lee) Hanelt; *B. rapa* subf. *depressa* M.Hiroe; *B. rapa* stirps *depressa* DC.; *B. rapa* var. *depressa* (DC.) Lej. & Courtois; *B. rapa* f. *depressa* (DC.) Prain; *B. rapa* subsp. *dichotoma* (Roxb. ex Fleming) Hanelt; *B. rapa* var. *dichotoma* (Roxb. ex Fleming) Kitam.; *B. rapa* var. *dissecta* (O.E.Schulz) Gladis; *B. rapa* var. *dubiosa* (L.H.Bailey) Gladis & K. Hammer; *B. rapa* var. *erythroplax* Alef.; *B. rapa* convar. *Europea* Shebalina; *B. rapa* subvar. *flavescens* DC.; *B. rapa* var. *frenosiana* Alef.; *B. rapa* f. *glabra* (Sinskaya) M.Hiroe; *B. rapa* var. *glabra* Regel; *B. rapa* var. *globosa* Peterm.; *B. rapa* subf. *globosa* M. Hiroe; *B. rapa* var. *hybrida* Shebalina; *B. rapa* var. *intermedia* Shebalina; *B. rapa* convar. *japonica* (Shebalina) Gladis & K.Hammer; *B. rapa* subsp. *japonica* Shebalina; *B. rapa* var. *jodopyrga* Alef.; *B. rapa* convar. *laxa* (Tsen & S.H.Lee) G.H.Loos; *B. rapa* var. *laxa* (Tsen & S.H.Lee) Hanelt; *B. rapa* var. *leptorhiza* Spach; *B. rapa* var. *leucoplax* Alef.; *B. rapa* var. *leucopyrga* Alef.; *B. rapa* var. *lorifolia* L.H. Bailey; *B. rapa* var. *lugdunensis* Alef.; *B. rapa* var. *meauxiana* Alef.; *B. rapa* var. *media* Alef.; *B. rapa* var. *melaina* Alef.; *B. rapa* convar. *mesopotamica* (Shebalina) Gladis & K.Hammer; *B. rapa* subsp. *mesopotamica* Shebalina; *B. rapa* var. *morignyana* Alef.; *B. rapa* f. *nagakabu* Kitam.; *B. rapa* subf. *nagakuba* (Kitam.) M.Hiroe; *B. rapa* subsp. *narinosa* (L.H.Bailey) Hanelt; *B. rapa* var. *narinosa* (L.H.Bailey) Kitam.; *B. rapa* subvar. *nigricans* DC.; *B. rapa* subsp. *nipposinica* (L.H.Bailey) Kitam.; *B. rapa* var. *nipposinica* (L.H.Bailey) Hanelt; *B. rapa* stirps *oblonga* (Mill.) DC.; *B. rapa* subf. *oblonga* M.Hiroe; *B. rapa* var. *oblonga* (Mill.) Lej. & Courtois; *B. rapa* subsp. *oleifera* (DC.) Metzg.; *B. rapa* stirps *oleifera* DC.; *B. rapa* var. *oleifera* (DC.) Lej. & Courtois; *B. rapa* convar.

oleifera (DC.) Alef.; *B. rapa* var. *oleronensis* (A.Sav. ex Foucaud) Sav.; *B. rapa* var. *orientalis* (Sinskaya) Gladis & K.Hammer; *B. rapa* convar. *orientalis* (Sinskaya) Gladis & K.Hammer; *B. rapa* var. *parachinensis* (L.H.Bailey) Hanelt; *B. rapa* convar. *parachinensis* (L.H.Bailey) G.H.Loos; *B. rapa* subvar. *pe-tsai* (L.H.Bailey) Kitam.; *B. rapa* subsp. *pekinensis* (Lour.) Kitam.; *B. rapa* var. *pekinensis* (Lour.) Hanelt; *B. rapa* var. *perviridis* L.H.Bailey; *B. rapa* var. *piaoh-tsai* Kitam.; *B. rapa* subvar. *punicea* DC.; *B. rapa* var. *purpuraria* (L.H.Bailey) Kitam.; *B. rapa* var. *rosularis* (Tsen & S.H.Lee) Hanelt; *B. rapa* convar. *rosularis* (Tsen & S.H.Lee) G.H.Loos; *B. rapa* var. *rubra* (Sinskaya) Shebalina; B. *rapa* var. *ruvo* (L.H.Bailey) Gladis & K.Hammer; *B. rapa* var. *sarcorhiza* Spach; *B. rapa* var. *sarson* (Prain) Mehrotra & Aswal; *B. rapa* subsp. *sarson* (Prain) Denford; *B. rapa* var. *sativa* Mérat; *B. rapa* var. *saulieua* Alef.; *B. rapa* var. *septiceps* L.H.Bailey; *B. rapa* var. *sinensis* Gladis & K.Hammer; *B. rapa* var. *sylvestris* (Lam.) Gilib.; *B. rapa* subsp. *sylvestris* (Lam.) Janch.; *B. rapa* var. *teltoviensis* Alef.; *B. rapa* subsp. *trilocularis* (Roxb.) Hanelt; *B. rapa* var. *variegata* Alef.; *B. rapa* var. *vilmorinii* Alef.; *B. rapa* var. *violascens* Shebalina; *B. rapa* subvar. *viridis* DC.; *B. rapa* var. *xanthocona* Alef.; *B. rapa* subvar. *youngtungtsai* Kitam.; *B. ruvo* L.H.Bailey; *B. saruna* Siebold; *B.sativa* var. *campestris* (L.) Clavaud; *B. septiceps* (L.H.Bailey) L.H.Bailey; *B. sphaerorhiza* var. *oblonga* Gray; *B. trilocularis* Hook.f. & Thomson; *B. trimestris* Boenn.; *B. tuberosa* Salisb.; *Napus rapa* (L.) K.F.Schimp. & Spenn.; *Barbarea derchiensis* S.S.Ying; *Caulanthus sulfureus* Payson; *Crucifera rapa* E.H.L.Krause; *Napus campestris* (L.) K.F.Schimp. & Spenn.; *Rapa oblonga* Mill.; *Raphanus rapa* (L.) Crantz; *R. amplexicaulis* Viv.; *R. campestris* (L.) Crantz; *R. chinensis* (L.) Crantz; *Sinapis auriculata* DC.; *S. campestris* Besser; *S. communis* Noronha; *S. dichotoma* Roxb.; *S. pekinensis* Lour.; *S. pekinensis* var. *cai-ben* Lour.; *S. pekinensis* var. *pe-tsai* Lour.; *S. rapa* Brot.; *S. trilocularis* Roxb.; *S. tuberosa* Poir. (www.powo.science.kew.org).

Description Annuals or biennials; roots fleshy or slender; green to slightly glaucous, glabrous or sparsely hairy. The stems are unbranched or branched distally, 3–10 dm., basal leaves possess a petiole (winged), (1–)2–10(−17) cm; blade ± lyrate-pinnatifid to pinnate to pinnatisect, (5–)10–40(−60) cm × 30–100(−200) mm, (margins sinuate-dentate, sometimes ciliate), lobes 2–4(−6) each side, (terminal lobe oblong-obovate, obtuse, large, blade surfaces usually setose). The cauline leaves (middle and distal) are sessile; base auriculate to amplexicaul, (margins subentire); racemes not paniculately branched, (with open flowers overtopping or equal to buds); fruiting pedicels ascending to spreading, (5–)10–25(−30) mm., flowers show sepals (3–)4–6.5(−8) × 1.5–2 mm; petals deep yellow to yellow, obovate, 6–11(−13) × (2.5–)3–6(−7) mm, claw 3–7 mm, apex rounded; filaments 4–6(−7) mm; anthers 1.5–2 mm. The fruits are ascending to somewhat spreading, torulose, terete, (2–)3–8(−11) cm × 2–4(−5) mm; valvular segment with 8–15 seeds per locule, (1.3–)2–5(−7.5) cm, terminal segment seedless, 8–22 mm; seeds are black, brown, or reddish, 1.1–2 mm diam.; seed coat very finely reticulate-lightly alveolate, not mucilaginous when wetted (Fig. 2.2) (Schulz 1919).

Fig. 2.2 *Brassica rapa*

2.1.31 Brassica repanda (Willd.) DC.

Synonyms *B. humilis* var. *repanda* (Willd.) Coss.; *B. saxatilis* subsp. *repanda* (Willd.) Maire; *Brassicaria repanda* (Willd.) Gillet & Magne; *Diplotaxis humilis* subsp. *repanda* (Willd.) P.Fourn.; *D. repanda* (Willd.) Gren. & Godr.; *D. saxatilis* subsp. *repanda* (Willd.) Braun-Blanq.; *Erucastrum repandum* (Willd.) Calest.; *Guenthera repanda* (Willd.) Gómez-Campo; *Sisymbrium repandum* Willd. (www.powo.science.kew.org).

Description The plants are very glabrous, low, turfy, root long, above remaining petioles dry; scape 2–10 cm, up to 2.5 cm long, simple; leaves often long-petiolate, spatulate or obovate, apex obtuse, decurrent at the base of the petiole, on both sides deeply 1–5-repeating-toothed, rarely entire, 0.5–3.5 cm, up to 6 cm long, rarely fleshy; raceme is dense corymbose below inflorescence, then slightly elongate, rarely 5–8, 12-flowered, up to the pedicels which are 4–5 mm long; petals yellow, 0.9–1.2 cm long; pistil minute, 4 mm long, narrowly ampullaceous; ovary 17-ovulate, ovules subserrate; the style is 1 mm long; stigma stylus evidently wider; pods 2–3 cm, radii up to 6.5 cm long, wider, 3–3.5 mm wide, suddenly contracted into a rostrum 2.5–3 mm long and seeds are large, 2 mm uniserial (Schulz 1919).

Brassica repanda also includes 16 subspecies (www.powo.science.kew.org).

(a) Brassica repanda **subsp.** africana **(O.E.Schulz) Greuter**
Synonyms: *B. nudicaulis* var. *africana* O.E.Schulz; *B. saxatilis* subsp. *africana* (O.E.Schulz) Maire; *Guenthera repanda* subsp. *africana* (O.E.Schulz) Gómez-Campo; *Guenthera repanda* var. *africana* (O.E.Schulz) Dobignard; *B. gravinae* var. *atlantica* (Batt.) Batt.; *B. humilis* Boiss.; *B. humilis* var. *atlantica* Batt.; *B. humilis* var. *nudicaulis* (Lag.) Coss.; *B. nudicaulis* Pomel; *B. nudicaulis* (Lag.) Gonz.Albo; *B. nudicaulis* var. *intermedia* Caball.; *B. nudicaulis* var. *lagascae* (Rouy) O.E.Schulz; *B. repanda* subsp. *nudicaulis* (Lag.) Heywood; *B. rouyana* Janka; *B. saxatilis* f. *maroccana* Maire; *B. saxatilis* var. *nudicaulis* (Lag.) Font Quer; *B. saxatilis* subsp. *nudicaulis* (Lag.) Maire; *B. saxatilis* f. *pomeliana* Maire; *Diplotaxis brassicoides* var. *lagascae* Rouy; *D. nudicaulis* (Lag.) Pau; *D. saxatilis* var. *lagascae* (Rouy) Willk.; *Erucastrum repandum* subsp. *rouyanum* (Janka) Bonnier; *Erucastrum rouyanum* (Janka) Bonnier; *Sinapis nudicaulis* Lag. (www.powo.science.kew.org).

(b) Brassica repanda **subsp.** almeriensis **Gómez-Campo**
Synonyms: *Guenthera repanda* subsp. *almeriensis* (Gómez-Campo) Gómez-Campo
(www.powo.science.kew.org).

(c) Brassica repanda **subsp.** blancoana **(Boiss.) Heywood**
Synonyms: *B. blancoana* Boiss.; *B. repanda* var. *blancoana* (Boiss.) O.Bolòs & Vigo; *B. saxatilis* subsp. *blancoana* (Boiss.) Maire; *B. foliata* Pau & Font Quer *Diplotaxis blancoana* (Boiss.) Giraudias; *D. brassicoides* var. *longifolia* Rouy; *D. saxatilis* var. *longifolia* (Rouy) Willk.; *Guenthera repanda* subsp. *blancoana* (Boiss.) Gómez-Campo; *D. brassicoides* Rouy (www.powo.science.kew.org).

(d) Brassica repanda **subsp.** cadevallii **(Font Quer) Heywood**
Synonyms: *B. repanda* var. *cadevallii* (Font Quer) O.Bolòs & Vigo; *B. saxatilis* var. *cadevallii* Font Quer; *B. nudicaulis* var. *intermedia* (Rouy) O.E.Schulz; *B. repanda* var. *tripontina* (Font Quer) O.Bolòs & Vigo; *B. saxatilis* var. *tripontina* Font Quer; *Diplotaxis brassicoides* var. *intermedia* Rouy; *D. saxatilis* var. *intermedia* (Rouy) Willk., *D. leucanthemifolia* Jord.; *Guenthera repanda* subsp. *cadevallii* (Font Quer) Gómez-Campo; (www.powo.science.kew.org).

(e) Brassica repanda **subsp.** cantabrica **(Font Quer) Heywood**

Synonyms: *B. saxatilis* var. *cantabrica* Font Quer; *Guenthera repanda* subsp. *cantabrica* (Font Quer) Gómez-Campo (www.powo.science.kew.org).

(f) Brassica repanda **subsp.** confusa **(Emb. & Maire) Heywood**

Synonyms: *B. repanda* var. *confusa* (Emb. & Maire) O.Bolòs & Vigo; *B. saxatilis* subsp. *confusa* Emb. & Maire; *B. saxatilis* var. *berberica* Litard. & Maire; *B. saxatilis* var. *brevisiliqua* Emb. & Maire; *B. saxatilis* var. *embergeri* Maire; *Guenthera repanda* subsp. *confusa* (Emb. & Maire) Gómez-Campo; *Diplotaxis humilis* var. *granatensis* Rouy; *D. subcuneata* Jord. (www.powo.science.kew.org).

(g) Brassica repanda **subsp.** dertosensis **Molero & Rovira**

Synonyms *Guenthera repanda* subsp. *dertosensis* (Molero & Rovira) Gómez-Campo

(www.powo.science.kew.org).

(h) Brassica repanda **subsp.** galissieri **(Girard) Heywood**

Synonyms: *B. saxatilis* var. *galissieri* (Girard) O.E.Schulz; *Diplotaxis galissieri* Girard; *Guenthera repanda* subsp. *galissieri* (Girard) Gómez-Campo; *Erucastrum repandum* subsp. *humile* Bonnier & Layens (www.powo.science.kew.org).

(i) Brassica repanda **subsp.** glabrescens **(Poldini) Gómez-Campo**

Synonyms: *B. glabrescens Poldini*; *Guenthera repanda* subsp. *glabrescens* (Poldini) Gómez-Campo

(www.powo.science.kew.org).

(j) Brassica repanda **subsp.** gypsicola **Gómez-Campo**

Synonyms: *Guenthera repanda* subsp. *gypsicola* (Gómez-Campo) Gómez-Campo (www.powo.science.kew.org).

(k) Brassica repanda **subsp.** latisiliqua **(Boiss. & Reut.) Heywood**

Synonyms: *B. humilis* var. *latisiliqua* (Boiss. & Reut.) Coss.; *B. latisiliqua* Boiss. & Reut.; *B. nudicaulis* var. *latisiliqua* (Boiss. & Reut.) O.E.Schulz; *B. saxatilis* var. *latisiliqua* (Boiss. & Reut.) Font Quer; *B. nudicaulis* var. *brevifolia* (Rouy) O.E.Schulz; *Diplotaxis saxatilis* var. *latisiliqua* (Boiss. & Reut.) Willk.; *D. brassicoides* var. *brevifolia* Rouy; *D. nevadensis* Jord.; *D. saxatilis* var. *brevifolia* (Rouy) Willk. *Guenthera repanda* subsp. *latisiliqua* (Boiss. & Reut.) Gómez-Campo; (www.powo.science.kew.org).

(l) Brassica repanda **subsp.** maritima **(Rouy ex Willk.) Heywood**

Synonyms: *B. maritima* (Rouy ex Willk.) Ballester, Figuerola, Peris & Stübing; *B. nudicaulis* var. *maritima* (Rouy ex Willk.) O.E.Schulz; *B. saxatilis* var. *maritima* (Rouy ex Willk.) Font Quer; *B. dianiae* Lacaita ex Willk.; *Diplotaxis brassicoides* var. *maritima* Rouy ex Willk.; *D. maritima* (Rouy ex Willk.) Rouy ex Willk.; *D. brassicoides* var. *maritima* Willk.; *Guenthera repanda* subsp. *maritima* (Rouy ex Willk.) Gómez-Campo; (www.powo.science.kew.org).

(m) Brassica repanda **subsp.** repanda

Synonyms: *B. viciosoi* Gonz.Albo; *Diplotaxis humilis* var. *delphinensis* Rouy (www.powo.science.kew.org).

(n) Brassica repanda **subsp.** saxatilis **(Lam.) Heywood**

Synonyms: *B. humilis* f. *saxatilis* (Lam.) Coss.; *B. saxatilis* (Lam.) Amo; *B. humilis* DC.; *B. repanda* subsp. *humilis* (DC.) O.Bolòs & Vigo; *B. saxatilis* subsp. *humilis* (DC.) O.Bolòs & Vigo; *Diplotaxis humilis* (DC.) Gren. & Godr; *D. humilis* subsp. *saxatilis* (Lam.) P.Fourn.; *D. saxatilis* (Lam.) DC.; *D. humilis* var. *provincialis* Rouy; *D. saxatilis* subsp. *humilis* (DC.) Braun-Blanq.; *Erucastrum humile* (DC.) Bonnier & Layens; *E. repandum* subsp. *saxatile* (Lam.) Bonnier & Layens; *E. saxatile* (Lam.) Bonnier & Layens; *Guenthera repanda* subsp. *saxatilis* (Lam.) Gómez-Campo; *G. repanda* subsp. *humilis* (DC.) Gómez-Campo; *Sisymbrium monense* L., *S. saxatile* Lam.; *Brassicaria humilis* (DC.) Gillet & Magne; *B. saxatilis* (Lam.) Gillet & Magne; (www.powo.science.kew.org).

(o) Brassica repanda **subsp.** silenifolia **(Emb.) Greuter & Burdet**

Synonyms: *B. saxatilis* subsp. *silenifolia* Emb.; *B. saxatilis* var. *silenifolia* (Emb.) Maire; *Guenthera repanda* subsp. *silenifolia* (Emb.) Gómez-Campo (www.powo.science.kew.org).

(p) Brassica repanda **subsp.** turbonis **(P.Monts.) J.M.Monts. & Romo**

Synonyms: *B. repanda* var. *turbonis* (P.Monts.) O.Bolòs & Vigo; *B. saxatilis* var. *turbonis* P.Monts.; *B. turbonis* (P.Monts.) Rivas Mart. & al.; *B. repanda* var. *montsicciana* Romo; *Guenthera repanda* subsp. *turbonis* (P.Monts.) Gómez-Campo; (www.powo.science.kew.org).

2.1.32 *Brassica rupestris Raf.*

Synonyms *B. oleracea* subvar. *rupestris* (Raf.) Coss.; *B. oleracea* var. *rupestris* (Raf.) Paol.

(www.powo.science.kew.org).

Description Suffrutescent perennial, usually 100–150 cm high, in cultivation often reaching 3–4 m, usually with a conspicuous main stem branching only apically to form a large inflorescence; young plant first forming a rosette of large leaves, soon prolonged and flowering after 1–5 years, often perennating for more flowering periods, in cultivation usually dying in fruit after first flowering. Stem strongly lignified with conspicuous ear rings, often 15–25 mm thick in basal parts; leaf scars are conspicuous and often remain visible on older stems. Plant glabrous or sparse indumentum, mainly on the veins and margins of young leaves. Hairs, when present, with a bulbous base. Leaves comparatively thin, herbaceous, losing their texture during drying, in one populations lightly greyish; basal and cauline leaves including petiole (15–)20–35(−45) × (8–)10–13(−17) cm, petiole (7–)10–25 cm, lamina broadly lanceolate to obovate or rarely+ orbicular, in basal part more or less deeply pinnatifid, 1(−3) pairs of usually very small lobes on upper part of petiole, petiole never winged or with any auricles on leaf foot; margin with prominent, usually acute teeth, midrib strong, side veins conspicuous, reaching the margin in the largest teeth, veins of higher order anastomosing; upper leaves gradually smaller and more short-petiolate, their margin varying among individuals, sometimes with few, large teeth, sometimes entire, bracts of inflorescence branches often linear and only 2–5 cm. First inflorescence usually very large, repeatedly branched, under favourable conditions about 1 m long and wide with thousands of flowers; if growth continues for more seasons, later inflorescences often also large; partial racemes many-flowered, buds usually densely placed, flowers thus opening close to the top. Pedicels (7–)9–20(−25) mm, at anthesis from 45° erecto-patent to quite patent. Flowers usually conspicuously protandrous with anthers ripening 1–2 days before the stigma. Sepals (7–)9–11(−12) mm, usually divergent at 20–40, rarely erect and then usually distant from the petal claws, the outer 2.5–3.5(−4) mm broad, boat-shaped, saccate at base, sometimes slightly cucullate, the inner 2–3 mm broad, flatter, conspicuously cucullate especially in bud. Petals at anthesis patent or up to 300 deflexed, rarely quite deflexed in male phase, light yellow; claw (6–)7–8(−9) mm long, usually tapering towards the base, limb (9–)11–16(−18) × (6–)7.5–11(−13) mm. Stamens exserted 2–4 mm above the petal disc, outer filaments (5–)6–8(−9) mm, inner (6–)7.5–9(−10) mm, ratio inner/outer filament (1.05–) 1.20–1.30; anthers (2–)2.5–3 mm, yellow, connective tip inconspicuous or up to 0.1 mm, whitish. Pistil exserted above anthers at maturity, ripe ovary 5–8 mm, style 2–4(−5) mm, stigma c. 1 mm broad. Fruiting pedicels 10–30(−40) mm, usually patent or almost so, rarely erecto-patent to 60°. Siliqua except rostrum (25–)35–65 mm long, 3–5.5(−6) mm broad, often at least equally thick, with a prominent, or on ripe siliquas sometimes inconspicuous, midrib on valves and side veins, veins more conspicuous in dry material. Rostrum 4–11 mm long, (1.5–)2–3 mm broad, usually seedless, rarely 1-seeded. Seeds 2.5–2.8(−3.3) mm, network obscure when dry, when fresh with a conspicuous pattern of comparatively small meshes, variable (Snogerup et al. 1990).

Brassica rupestris also includes four subspecies (www.powo.science.kew.org).

(a) Brassica rupestris **subsp.** hispida **Raimondo & Mazzola**
Synonyms: None reported.
(b) Brassica rupestris **subsp.** monilicarpa **Raimondo & Spadaro**
Synonyms: None reported.
(c) **Brassica rupestris subsp.** rupestris
Synonyms: *B. crispa* Raf.; *B. montana* Raf.; *B. sempervirens* Ten. ex Schrank (www.powo.science.kew.org).
(d) Brassica rupestris **subsp.** tardarae **(Ilardi, Geraci & Troìa) Raimondo**
Synonyms: *Brassica tardarae* Ilardi, Geraci & Troìa (www.powo.science.kew.org).

2.1.33 Brassica setulosa (Boiss. & Reut.) Coss.

Synonyms *Brassica munbyana* Coss. & Durieu; *Eruca setulosa* Boiss. & Reut.; *Guenthera setulosa* (Boiss. & Reut.) Gómez-Campo; *Velleruca setulosa* (Boiss. & Reut.) Pomel (www.powo.science.kew.org).

Description Perennial rhizome short, with scaly neck and rosy petioles, leaves all radical rosulate, petiolate, pinnate, oblong, entire, obtuse segments along the dilated midrib, often 1–2 dentate, all terminal, submajor, margin setose-ciliate, stems naked, erect, in racemes, elongate, laxly ascending, calyx scarcely saccate, sepals oblong-linear, red, deciduous, setose at apex, petals all vee laced, nail-like, inserted into oblong-linear lamina slightly dilated, silicules, pedicel thick, striate, scarcely shorter, supported by oblong, inflated, scarcely compressed, obtuse, rostrum much thinner, ensiform, rough, half or three times shorter, valves convex, reticulate, glabrous, with prominent nerve (Boissier 1867).

2.1.34 *Brassica somalensis Hedge & A.G.Mill.*

Synonyms None reported.

Description Perennial herb, glabrous or slightly strigose, squarrosoramorous below, grey bark; stems erect, to 60 cm, simple, herbaceous, terete, slightly striate, glabrous or with short hairs on the back, ± copiously foliated below, finally forming a tight raceme; leaves all narrowly oblong-linear, (10–)20–55(−60) × (2–)5–12(−15) mm, base narrowly attenuated cuneate, glabrous or with short hairs on the front, rigidly leathery (when dry), rarely sinuate-dentate, apex obtuse, venation indistinctly reticulate-pinnate; petiole subnullus. The racemes lie below the anthers, corymbose, dense, very elongate in fruit; pedicels erect-open 5–11 mm long, elongated to 14(−2o) mm in fruit, strong; sepals are linear-elliptic, (4–)5 × (0–8–)1.2–1.6 mm, glabrous or with two anterior hairs, erect-open, hooded at apex; inner sepals slightly wider than outer ones, base subsaccate; petals pale yellow (when dry), c. 8–10 mm; lamina obovate, or rarely narrowly elliptic c. 6 × 3.5–4.5 mm, attenuated into a perangustum 3–4 mm. The stamens subequal; inner filaments 3.5–4.8 mm; outer filaments 4–6 mm; anthers oblong c. i mm, finally recurved; median glands oblong, lateral semi-lunate, smaller; ovary c. 16-ovulate, not stipitate; stigma is capitate, barely bilobate. The pods are erect, somewhat torulose, attenuated into a rough (1–)2–2.5 mm long rostrum; submembranous valves between seeds, barely constricted, with a prominent median nerve, blunt at the apex; seeds ovoid, 1–1.25 × 0.7–0.8 mm, slightly mucous, brownish-yellowish sublaevate (Hedge and Miller 1977).

2.1.35 *Brassica souliei (Batt.) Batt.*

Synonyms *B. amplexicaulis* var. *souliei* (Batt.) O.E.Schulz; *B. amplexicaulis* subsp. *souliei* (Batt.) Maire & Weiller; *Camelina souliei* Batt.; *Guenthera amplexicaulis* subsp. souliei (Batt.) Gómez-Campo (www.powo.science.kew.org).

Description Annuals; stems glabrous, flowers yellow, petals grooved at apex, obcordate to rounded; leaves entire but toothed on margin, glossy green; upper leaves narrower, oblong, coarsely toothed on the crown; petiole very short (Battandier 1893).

Brassica souliei also includes two subspecies (www.powo.science.kew.org).

(a) Brassica souliei **subsp.** amplexicaulis **(Desf.) Greuter & Burdet**
Synonyms: *B. amplexicaulis* (Desf.) Pomel; *B. amplexicaulis* f. *cedrorum* O.E.Schulz; *B. amplexicaulis* f. *cordifolia* O.E.Schulz; *B. amplexicaulis* var. *hirtula* Pau; *B. amplexicaulis* var. *maroccana* O.E.Schulz; *B. amplexicaulis* var. *occidentalis* Pau *Guenthera amplexicaulis* (Desf.) Gómez-Campo; *Melanosinapis amplexicaulis* (Desf.) Calest.; *Sinapis amplexicaulis* (Desf.) DC.; *Sisymbrium amplexicaule* Desf.; (www.powo.science.kew.org).
(b) Brassica souliei **subsp.** souliei
Synonyms: *Erucastrum varium* subsp. *barbei* Vindt (www.powo.science.kew.org).

2.1.36 Brassica spinescens Pomel

Synonyms *B. scopulorum* var. *spinescens* (Pomel) Batt.; *B. scopulorum* Coss. & Durieu; *B. spinescens* var. *scopulorum* (Coss. & Durieu) Maire (www.powo.science.kew.org).

Description Flowers bloom in corymbs, the sepals are membraneous at the edges, erect; filaments unequal; very loose fruiting raceme with spreading peduncles, as well as the siliques, which are long 30–40°, cylindrical, slightly compressed, dented, with valves slightly keeled by the dorsal vein and clearly veined in anastomosis on the sides. The style is 6–7 cm long, subconical, striated, often ovuliferous; seeds are suburicate, globose, tawny, tuberous. The leaves are small, fleshy, thick, glaucous, obovate or oblong, entire or crenate, attenuated into a petiole with or without a pair of small lobules detached from the base of the blade; stems fruitfully branched, knobbly, bearing, intermingled with the flowering branches: hardened and spinescent remains of the previous inflorescences and aborted twigs; the plants are entirely glabrous, with very pale purple flowers (Pomel 1874).

2.1.37 Brassica taiwanensis S.S.Ying

Synonyms None reported.

Description Annual herbs about 120 cm tall; taproots somewhat fleshy, 2–3.5 cm long, 5–6 mm across; fibrous roots many. slender; stem erect to ascending, simple to branched, terete, hollow, green, glabrous; branches slender, erect to ascending, green, glabrous. Leaves radical and cauline; the radical ones simple, alternate, oblong to oblong-elliptic, 22–32 cm long, 6.5–8.5 cm wide, obtuse at apec, attenuate to the petioles at base, chartaceous, deeply clefted to double serrate at margins, glabrous on both surfaces, midrib and lateral veins (5–7 per side) slightly impressed surface and elevated beneath; the petioles 6–9 cm long, grooved, glabrous; cauline leaves oblong, oblong-lanceolate to lanceolate, 10–18 cm long, 2.5–4 cm wide, obtuse at apex, attenuate to the petioles at base, double serrate at margins, midrib and lateral veins (3–5 per side) slightly impressed surface and elevated beneath; the petioles 2.5–5 cm long, grooved, glabrous. Flowers numerous, yellow, 6–8 mm in diameter and somewhat capitate-like when opened, in a terminal 30–60 cm long panicle; the panicles often with 4–7 racemes; the racemes ascending, 15–30 cm long, glabrous, with 10–35 flowers; bracts absent; pedicels 5–8 mm long, green, glabrous; sepals 4, lanceolate, 4.5–6 mm long, 2–3 mm wide, obtuse at apex, involute, pale green to pale yellow, glabrous; petals 4, oblong to ovate-oblong, 5–6.5 mm long, 3.5–4 mm wide, obtuse at apex, slightly clawed at base, yellow, glabrous; stamens 6, tetradynamous, filaments slender, 2.5–3.5 mm long, yellow; anthers oblong, 2 mm long, obtuse at apex; nectar glands confluent; median glands present; ovary cylindric, 2.5–3 mm long, glabrous, green; style short, 0.6–0.8 mm long; stigma capitate. Fruits dehiscent siliques, slender, terete, 3.5–5.5 cm long, 2–2.5 mm across, stipitate, smooth or slightly torulose, dehiscent 2-valved when mature; valves with 1 midvein, latral veins inconspicuous, glabrous; seeds 6–15 in per fruit, 1-seriated, rounded, 1.5–2.0 mm in diameter, glabrous, light brownish; the pedicels slender 1.5–2 cm long, glabrous (Ying 2021).

2.1.38 *Brassica taurica (Tzvelev) Tzvelev*

Synonyms *B. sylvestris* subsp. *taurica* Tzvelev (www.powo.science.kew.org).

Description Perennial plants; stems grow upright, branching abundantly from the base; leaves are broad and clustered at the base, resembling typical cabbage leaves, and are glabrous; entire to slightly segmented. Inflorescence is profusely flowered, much-branched; flowers are numerous and reach the tips of the branches. The sepals are calyx-shaped and glabrous; colour is pale yellow to dark yellow and long-stalked. The siliques are borne at the ends of the ascending stems, thickly petiolated, glabrous, having 10–12 seeds; pedicels are thickened at the tip; seeds are dark, round, and smooth (www.redbook-ua.org).

2.1.39 *Brassica trichocarpa* C.Brullo, Brullo, Giusso & Ilardi

Synonyms None reported.

Description Suffructescent hairy, 70–120 m tall, with a robust main stem, branching often from the base. The stems are woody, up to 25 mm in diameter at the base, green-purplish to purple mainly in the inflorescence; leaves are rigid, lightly thickened, green-glaucescent, more whitened in the abaxial face, often purplish at the margin, densely covered by rigid, simple hairs; basal and lower cauline leaves 10–30 cm long (including petiole), with petiole 3–9 cm long and lamina ovate to ovate-lanceolate, 7–20 × 6–13 cm, lobate or basally pinnatifid and pinnatisect, with lobes rounded or slightly obtuse at apex and petiole in upper part often with 1–2 pairs of small lobes, never winged or auriculate at base; margin irregularly and coarsely dentate; midrib strong with conspicuous lateral veins; upper cauline leaves gradually smaller, shortly petiolate to sessile, with lamina often linear, entire, only 2–3 cm long. Inflorescence loosely paniculate, branched from the base, pyramidate. The pedicels are 10–19 mm long, hairy, patent or subpatent at anthesis; sepals dark yellow to yellow-greenish, hairy outside, the outer ones ovate-lanceolate, 11.5–12.5 × 3.2–3.8 mm, the inner ones linear to linear-lanceolate, 11–12 × 2–2.4 mm, cucullate, slightly saccate at base; the petals are spathulate, yellow, 20–22-mm; claw 6–7 mm long; limb 8–8.5 mm wide, rounded; stamens are exserted 1–2 mm above the petal disc; outer filaments whitish, 6–7.5 mm long; inner filaments 6.5–8 mm long; anthers yellow, 4.5–5 mm long, with a conspicuous connective tip. The pistil is shorter or subequalling the stamens; ovary 2.2–3.7 mm long, white greenish, densely ciliate along the sutures and midribs and pubescent on the valves; style green, often tinged with purple, 2–2.5 mm long, subcylindric; stigma 0.7–0.8 × 1–1.2 mm, subcapitate, minutely papillose, yellow greenish; basal glands are dark green, 0.8–0.9 mm in diameter; fruiting pedicels are 1.2–2.2 cm long, patent to erect-patent, usually purplish. The fruits dehiscence is difficult, with corpus subglobose to ellipsoid, smooth not torulose, hairy, tinged with dark purple, 8–18 × 8–11 mm, with valves thickened, hardened, spongy, smooth and furrowed along the sutures at maturity, while it shows 4 ribs in correspondence of midribs and sutures when dry; rostrum linear-subulate, 57 mm long, seedless. Seeds 4–16, subglobose, 2–2.5 mm in diameter, brown-reddish, minutely reticulate, arranged in 2 rows in each loculus (Brullo et al. 2013).

2.1.40 *Brassica tyrrhena* Giotta, Piccitto & Arrigoni

Synonyms None reported.

Description A deciduous plant, 30–40(80) cm tall. The stems are erect-ascending, with a woody axis above ground, 4–6 cm and completely glabrous, branching annual stems. Lower leaves of the stem possess a 2–10 cm petiole, sub-elliptical blade, 10–15 × 4–8 cm, glossy green in colour, crenulate-dentate at the margin, often briefly lobed-hastate at the base; upper leaves lanceolate-cuneate to linear, 2–12 × 1–3 cm, rounded at the apex. The terminal racemose inflorescences is beset with sweet-scented flowers and patent pedicels of 0.5–2 cm; the sepals oblong-elliptical, 0.8–1 × 0.3–0.5 cm, yellowish-white; whereas the petals are yellow, obovular-unguiculate, 1.6–1.8 cm, with a developed 0.5–0.7 cm claw; stamens approximately 1 cm long, with glabrous filaments and yellow anthers approximately 2 mm long; style is longer than the stamens, with a capitate stigma; and silique is quadrangular in cross-section, 3–9 cm, with a terminal cone of 6–12 mm (Giotta et al. 2002).

2.1.41 Brassica villosa Biv.

Synonyms *B. oleracea* var. *villosa* (Biv.) Coss.; *B. rupestris* subsp. *villosa* (Biv.) Snogerup ex Hanelt (www.powo.science.kew.org).

Description Suffrutescent perennials, usually 80–150 cm high, in cultivation often reaching 3–4 m, usually with a conspicuous main stem, branching only apically to form a large inflorescence; young plant first forming a rosette of large leaves, soon prolonged and after 1–5 years flowering, in nature often forming new stems during more flowering seasons, in cultivation usually withering in fruit. The stems are strongly lignified with conspicuous year rings, often 15–25 mm thick in the basal part; leaf scars are conspicuous and often remain visible on older stems. Indumentum is moderate, but usually dense, simple hairs without a bulbous base. Leaves are comparatively thin, herbaceous, losing their texture during drying; basal and cauline leaves, including petiole (16–)19–40(−45) × (7–)9–15(−18) cm, petiole 6–21 cm, lamina broadly lanceolate to obovoid, undivided or basally pinnatifid or pinnatisect, petiole in upper part often with 1–2(−4) pairs of small lobes, never winged or with any auricles on leaf foot; margin usually with prominent and acute teeth, rarely more obtusely dentate, midrib strong, side veins conspicuous, reaching the margin in the largest teeth; upper cauline leaves gradually smaller, more short-petiolate to sessile, variable in shape and margin, often some with few but prominent teeth, bracts of inflorescence branches often linear and only 2–5 cm. The first inflorescence is usually very large, repeatedly branching, under favourable conditions about 1 m long and wide with thousands of flowers, if growth continues for more seasons, later inflorescences are often also large; partial racemes are many-flowered, buds densely placed, and flowers thus opening close to the top. The pedicels are (6–)8–17(−28) mm; flowers usually conspicuously protandrous, anthers ripening 1–2 days before the stigma. Sepals (6–)8–12 mm, at anthesis usually divergent at c. 20–45°, or erect, adpressed or distant from the petal claws, the outer (2–)3–4 mm broad, boat-shaped, saccate at base, often slightly cucullate, the inner (1.5–)2–2.5(−3.5) mm broad, flatter, conspicuously cucullate especially in bud. The petals light yellow; claw 6–9(−10) mm, limb (10–)12–18(−20) × (6–)8–11(−13) mm; stamens exserted 2–4 mm above the petal disc; outer filaments 5–8(−10.5) mm, inner (5.5–)6–10(−12) mm; anthers (2–)2.5–3(4) mm, yellow, connective tip lacking or inconspicuous, a very small dark tip observed only in one individual. The pistil is exserted above anthers at maturity, ripe ovary 5–8(−9) mm, style (1.5–)2–4 mm, stigma c. 1 mm broad; fruiting pedicels (12–)15–25 mm, erecto-patent at 45° to patent; siliqua except rostrum (25–)35–65 mm long, 3–5 mm broad, varying in thickness at least between 3 and 7 mm when ripe and dry, if well filled with (12–)20–30 seeds, with a conspicuous midrib of valves and side veins. The rostrum is 6–12 mm, usually 1.5–2 mm broad, seedless; the seeds are (2.1)2.4–2.8(−3) mm, surface pattern obscure when dry, when wet a conspicuous reticulum is observed (Snogerup et al. 1990).

Brassica villosa also includes five subspecies (www.powo.science.kew.org).

(a) B. villosa **subsp.** bivoniana **(Mazzola & Raimondo) Raimondo & Mazzola**
Synonyms *B. bivoniana* Mazzola & Raimondo (www.powo.science.kew.org).
(b) B. villosa **subsp.** brevisiliqua **(Raimondo & Mazzola) Raimondo & Geraci**
Synonyms: *B. rupestris* subsp. *brevisiliqua* Raimondo & Mazzola (www.powo.science.kew.org).
(c) B. villosa **subsp.** drepanensis **(Caruel) Raimondo & Mazzola**
Synonyms: *B. drepanensis* (Caruel) Damanti; *B. macrocarpa* var. *drepanensis* (Caruel) Paol.; *Eruca drepanensis* Caruel (www.powo.science.kew.org).
(d) B. villosa **subsp.** tineoi **(Lojac.) Raimondo & Mazzola**
Synonyms: *B. tineoi* Lojac.; *B. villosa* var. *tineoi* (Lojac.) O.E.Schulz (www.powo.science.kew.org).
(e) B. villosa **subsp.** villosa
Synonyms None reported.

2.1.42 *Brassica × harmsiana O.E.Schulz*

Synonyms None reported.

Description This hybrid has not been observed spontaneously as yet, but has been produced artificially. It should be emphasised at this point that *B. campestris* and *B. napus* are often very similar. The main difference between these two species is that *B. napus* always has elongated flower buds that more or less extend beyond the open flowers. *B. campestris*, on the other hand, has shorter, egg-shaped flower buds that are far exceeded by the open flowers (Schulz 1919).

2.2 Distribution in the World

Inferences from antiquity, excavated plant fragments, references to the genus in ancient literature from various civilisations, comparative taxonomy, the natural distribution of the genus, cytogenetic and biochemical evidence, and, more recently, molecular markers have all helped to reinterpret past events concerning the origin of the genus *Brassica* (Gomez-Campo 1999a).

An evaluation of Table 2.1 reveals that most of the *Brassica* species are naturally distributed in the world. Looking at the data presented in Table 2.1, we find a total of 42 wild *Brassica* species in the world and natural distribution areas of species like *B. oleracea*, *B. rapa* and *B. napus*; the introduced areas in different geographical regions of the world are shown in Table 2.2.

Table 2.1 Natural distribution areas of wild *Brassica* species in the world

No	Species	Country–region
1	*Brassica assyriaca*	Syria
2	*Brassica aucheri*	Iran, Iraq, Lebanon-Syria, Palestine, Jordan
3	*Brassica baldensis*	Italy
4	*Brassica balearica*	Baleares
5	*Brassica barrelieri*	Algeria, Morocco, Portugal, Spain
6	*Brassica beytepeensis*	Türkiye
7	*Brassica bourgeaui*	Canary Is.
8	*Brassica cadmea*	Greece
9	*Brassica cretica*	Albania, East Aegean Is., Greece, Kriti, Lebanon-Syria, Palestine, Türkiye
10	*Brassica deflexa*	Afghanistan, Iran, Iraq, Kuwait, Lebanon-Syria, Pakistan, Saudi Arabia, Türkiye
11	*Brassica deserti*	Sinai
12	*Brassica desnottesii*	Morocco
13	*Brassica dimorpha*	Algeria, Tunisia
14	*Brassica elongata*	Afghanistan, Austria, Bulgaria, Russia, Czechia-Slovakia, Hungary, Iran, Kazakhstan, Krym, Lebanon-Syria, Mongolia, Morocco, NW. Balkan Pen., Poland, Romania, Tajikistan, Transcaucasia, Turkmenistan, Türkiye, Ukraine, Uzbekistan, West Siberia, Xinjiang
15	*Brassica fruticulosa*	Algeria, Corse, France, Italy, Morocco, NW. Balkan Pen., Sardegna, Sicilia, Spain
16	*Brassica gravinae*	Algeria, Italy, Libya, Morocco, Tunisia
17	*Brassica hilarionis*	Cyprus, Lebanon-Syria
18	*Brassica huseyinduralii*	Türkiye
19	*Brassica incana*	Albania, Greece, Italy, NW. Balkan Pen., Sicilia
20	*Brassica insularis*	Algeria, Corse, Sardegna, Sicilia, Tunisia
21	*Brassica loncholoma*	Algeria, Tunisia
22	*Brassica macrocarpa*	Sicilia
23	*Brassica maurorum*	Algeria, Morocco
24	*Brassica montana*	France, Italy, Spain
25	*Brassica napus*	France
26	*Brassica nivalis*	Bulgaria, Greece
27	*Brassica oleracea*	France, Great Britain, Spain

(continued)

Table 2.1 (continued)

No	Species	Country–region
28	*Brassica oxyrrhina*	Morocco, Portugal, Spain
29	*Brassica procumbens*	Algeria, Tunisia
30	*Brassica rapa*	Albania, Algeria, Bulgaria, Eritrea, Ethiopia, Greece, Iran, Iraq, Italy, Kriti, Lebanon-Syria, Libya, Morocco, North Caucasus, NW. Balkan Pen., Oman, Saudi Arabia, Transcaucasia, Tunisia, Türkiye
31	*Brassica repanda*	Algeria, France, Italy, Morocco, Spain
32	*Brassica rupestris*	Italy, Sicily
33	*Brassica setulosa*	Algeria, Morocco
34	*Brassica somalensis*	Somalia
35	*Brassica souliei*	Algeria, Morocco, Sicily, Tunisia
36	*Brassica spinescens*	Algeria
37	*Brassica taiwanensis*	Taiwan
38	*Brassica taurica*	Krym
39	*Brassica trichocarpa*	Italy
40	*Brassica tyrrhena*	Sardegna
41	*Brassica villosa*	Sicilia
42	*Brassica* × *harmsiana*	Great Britain

Modified from www.powo.science.kew.org

Table 2.2 Distribution areas of introduced *Brassica* species in the world

No	Species	Country–region
1	*Brassica barrelieri*	Western Australia
2	*Brassica cretica*	Bangladesh, East Himalaya, Vietnam
3	*Brassica elongata*	Baltic States, Belgium, Botswana, Cape Provinces, Denmark, France, Germany, Great Britain, Italy, Kirgizstan, Netherlands, Norway, Russia, South Australia, United States
4	*Brassica fruticulosa*	Australia, Germany, New Zealand, Spain, Tasmania, United States
5	*Brassica napus*	Afghanistan, Albania, Algeria, Argentina, Australia, Baltic States, Bangladesh, Belgium, Bolivia, Bulgaria, Canada, Chile, China, Colombia, Costa Rica, Cuba, Czechia-Slovakia, Dominican Republic, Ecuador, Ethiopia, France, Georgia, Germany, Great Britain, Greece, Guatemala, Haiti, Holland, Indiana, Inner Mongolia, Iran, Iraq, Ireland, Italy, Japan, Kenya, Korea, Krym, Libya, Mali, Marshall Is., Mauritius, Mexico, Morocco, New Zealand, North Caucasus, Russia, Norway, Pakistan, Palestine, Portugal, Puerto Rico, Réunion, Saudi Arabia, Sicilia, Spain, Tasmania, Transcaucasia, Tunisia, Ukraine, United States, West Himalaya, West Siberia, Yemen

(continued)

Table 2.2 (continued)

No	Species	Country–region
6	*Brassica oleracea*	Albania, Algeria, Australia, Baltic States, Bangladesh, Belarus, Belgium, Bulgaria, Canada, Canary Is., Caroline Is., Chile, China, Costa Rica, Czechia-Slovakia, DR Congo, East Himalaya, Easter Is., Ecuador, El Salvador, Ethiopia, Falkland Is., Germany, Greece, Guatemala, India, Iraq, Ireland, Juan Fernández Is., Kazakhstan, Kenya, Korea, Krym, Lebanon-Syria, Libya, Marianas, Marshall Is., Morocco, Netherlands, New Caledonia, New Zealand, Portugal, Russia, Sardegna, Saudi Arabia, Tajikistan, Tanzania, Tasmania, Trinidad-Tobago, Tunisia, Turkmenistan, Ukraine, United States, Vietnam
7	*Brassica oxyrrhina*	New Zealand North
8	*Brassica procumbens*	Corsica-France
9	*Brassica rapa*	Andaman Is., Argentina, Australia, Baltic States, Bangladesh, Belarus, Belgium, Bolivia, Brazil, Cambodia, Canada, Cape Provinces, Cayman Is., Chile, China, Colombia, Comoros, Cook Is., Corse, Costa Rica, Cuba, Czechia-Slovakia, Denmark, Dominican Republic, East Himalaya, Ecuador, Egypt, El Salvador, Falkland Is., Finland, France, Free State, Føroyar, Galápagos, Georgia, Germany, Gilbert Is., Great Britain, Guatemala, Haiti, Hawaii, Honduras, Hungary, India, Ireland, Juan Fernández Is., Kazakhstan, Kenya, Kirgizstan, Korea, Krym, Kuwait, Leeward Is., Line Is., Marianas, Marshall Is., Mauritius, Mexico, Mozambique, Nepal, Netherlands, New Caledonia, New Zealand, Nicaragua, Nicobar Is., Norway, Pakistan, Palestine, Panamá, Paraguay, Peru, Poland, Portugal, Puerto Rico, Romania, Rwanda, Réunion, Russia, Sardegna, Sicilia, Society Is., Solomon Is., Spain, Sweden, Switzerland, Tajikistan, Taiwan, Tanzania, Tasmania, Trinidad-Tobago, Tristan da Cunha, Tuamotu, Turkmenistan, Uganda, Ukraine, Unites States, Uruguay, Uzbekistan, Vietnam, Virginia, West Himalaya, West Siberia, Windward Is., Zimbabwe
10	*Brassica* × *harmsiana*	Great Britain

Modified from www.powo.science.kew.org

Wild forms of *B. oleracea* have originated in the Mediterranean region and can be found along the coasts of the Mediterranean Sea and the Atlantic Ocean, spanning from Greece to Britain. The cultivated forms are widely grown across the globe, except in tropical regions (Warwick 2010; OECD 2016; Jabeen 2020). *B. rapa*, also known as turnip rape, grows naturally in mountainous regions close to the Mediterranean Sea. It is believed to have originated in the Afghanistan-Pakistan region. The cultivated forms are widely grown in Europe, Africa, north-west China, Asia and the Indian subcontinent (OECD 2016; Jabeen 2020); these can be categorised into three main groups based on their morphological characteristics:

1. Oleiferous or oilseed type rapeseed (canola)

2. Leafy types (Chinensis group—celery mustard; Pekinensis group—Chinese cabbage; Perviridis group—soft green)
3. The Rapifera group (turnip) and the Rufo group (turnip broccoli and Italian turnip) (Jabeen 2020)

Brassica napus, commonly known as canola or rapeseed, is believed to have originated in the Mediterranean coastal region or southern Europe, although more than one centre of origin is generally accepted (Rakow 2004; OECD 2012; Jabeen 2020). Cultivated forms of this species are widely grown in Asia, Africa, Australia, Canada, Europe, Mexico, New Zealand, Russia and the Americas (USDA-ARS 2017). *B. napus* is an amphidiploid species resulting from hybridisation between *B. oleracea* and *B. rapa* (OECD 2012).

Cultivars of the *Brassica* genus are among the world's most economically important vegetables (Fig. 2.3). Although their ancestral species are believed to have originated in the Mediterranean region, these vegetables are now found all over the world. Different varieties have readily adapted to diverse production regions around the globe, ranging from the tropics to the Arctic Circle. Consequently, these cultivars have been cultivated for centuries and extensively crossed and hybridised (Fahey 2016). Breeders have developed viable hybrids by crossing genetically distinct parents using techniques such as ovule, ovary and embryo culture, as well as protoplast fusion. All these techniques have improved the yield, agronomic performance and potential disease resistance of *Brassica* vegetables. It has also enabled the development of new hybrid varieties with beneficial agronomic traits (Jabeen

Fig. 2.3 *Brassica rapa* cultivation area (Kashmir-India)

2020). Important criteria for obtaining these hybrid varieties include characteristics such as hybridisation potential, flowering period and the presence, frequency, activity, and behaviour of pollinators (Jabeen 2020).

Successful hybridisation between *B. oleracea* and other *Brassica* species has been reported, particularly with *B. bourgeaui*, *B. cretica*, *B. incana*, *B. insularis*, *B. macrocarpa*, *B. montana*, *B. maurorum*, *B. rupestris* and *B. villosa* (Chrungu et al. 1999; Warwick et al. 2000). Moreover, successful interspecific hybridisation has been reported between various *B. oleracea*, including *Moricandia*, *Raphanus* and *Sinapis* (Mizushima 1950; Eenink 1974; Ellerstrom 1978; McCollum 1979; Sarashima and Matsuzawa 1979; Namai 1980; Apel et al. 1984; Quiros et al. 1988; Sarmah and Sarla 1998; Jabeen 2020). Hybrids with the potential to produce viable seeds have also been obtained between *B. napus* and *B. rapa* (Bing et al. 1991).

2.3 Fossil Records and Pollen Studies

A large number of studies have demonstrated that the pollen morphology of the Brassicaceae family is usually uniform (Pokrovskaya et al. 1956; Lan and Cheo 1983; Deng and Hu 1995; Wang et al. 1995; Tang et al. 2005). While identification at the family level is generally straightforward, identification at the genus and species level is reportedly challenging (Yang et al. 2018).

The identification of pollen grains from cultivated plants, particularly at the genus and species level, is crucial in archaeobotanical investigations. A recent study presented in 2018 by Yang et al. in China has revealed significant differences in the equatorial diameter and polar axis values of *B. rapa*, *B. oleracea* var. *alboglabra*, *B. rapa* var. *parachinensis* and *B. rapa* ssp. *chinensis*. Some moderate differences in reticulum lumina length along the long axis have been recorded. These findings are expected to contribute significantly to species identification based on pollen grain micromorphological features (Yang et al. 2018). The same authors have at the same time compared the pollen grain values and long-axis length of lumina in modern Brassicaceae pollen samples with those in fossil pollen obtained from the Ming-Qing culture layer at the Fupingkou site in Chongqing, China. They concluded that the fossil pollen probably belongs to *B. rapa* var. *parachinensis* and related vegetable plants (Yang et al. 2018). Some of the authors have investigated the micromorphological pollen features in the study area. There findings are presented below:

Brassica rapa (oilseed rape, Chinese cabbage): The pollen grains are subspheroidal, trilobate (having 3 lobes or parts, when there are 3 colpi or furrows dividing the grain), circular in polar view, P22 (25) 28 × E18 (23) 28 μm minimum, (average) and maximum dimensions, tricolpate, colpi slender and long, extending to poles, sculpture reticulate, lumina uniformly distributed and irregular in size and shape, muri (ridges or walls between lumina of reticulum) granular, sexine (the outer sculptured part of the exine, the pollen wall) formed from columellae, exine ca. 1–2 μm thick (Yang et al. 2018).

B. oleracea var. *alboglabra* (Chinese kale): Pollen grains subspheroidal or prolate-spheroidal, P16 (24) 34 × E16 (20) 27 μm, morphology similar to *B. rapa* except in size (Yang et al. 2018).

B. rapa ssp. *parachinensis* (Chinese flowering cabbage): Pollen grains spheroidal, trilobate, circular in polar view, P19 (24) 29 × E21 (25) 28 μm, morphology similar to *B. rapa* and *B. oleracea* var. *alboglabra* except in size (Yang et al. 2018).

B. rapa ssp. *chinensis* (pak choi): Pollen grains prolate-spheroidal, P23 (28) 35 × E19 (25) 35 μm, morphology similar to the above three species (Yang et al. 2018).

References

Al-Shehbaz IA (2012) A generic and tribal synopsis of the Brassicaceae (Cruciferae). Taxon 61(5):931–954

Anjum NA, Gill SS, Ahmad I et al (2012) The plant family Brassicaceae: an introduction. The plant family Brassicaceae: contribution towards phytoremediation. Springer, pp 1–33

Apel P, Bauwe H, Ohle H (1984) Hybrids between *Brassica alboglabra* and *Moricandia arvensis* and their photosynthetic properties. Biochem Physiol Pflanz 179:793–797

Bailey CD, Koch MA, Mayer M et al (2006) A global nrDNA ITS phylogeny of the Brassicaceae. Mol Biol Evol 23:2142–2160

Başköse İ, Özüdoğru B, Savran A, Keskin A (2023) A new species of *Brassica* (Brassicaceae) from Bolkar Mountains (Türkiye) with morphological and molecular evidence. Phytotaxa 632(2):101–117

Battandier MA (1893) Botanique dans la region de L'oguar. Bulletin Societe Botanique Paris 1893:259–264

Beilstein MA, Al-Shehbaz IA, Kellogg EA (2006) Brassicaceae phylogeny and trichome evolution. Am J Bot 93:607–619

Beilstein MA, Al-Shehbaz IA, Mathews S, Kellogg EA (2008) Brassicaceae phylogeny inferred from phytochrome A and ndhF sequence data: tribes and trichomes revisited. Am J Bot 95:1307–1327

Bing DJ, Downey RK, Rakow GFW (1991) Potential of gene transfer among oilseed Brassica and their weedy relatives. In: GCIRC 1991 Congress, pp 1022–1027

Boissier E (1867) Flora orientalis: Brassicaceae, vol 1. Genevae, pp 390–394

Bowers JE, Chapman BA, Rong J, Paterson AH (2003) Unravelling angiosperm genome evolution by phylogenetic analysis of chromosomal duplication events. Nature 422:433–438

Brullo C, Brullo S, Del Galdo GG, Ilardi V (2013) Brassica trichocarpa (Brassicaceae), a new species from Sicily. Phytotaxa 122(1):45–60

Bruno C, Christophe G (2002) *Brassica barrelieri* (L.) Janka (Brassicaceae), plante nouvelle pour la France naturalisée dans le département de la Loire. Bull Mens Soc Linn Lyon 71(2):75–88

Chrungu B, Verma N, Mohanty A et al (1999) Production and characterization of interspecific hybrids between Brassica maurorum and crop brassicas. Theor Appl Genet 98:608–613

Cosson MME, Durieu M (1955) Notes sur quelgues especes nouveles DÂlgeria. Bulletin Societe Botanique De France 1955:303–314

Danin A, Hedge IC (1973) Contributions to the Flora of Sinai: New and confused taxa. Flora of Israel 1973:259–260

Davis PH (1965) Flora of Turkey and the East Aegean Islands, vol 1, p 264

Deng YB, Hu ZH (1995) The comparative morphology of the floral nectaries of Brassicaceae. Acta Phytotaxon Sin 33:209–220

Eastwood A (1996) The conservation and utilization of *Brassica incana* Ten. in Croatia, with particular reference to leaf trichomes as a potential source of insect resistance. MSc. thesis. Faculty of Science. Univ. Birmingham

Eenink AH (1974) Matromorphy in *Brassica oleracea* L. III. The influence of temperature, delayed prickle pollination and growth regulators on the number ofmatromorphic seeds formed. Euphytica 23:711–718

Ellerstrom S (1978) Species crosses and sterility in *Brassica* and *Raphanus*. Cruciferae Newslett 3:16–17

Emberger L, Maire R (1930) Contrubutions a Letude de la flore de L'Afrique Du Nord. Materiaux Pour La Flore Marocanie 15:21–33

Fahey JW (2016) Brassica: characteristics and properties. In: Encyclopedia of food and health. Elsevier Inc., pp 469–477

Giotta C, Picitto M, Arrigoni PV (2002) Un nuovo endemismo della Sardegna: *Brassica tyrrhena* sp. nov. (Brassicaceae). Webbia 57(1):1–5

Gladis T (1989) Die Gattung *Brassica* L. und die Reproduktion entomophiler Pjanzensippen in Genbanken. Dissertation zur Erlangung des Akademischen Grades. Doktor der Naturwissenschaften. Gatersleben

Gomez-Campo C (1980) Morphology and morphotaxonomy of the tribe Brassiceae. In: *Brassica* crops and wild allies. Biology and breeding. Japan Scientific Societies Press, Tokyo, pp 3–31

Gomez-Campo C (1999a) Biology of Brassica Coenospecies. Elsevier Science B.V

Gomez-Campo C (1999b) Seedless and seeded beaks in the tribe Brassiceae (Cruciferae). Cruciferae Newsl 21:1

Hedge IC, Miller AG (1977) New and interesting taxa from N.E. Tropical Africa. Notes R Botan Garden Edinburg 35(2):179–193

Huang CH, Sun R, Hu Y et al (2015) Resolution of Brassicaceae phylogeny using nuclear genes uncovers nested radiations and supports convergent morphological evolution. Mol Biol Evol 33(2):394–412

Jabeen N (2020) Agricultural, economic and societal importance of Brassicaceae plants. In: The plant family Brassicaceae: Biology and physiological responses to environmental stresses. Springer Nature Singapore, pp 45–128

Koch M, Kiefer C (2006) Molecules and migration: biogeographical studies in cruciferous plants. Plant Syst Evol 259:121–142

Koch MA, Mummenhoff K (2006) Evolution and phylogeny of the Brassicaceae. Plant Syst Evol 259:81–83

Lan YZ, Cheo TY (1983) Pollen morphology of Loxostermon (Brassicaceae) in China. Acta Phytotaxon Sin 21:436–440

Lanner C (1997) Genetic relationships within the *Brassica* cytodeme. Comparison of molecular marker systems. Acta Universitatis Agriculturae Sueciae Agraria 39:1–57

Lazaro A (1997) Evaluacion de la diversidad gentica en *Brassica oleracea* L. (Cruciferae) y espontdneas relacionadas (n = 9) mediante marcadores moleculares. Tesis Doctoral. Univ. Politcnica, Madrid

McCollum GD (1979) Sterility in successive backcrosses of Raphanobrassica (2n = 4x = 36) with recurrent *Brassica oleracea* (2n = 2x = 18). Can J Genet Cytol 21:479–485

Mizushima U (1950) Karyogenetic studies of species and genus hybrids in the tribe Brassiceae of Cruciferae. Tohoku J Agric Res 1:1–14

Namai H (1980) Effect of flower bud age at bud pollination on cross fertility in reciprocal intergeneric crosses, *Raphanus sativus, Brassica oleracea* and *B. campestris*. Cruciferae Newslett 5:26–27

Nikolov LA, Shushkov P, Nevado B et al (2019) Resolving the backbone of the Brassicaceae phylogeny for investigating trait diversity. New Phytol 222:1638–1651

OECD (2012) Organisation for Economic Co-operation and Development. Consensus document on the biology of the brassica crops (Brassica spp.). In: Series on Harmonisation of regulatory oversight of biotechnology, vol 54, p 142

OECD (2016) *Brassica* crops (*Brassica* species). In: Safety assessment of transgenic organisms in the environment, Volume 5: OECD consensus documents. OECD Publishing, Paris

Pokrovskaya IM et al (1956) Pollen analysis (Chinese translation by Wang. FX et al.). Science Press, Beijing

Pomel AN (1874) Nouveaux materiaux pour la flore Atlantique. Nou Mat Fl Atlan 1874:1–399

Prosser F, Bertolli A (2007) A new subspecies of *Guenthera repanda* (Brassicaceae) from Mt Baldo (SE Prealps, Italy). Willdenowia 37(1):191–198

Quiros CF, Ochoa O, Douches DS (1988) Exploring the role of x = 7 species in *Brassica* evolution: hybridization with *B. nigra* and *B. oleracea*. J Hered 79:351–358

Rakow G (2004) Species origin and economic importance of *Brassica*. In: *Brassica* biotechnology in agriculture and forestry, Berlin, pp 3–11

Salmeen O (1979) A systematic revision of the genus *Brassica* L. in the Mediterranean region. PhD. thesis. Univ. of Reading

Sanchez-Yelamo MD (1990) Relaciones quimiotaxonmicas y evolutivas en el complejo *Diplotaxis-Erucastrum-Brassica* (Cruciferae-Brassiceae) mediante anddisis cromatogr (~ficos y electrofor~ticos). Tesis Doctoral. Universidad Politdcnica de Madrid

Sarashima M, Matsuzawa Y (1979) Possibility to breed cabbage lines with the cytoplasm of radish. Cruciferae Newslett 4:32

Sarmah BK, Sarla N (1998) Erucastrum abyssinicum, Brassica oleracea hybrids obtained by ovary and ovule culture. Euphytica 102:37–45

Schulz OE (1919) Cruciferae-Brassiceae. Pars prima. Subtribus I. Brassicinae et II. Raphaninae. In: Engler, Das Pflanzenreich [...][Heft 70] IV. 105

Snogerup S, Persson D (1983) Hybridization between *Brassica insularis* Moris and *B. balearica* Pers. Hereditas 99:187–190

Snogerup S, Gustafsson M, Von Bothmer R (1990) Brassica sect. *Brassica* (Brassicaceae) I. Taxonomy and variation. Willdenowia 19:271–365

Takahata Y, Hinata K (1980) A variation study of subtribe Brassicinae by principal component analysis. In: *Brassica* crops and wild allies. Biology and breeding. Japan Scientific Societies Press, Tokyo, pp 33–49

Takahata Y, Hinata K (1986) A consideration of the species relationships in subtribe Brassicinae (Cruciferae) in view of cluster analysis of morphological characters. Pl Sp Biol 1:79–88

Tang GY, Sun ZY, Li FZ (2005) Observation on the pollen morphology in 9 species of *Lepidium* (Brassicaceae) in China by SEM. J Wuhan Bot Res 23:432–436

Tsunoda S, Hinata K, Gomez-Campo C (eds) (1980) *Brassica* crops and wild allies. Biology and Breeding. Japan Scientific Societies Press, Tokyo, pp 1–354

USDA-ARS (2017) Germplasm resources information network (GRIN) taxonomy for plants. Taxon: *Brassica napus* L. United States Department of Agriculture. Agricultural Research Service, Beltsville Area

Wang FX, Qian NF, Zhang YL, Yang HQ (1995) Pollen flora of China, 2nd edn. Science Press, Beijing

Warwick SI (2010) Brassicaceae in agriculture, in genetics and genomics of the Brassicaceae. In: Plants genetics and genomics: crops and models series, vol 9, pp 33–65

Warwick SI, Francis A, La Fleche J (2000) Guide to the wild germplasm of *Brassica* and allied crops (tribe Brassiceae, Brassicaceae), 2nd edn. Agriculture and Agri-Food Canada Eastern Cereal and Oilseeds Research Centre, Ottowa

Willkomni M, Lange J (1880) Prodromus florae hispanicae, seu synopsis methodica omnium plantarum in Hispanicae, vol 3. Stutgartiae, pp 853–860

www.efloras.org. Accessed 15 Apr 2025

www.meteofunghi.it. Accessed 15 Apr 2025

www.powo.science.kew.org. Accessed 15 Apr 2025

www.redbook-ua.org. Accessed 15 Apr 2025

Yang YW, Lai KN, Tai PY, Li WH (1999) Rates of nucleotide substitution in angiosperm mitochondrial DNA sequences and dates of diversification between Brassica and other angiosperm lineages. J Mol Evol 48:597–604

Yang S, Zheng Z, Mao L et al (2018) Pollen morphology of selected crop plants from southern China and testing pollen morphological data in an archaeobotanical study. Veg Hist Archaeobotany 27(6):781–799

Yıldırımlı Ş (2010) Some new taxa record and taxonomic treatment from Turkey. Ot Sistematik Botanik Dergisi 17(2):114

Ying SS (2021) New taxa and new names. Research Center of Flora of Taiwan

Chapter 3
Ecological and Ecophysiological Aspects

3.1 Ecological Features

3.1.1 Habitat Preferences of the Brassicas

Although *Brassicas* can be cultivated in a variety of climates, they are generally considered a winter crop. The ideal temperature range is 14–21 °C, with minimum and maximum temperatures ranging from 4 to 30 °C depending on the variety, thus grown in temperate climates as well as in tropical and subtropical regions at higher elevations or during the winter months (Ahmad et al. 2012). They generally prefer deep, well-drained, rich, friable soils with a neutral pH, such as sandy or silty loams. The optimum pH for soils rich in organic matter is reported to be 5.8, but for mineral-rich soils, it can be as high as 6.5 (Björkman et al. 2011).

Many studies have evaluated the ecological tolerance levels of certain *Brassica* species. The plants in the *B. oleracea* group can tolerate hard frosts down to −10 °C during their growth stages, but they can typically withstand frosts down to −6 °C on average. The *Brassica* vegetables grow and develop best within a temperature range of 10–24 °C, and water requirements generally vary between 380 and 500 mm, depending on ecological conditions and the growth stage of the plant. Furthermore, soil pH values between 6.0 and 6.8 are reported to be ideal for their growth, depending on the soil structure (Bradley et al. 2009; FAO 2019).

The ideal temperature range for the growth and development is 10–27.5 °C for *B. rapa*, which grows in the areas with an annual rainfall of 350–4100 mm and on soils with a pH level of 4.8–8.5, shows high ecological tolerance and can tolerate drought conditions, moderate temperatures and low-fertility soils (Duke 1983; Hannaway and Larson 2004; Clark 2007).

The ideal temperature range for the growth and development of *B. napus* varies by season; average summer temperature ranges from 13.0 to 16.0 °C, whereas average winter temperatures range from −14.5 to −8.0 °C. The regions with an annual

M. Ozturk et al., *Climate-Resilient* Brassicas,
https://doi.org/10.1007/978-3-032-12709-9_3

rainfall of 325–700 mm are ideal for its growth and development. It thrives on soils with a pH generally ranging from 5.5 to 8.3 (OGTR 2008; USDA-ERS 2014; USDA-ARS 2017).

The habitat preferences, climate and edaphic characteristics of some popular global *Brassica* vegetables have been evaluated together with their agro-ecological characteristics.

3.1.1.1 Climate Requirements

The *Brassica* plants in general require a dormancy period of 3–4 weeks at a temperature of 1–3 °C in order to produce flower shoots and seedlings in their second year of growth; any brief drops in air temperature, even below 0 °C, do not harm these plants. However, a lack of light during the seedling stage can cause wilting (Salehi et al. 2021). *Brassica*s are hardy, cool-season vegetables growing best at temperatures between 15 and 20 °C. The hardened cabbage plants can withstand short-term temperatures as low as 4 °C, while broccoli and cauliflower can tolerate light frosts. The quality of almost all cabbage plants deteriorates when temperatures exceed 26–27 °C. Depending on the temperature and climate, broccoli plants can be harvested 10–15 weeks after direct sowing, cabbage after 13–17 weeks and cauliflower approximately after 13–16 weeks (Fahey 2016).

(a) ***Brassica oleracea* var. *capitata*—cabbage:** Following a vernalisation at low temperatures, cabbages enter the generative stage in their second year when temperatures rise above 15 °C. The inflorescences are then formed at the tips of the flower shoots, emerging from the leaf axils of the plants. If climatic conditions are unsuitable, the plants do not flower and enter a dormancy stage. The flowering period for cabbage generally varies between 20 and 60 days depending on the climatic conditions (Eşiyok 2012).

Under normal conditions, cabbage seeds retain their germination potential for 4–6 years. The optimal germination temperature is 20–25 °C, but they germinate within a temperature range of 10–30 °C; low soil temperature and excess moisture delay the germination. Under normal conditions, seeds will germinate within 3–4 days (Eşiyok 2012).

Cabbage is a vegetable that has specific climate requirements. The temperature is particularly important during flowering; high temperatures degrade the quality of the leaves. The optimum temperature for cabbage heads to form is between 15 and 20 °C. They require low temperatures (below 10 °C) to flower; if these requirements are not met, the plants remain in the vegetative phase. It has been observed that plants do not flower when temperatures consistently exceed 15 °C (Eşiyok 2012). The cabbages require plenty of light, but if grown in shaded areas or under trees, where incoming light is weak, they grow more slowly and produce lower yields (Eşiyok 2012).

The cabbages flourish well in rainy conditions; they prefer in particular wet conditions; thrive well in humid environments such as around the seashores and lakesides; however, hail damages their leaves. The rainfall stimulates cabbage growth and development, improving the quality; cabbages grown in dry regions with little rainfall have a lower yield and are low in quality (Eşiyok 2012).

Wind can have a negative impact on cabbage plants, causing water to evaporate from the leaves, which can lead the plants to topple over and bend up to the soil. Any water loss degrades the quality of the leaves, the plants toppled by strong winds grow more slowly, and leaves touching the ground rot (Eşiyok 2012).

(b) ***Brassica oleracea* var. *gemmifera*—Brussels sprouts:** Under normal conditions, remain viable for 4–5 years. The seeds germinate within 2–3 months if the temperature is above 7 °C. Brussels sprouts are a cool-season vegetable. The winter months are the most suitable time for cultivation, provided there is no threat of frost. Plants produce high-quality mini cabbages at temperatures as low as 15 °C. However, higher temperatures may prevent mini cabbages from forming or cause them to wilt. This vegetable can be grown very successfully in regions with regular rainfall (Eşiyok 2012). These are the most tolerant to low temperatures, can withstand drops in temperature up to at least −12 °C, but require well-moistened soil and high air humidity, particularly during the autumn head formation period (Babik 2006; Dhaliwal 2017; Salehi et al. 2021).

(c) ***Brassica oleracea* var. *botrytis*—cauliflower:** The cauliflower is a winter vegetable when evaluated climatically. The temperature is an important factor in its cultivation than light or humidity. These plants are generally considered day-neutral, requiring ample soil moisture, particularly during the crown formation phase. The market-ripened crowns and flowering cauliflower plants both dislike excessive rainfall and windy weather (Eşiyok 2012).

Under normal conditions, the cauliflower seeds retain their germination potential for 4–6 years. The optimal germination temperature is 20–25 °C, the range being 10–30 °C. A delayed germination and emergence occur when the soil temperature and the moisture level are too high or too low. Under normal conditions, the seeds germinate and emerge from the soil surface within 3–4 days (Eşiyok 2012).

Cauliflower cultivars are highly sensitive to temperature and photoperiod. Early cultivars require higher temperatures and longer days, whereas late cultivars respond to lower temperatures and shorter days (Salehi et al. 2021). The different varieties require different temperatures to enter the generative stage. Crown quality deteriorates at temperatures above 20 °C, and the optimum temperature for crown formation is around 15–17 °C, but it occurs at temperatures between 10 and 20 °C. Once the crown is formed, an increase in temperatures accelerates the vegetative growth of these plants; the bracts grow rapidly between the immature flower buds, and leaves emerge from between the lobes of the crown, resulting in leafy crowns that have no market value (Eşiyok 2012).

The temperature fluctuations during the seedling stage can affect the quality of vegetable crowns. Those grown under fluctuating conditions exhibit slower growth, develop smaller leaves and have a more scattered crown. If temperatures drop below 0 °C, the growing tip of the vegetable is damaged, and plants produce only leaves and no crowns. These leaves are thicker and stiffer than those of healthy plants. The plants reaching harvest time are more sensitive to low temperatures than seedlings, and when the crowns reach market maturity, temperatures below 0 °C cause a fuzzy structure to form on their surface. Any persistent low temperatures cause the crowns to turn purple, decreasing their market value (Eşiyok 2012).

(d) ***Brassica oleracea* var. *italica*—broccoli:** The characteristics of seed germination of this plant are similar. Under normal conditions, the seeds retain their germination capacity for up to 5 years. The optimum germination temperature is 20 °C; the seeds germinate and emerge above the soil surface within 3–4 days under optimal conditions (Eşiyok 2012).

Immature broccoli flower buds are damaged in hot, dry climates, as well as in areas with low temperatures and excessive rainfall. The plants grown under these conditions produce poor-quality vegetables of little value. Even if the plant develops normally in hot, dry conditions, it produces loosely structured crowns and shoots (Eşiyok 2012). The broccoli is more tolerant of cold weather than cauliflower. In temperate climates, it can withstand severe autumn frosts, and frozen broccoli florets have a particularly tasty flavour; optimum growth temperature is 16–18 °C, but it can tolerate temperatures ranging from 4 to 30 °C (Björkman et al. 2011; Salehi et al. 2021).

(e) ***Brassica oleracea* var. *gongylodes*–kohlrabi:** It is a temperature-sensitive plant. The climate requirements of different varieties can vary considerably. Under favourable conditions, the seeds remain viable for 4–6 years, but under normal conditions, they germinate within 3–4 days. The optimum germination temperature requirement lies between 20 and 30 °C (Eşiyok 2012). If temperatures remain below 10 °C for 7 days after germination, the plants enter the generative stage without forming kohlrabi or flowers. The early-maturing varieties are sensitive to low temperatures and require an optimum temperature of around 22 °C, but insensitive and late-maturing varieties do not enter the generative stage and may flower even at temperatures around 5 °C (Eşiyok 2012).

Kohlrabi is also sensitive to day length because long-day conditions accelerate flowering. The humidity is crucial for cultivating kohlrabi. In dry weather, development slows, cracking occurs, and flowering accelerates. Its yield and quality as a vegetable increase under high humidity conditions (Eşiyok 2012).

(f) ***Brassica rapa* var. *rapa*—turnip:** Under normal conditions, seeds retain their germination capacity for 3–5 years, optimum germination temperature being 15–20 °C. If soil temperature and moisture levels are favourable, seeds germinate and emerge from the soil surface within 2–4 days following their sowing (Eşiyok 2012).

The turnips are classified as both cool- and warm-season vegetables. They are more resistant to the cold than other vegetables and do not prefer high temperatures. They are planted in spring in cold regions and in autumn in warmer regions. The growth slows down when temperatures drop to 0 °C, and flowering begins when temperatures rise. Turnips remain small when exposed to high temperatures and decreased humidity during their development (Eşiyok 2012).

(g) ***Brassica rapa* var. *pekinensis*—Chinese cabbage:** It thrives in cool and temperate climates. The seeds remain viable for 3–4 years if stored under favourable conditions. Under optimal conditions, germination occurs within 3–4 days. The Chinese cabbage varies in terms of temperature requirements; most varieties require temperatures between 13 and 20 °C. A temperature of around 22 °C is required for seed germination (Eşiyok 2012; Golian et al. 2016; Salehi et al. 2021). A significant difference between day and night temperatures positively influences the bulb formation. Some early-maturing varieties can form bulbs at temperatures above 25 °C; however, the higher temperatures cause leaves to shrink and delay bulb formation. The temperatures above 25 °C lead to a decrease in leaf calcium and an inability to photosynthesise sufficiently, resulting in the development of diseases and poor bulb formation (Eşiyok 2012). Light intensity affects both core formation and leaf size. High light intensity encourages core formation and leaf development. However, low light intensity results in narrower leaves; when light intensity decreases, the yield of leaves in this vegetable also decreases (Eşiyok 2012).

3.1.1.2 Soil Requirements

Brassica vegetables can be grown on a wide variety of soil types. However, those with a high humus content, a neutral pH (6.2–7.0) and a loose texture are generally considered the most suitable (Fahey 2016; Salehi et al. 2021; Ozturk et al. 2022).

(a) ***Brassica oleracea* var. *capitata*—cabbage:** These are generally not particular about their soil requirements, thriving well in soil with a high water-holding capacity. The heavy soils produce a higher yield, whereas light soils produce a lower yield but more quickly. For autumn and winter cultivation, well-drained soil is preferable. The optimum pH for cabbage plants is between 6.0 and 6.5. Growth is generally poor in acidic and saline soils. The cabbage plants should not be grown in the same soil year after year; these need a rotation with vegetables from a different family. Otherwise, loss of soil fertility will prevent them from reaching their full potential (Eşiyok 2012).

(b) ***Brassica oleracea* var. *gemmifera*—Brussels sprouts:** These do not require any particular type of soil for their growth and can be grown in all soil types. However, their growth is more consistent and higher yields are recorded in sandy soils containing clay. Early varieties produce better yields in light soils, whereas late varieties produce better yields in heavy soils. A deep tillage of soil increases the yield (Eşiyok 2012). These plants have lower soil requirements than white cabbage and can be easily grown in soils which are less fertile than those used for growing broccoli or cabbage (Babik 2006; Dhaliwal 2017; Salehi et al. 2021).

(c) ***Brassica oleracea* var. *botrytis*—cauliflower:** The varieties of this vegetable are successfully cultivated in deep, sandy loam soils with high water-holding capacity. In heavy-textured soils, particularly during winter cultivation, it is important that any rainfall drains effectively. As cauliflower is susceptible to drought, it should not be grown in light-textured soils (Eşiyok 2012; Salehi et al. 2021). These plants thrive on soils with a pH varying between 6.0 and 6.5; growth is poor in acidic soils. This vegetable is susceptible to overused soil and should not be grown in the same soil repeatedly. There should always be a rotation with vegetables from different families (Eşiyok 2012).

(d) ***Brassica oleracea* var. *italica*—broccoli:** These plants thrive in soils with a pH level between 6.0 and 6.5; growth is poor in acidic soils. These plants are susceptible to overused soils, so these should not be grown repeatedly in the same soil. There should be a rotation with vegetables from different families (Eşiyok 2012).

This vegetable has fairly high soil requirements (Babik 2005; Salehi et al. 2021). The best soils for growing these plants are deep, fertile and rich in humus and calcium. Early varieties can be cultivated in fertile sandy or sandy loam soils, whereas late and overwintering varieties can be grown in heavier soils in temperate climates (Babik 2006; Salehi et al. 2021).

(e) ***Brassica oleracea* var. *gongylodes*—kohlrabi:** It is noted as a vegetable grown above ground and cultivated in any type of soil. However, in order to thrive, the plants require well-cultivated, nutrient-rich soils with an ideal soil pH around 5.6–7.0 (Eşiyok 2012).

(f) ***Brassica rapa* var. *rapa*—turnip:** They prefer deep, permeable soils such as loam, loam-clay and sandy loam, which are rich in organic matter and nutrients. It can be grown in most soil types, provided these are deep-textured but do not prefer very light or very heavy. A successful production is observed in sandy and sandy loam soils if adequate fertilisation is done (Eşiyok 2012).

(g) ***Brassica rapa* var. *pekinensis*—Chinese cabbage:** These plants possess a weak root system, typically growing within the top 20–30 cm of soil. It is included among the vegetables requiring good soil, fertilisation and water requirements. The fertile, well-cultivated soil, rich in humus and nutrients, not forming a crust, and retaining moisture effectively is most suitable for its growth (Golian et al. 2016; Salehi et al. 2021). It thrives particularly well in loamy-sandy soils rich in organic matter and even better in nutrients (Eşiyok 2012).

3.1.1.3 Some Agro-Ecological Characteristics of *Brassica* Vegetables

These features of the *Brassica* vegetables enlighten in particular their cultivation, planting and maintenance (especially hoeing, irrigation and fertilisation) and are summarised as follows.

(a) ***Brassica oleracea* var. *capitata*—cabbage:**

It flourishes in regions with regular rainfall, being grown by direct seeding if the ecological conditions are suitable. In some countries, it is grown from seedlings. The seedling nurseries are tilled with well-rotted farm manure at least 1 month before planting, and under favourable weather conditions, the cabbage seeds germinate very quickly; the young plants emerge from the soil 3–5 days after sowing, and seedlings reach a suitable size for planting approximately 1 month after sowing (Eşiyok 2012).

Overwatering can cause soil erosion around the roots of seedlings in nurseries; as such, the nurseries should initially be watered using a filtered mini-sprinkler system until the seeds have emerged from the soil. Later, irrigation should be carried out using a low-flow flood system. The seedlings that have reached the appropriate size for planting should be watered in the evening or the day before, depending on weather conditions, to facilitate easy removal (Eşiyok 2012).

Depending on the soil type, a hoe should be used once or twice, or after each irrigation. If the cabbage seedlings in the growing area are not growing as desired, water them until they reach the appropriate size. Once they have reached the required size, superficial hoeing can be done as their hairy roots develop close to the soil surface. After this, the plants will begin to grow rapidly. The watering should be done regularly, depending on their development. A second hoeing should be carried out when the cabbage leaves reach a diameter of 30–35 cm. After this, apply the reserved nitrogenous fertiliser to the leaf projections of the plants, taking care to avoid the roots (Eşiyok 2012).

Following the planting of seedlings, water the soil to prevent cracks from forming around their roots. Generally, regular irrigation is required after the initial watering, taking into account the soil structure, the environment and the plant developmental status. The cabbages require plenty of water; inadequate watering results in stunted growth and lower-quality produce over time. Therefore, irrigation should be performed regularly and at regular intervals (Eşiyok 2012).

When cultivating cabbages, if fertilisation with farm manure is not possible, at least 15–16 kg of nitrogen (N) should be applied per decare. For larger varieties, depending on the structure of the soil, 18 kg of nitrogen (N), 8–10 kg of phosphorus (P_2O_5) and 12–16 kg of potassium (K_2O) should be applied. Half of the nitrogen, along with all of the phosphorus and potassium, should be applied as $(NH_4)_2SO_4$ during the final tillage before planting. The remaining nitrogen should be applied as NO_3 to the leaf projections of the plants during the final hoeing (Eşiyok 2012).

(b) ***Brassica oleracea* var. *gemmifera*—Brussels sprouts:**

The processes involved in the preparation of the growing area, raising seedlings and planting them are very similar to those for head cabbage. The method of cultivating seedlings is the same. As with cabbage, Brussels sprout seedlings should be planted carefully to avoid damaging the growing tips. When planting in tubers, the spacing between and within rows varies depending on the growth characteristics and earliness of the variety. The seedlings are planted in the required field with 70–80 cm between rows and 60–70 cm within rows. The water is applied before planting in tubers. The seedlings are then manually placed in the mud at the neck of the tubers before the water has drained away, with the growing tips left above the water level. Watering should be repeated 4–5 days after planting, before the soil cracks (Eşiyok 2012).

Depending on the ecological conditions, hoeing should be carried out promptly after planting, provided that the climate and soil conditions are suitable. An application of the same amount of water and fertiliser regularly and without interruption is needed (Eşiyok 2012).

(c) ***Brassica oleracea* var. *botrytis*—cauliflower:**

The general procedure here follows: preparation of areas for seedling cultivation, the planting of seeds in these areas, the cultivation requirements in the seedling beds, the preparation of the main growing areas, the planting of seedlings in these areas and the cultural processes until harvest are all the same as for cabbage plants (Eşiyok 2012).

The spacing between cauliflower plants depends on whether they are an early or late variety. Interplant spacing is reduced for early varieties with a small habit and increased for late varieties with a larger habit. Under normal conditions, cauliflower seedlings are planted at a spacing of 60–70 cm between rows and 40–50 cm within rows in their primary production areas (Eşiyok 2012).

When growing cauliflower, the production area should be fertilised with farm manure 2–3 months before planting. The soil nutrients need to be incorporated before planting, taking into account the previous harvest. If there is insufficient organic matter, 3–6 tonnes of farm manure needs to be applied per decare. Additionally, deficiencies in boron and molybdenum in the soil during cauliflower cultivation can negatively impact plant growth. The areas experiencing these deficiencies must be supplemented with the necessary nutrients (Eşiyok 2012).

(d) ***Brassica oleracea* var. *italica*—broccoli:**

The production method for broccoli is very similar to that for cabbage and cauliflower. The preparation of seedling nurseries, sowing of seeds, cultural processes, preparation of main growing areas, planting of seedlings and cultural processes up to harvest are all the same as in the other plants (Eşiyok 2012).

The soil analysis results show that applying fertiliser positively affects broccoli yield and quality. This vegetable removes more nutrients from the soil than cauliflower. Its cultivation requires an application of 15–20 kg of nitrogen (N), 20 kg of phosphorus (P_2O_5) and 20–22 kg of potassium (K_2O) per decare. Half of the nitrogen fertiliser, along with all the phosphorus and potassium, is applied at planting time; the rest should be applied to the leaves after harvesting the main flower buds. Boron and molybdenum deficiencies negatively affect plant development in particular the molybdenum deficiency causes leaf blade shrinkage (Eşiyok 2012).

(e) ***Brassica oleracea*** **var.** ***gongylodes*****—kohlrabi:**

This vegetable is cultivated by either sowing the seeds directly or planting seedlings. The latter method is more commonly used in cold regions and for greenhouse cultivation (Eşiyok 2012).

The production areas are carefully cultivated; once fertilised, the soil is ready for sowing and planting. The tubers for sowing or transplanting are spaced 60–70 cm apart, and for direct sowing, two or three seeds are sown at a depth of 1–2 cm on either side of the tubers at the neck point, 30–35 cm apart within the rows in spring in warmer regions. The seeds germinate in 4–6 days after sowing, and plants emerge from the soil. Once the plants have reached a height of 3–5 cm, thinning needs to be carried out to leave one plant per planting area. The cultural practices need to be carried out regularly and without interruption until harvest (Eşiyok 2012).

The second cultivation method is seedling cultivation, which is used in cold regions. In this method, warm or humid beds are prepared. Seeds are sown in these beds to produce 700–1000 seedlings per square metre. After 6–8 weeks, the seedlings reach a suitable size for planting. The seedlings are then carefully uprooted and planted in mud at the neck of irrigated furrows, spacing is maintained between and within the rows, as in the direct seeding method. The post-planting cultural practices need to be carried out promptly; the plants sensitive to low temperatures should not be kept in such conditions during direct seed sowing or seedling cultivation (Eşiyok 2012).

Another cultivation method used in colder regions is greenhouse cultivation. Once they have reached the appropriate size, kohlrabi seedlings are transplanted to their primary growing areas at a spacing of either 20 × 20 cm or 15 × 15 cm between and within the rows. As the number of plants per unit area increases—that is, as the distance between plants decreases, the size of the kohlrabi decreases. However, the amount of produce harvested per unit area increases. Increasing the number of plants per unit area also promotes earlier harvesting (Eşiyok 2012).

After sowing seeds or planting seedlings, it is important to regularly weed the soil. This vegetable requires a regular and consistent supply of water, as inconsistent watering can result in kohlrabi remaining small and poor in quality. An application of excessive amounts of nitrogen fertiliser to kohlrabi plants accelerates vegetative growth, thus delaying the harvest and causing it to split. Generally, 12–14 kg/da of nitrogen (N), 10–15 kg/da of phosphorus (P_2O_5) and 20–25 kg/da of potassium (K_2O) needs to be applied to the production area before sowing and planting. The potassium fertiliser should be applied in the form of potassium sulphate (Eşiyok 2012).

(f) ***Brassica rapa*** **var.** ***rapa*****—turnip:**

In order to prepare the soil for sowing seeds, it is ploughed to a depth of 10–20 cm, then cleared of stones and other debris using a harrow before being levelled. Good soil preparation is essential to ensure proper seed germination and uniform plant emergence. The pans are prepared according to the slope of the growing area, and seeds are sown either by broadcasting or in rows. A row planting is preferred for easier cultural practices. If sprinkler irrigation is used, it is better to plant in a flat field than in pans. The seeds can be sown by hand, and 2–4 days after sowing, they germinate and the plants begin to emerge from the soil surface (Eşiyok 2012).

Seed spacing depends on plant development, the length of the growing season and the type of soil. The seeds are typically sown 30–40 cm apart within a row and 5–10 cm between rows, or broadcast at a depth of 2–3 cm. In large areas and on flatbeds, seeding is usually done mechanically. The machine is set to sow seeds either 30–40 cm apart or at 5–10 cm intervals; sowing is done at a depth of 2–3 cm in tempered soil (Eşiyok 2012).

The hoeing method is not used for cultivating turnips, as such a method of weeding can damage turnips and reduce the yield. To ensure high-quality, uniform growth, turnips must be watered regularly and consistently, and the amount of water applied depends on the developmental stage of the plant and the climatic conditions. The plants are watered 3–6 times during the production period; regular watering improves the quality of turnip radishes and prevents cracking. This vegetable benefits much from manure; as such, applying 4–6 tonnes of decomposed manure per decare before seeding has a positive impact on its yield. Additionally, 10–15 kg of nitrogen, 8–10 kg of phosphorus, 20–30 kg of potassium and 10–15 kg of calcium are needed to be applied per decare. If turnips are to be grown in sandy soil, the fertiliser should be applied in 3 stages: 1/3 before planting, 1/3 when the radishes begin to develop and the final third when they begin to grow; for better yields and quality, fertiliser should be applied according to the results of a soil analysis (Eşiyok 2012).

(g) ***Brassica rapa* var. *pekinensis*—Chinese cabbage:**

This vegetable lends itself better to autumn cultivation, and in irrigated areas, it is successfully grown as a second crop after the grain harvest. The seedlings are cultivated first to reduce the amount of time the plants spend in the field and to simplify care, and once they reach the required size, the seedlings are transplanted to their final growing location; the distance between plants varies depending on the variety (Eşiyok 2012).

Direct seeding is not used in Chinese cabbage cultivation. As its growing season is short, hoeing is only done once or twice, depending on the soil type; the first hoeing takes place around 10 days after planting, and the second hoeing is done around 20 days later. This cabbage requires plenty of water after planting and during the bud formation period because frequent changes in soil moisture can cause abnormal plant development. Therefore, watering needs to be frequent and regular to meet the needs of these plants. Before planting, 6 kg of nitrogen (N), 9 kg of phosphorus (P_2O_5) and 6 kg of potassium (K_2O) per decare should be applied in general as a base fertiliser during the final tillage (Eşiyok 2012).

3.1.2 Germination Behaviour and Role of Hormones

In addition to external stimuli, internal growth regulators like hormones play an important role in the growth and development regulation (Ozturk et al. 2023a). The production and perception of plant hormones are influenced by the environmental signals, pointing out to the fact that hormones not only coordinate intrinsic developmental processes but also transmit environmental inputs (Shahrasbi et al. 2021). Auxins (IAA), abscisic acid (ABA), gibberellic acid (GA), ethylene (ET), cytokinins (CK), salicylic acid (SA), jasmonic acid (JA), brassinosteroids (BR) and strigolactones (SLT) are hormones which play major role in growth and development along with stress management in plants (Ozturk et al. 1992). The phytohormones work in collaborative or antagonistic connections with each other, resulting in an ultimate growth or development that reflects their combined impact. IAA signalling pathways majorly help in growth and organ architecture. A development of adventitious roots (AR) and lateral roots (LR) is regulated by IAA; higher levels of IAA promote more AR and LR formation; it also helps in shoot growth by stimulating strigolactones, which help in increased expression of TB1 (teosinte branched 1)/ BRC1 (branched 1), which act on axillary buds to inhibit lateral shoot growth. The flower and fruit development are also regulated by IAA. Through the regulation of PIN and YUC gene expression trends and stages, certain TFs can modify the regional IAA distribution and regulate the growth of floral organs. In the fruit development and setting times, IAA signalling is strictly controlled. IAA control is among the most important plant hormones controlling the complicated procedure of the ovary-to-fruit transition. It works by regulating the expression of several genes involved in the production and distribution of IAA, cell division and proliferation and the development of flowers and fruits (Zhang et al. 2022). The cytokinins (CK) are compounds based on adenine or phenyl urea and are well known for controlling a variety of plant growth and developmental processes in addition to helping plants cope with stress. CKs play an important role in controlling chloroplast growth and efficiency and postpone the senescence of leaves. The first phase of the development of leaves is when the highest CK concentrations are observed, and this is explained by cytokinesis stimulation, membrane construction, plastid division and, at that time, intense protein synthesis. By preventing the depletion of chlorophyll as well as preserving the green colour, CKs prevent or slow down the ageing process of plants. The gibberellic acid (GA) is a pentacyclic diterpene which plays a vital role in the germination of seeds, dormancy, elongation of cells, reproductive growth, senescence process, yield, quality traits of different plants and tolerance against various stresses. ABA/GA ratio is the primary factor for germination and dormancy of seeds. In contrast to ABA, which keeps seeds dormant, GA positively controls seed germination. GA enhances plant growth by initiating division and elongation of cells, facilitating the meristem to shoot growth transition, vegetative phase to blooming. Numerous findings have demonstrated that GA_3 is essential for regulating a range of plant development traits, including vegetative growth, plant biomass, root and stem elongation and leaf area enlargement. It is an essential messenger which allows plants to become more resilient to abiotic stressors by modifying a number of cellular functions. Several investigators have hypothesised that by significantly improving the physio-biochemical characteristics of numerous crop plants, external GA_3 treatment enhances the physio-biochemistry of plants. The gibberellins have been shown to increase physiological processes by enhancing nutrient-use efficiency, limiting lipid peroxidation and ROS production while improving mineral nutrient absorption, chlorophyll pigment content, membrane permeability, seed germination, ion homeostasis, antioxidant activity, OP accumulation, fruit development, forced blooming and inter-nodal elongation, thereby mitigating the negative effects of stress conditions. GA_3 increases the amount of soluble sugars, membrane permeability, fatty acids, OP accumulation and antioxidant enzyme activities in crop plants, reducing the impacts of water stress. Through increased levels of endogenous hormones, OP accumulation, and antioxidant enzyme activity and ion homeostasis, GA_3 protects crop plants from the negative effects of temperature fluctuations and toxicity of HMs in agricultural plants (Shah et al. 2023). ABA is a 15-carbon sesquiterpenoid hormone that controls many aspects of plant growth and development, such as stomatal closure, abscission, seed dormancy, fruit ripening, germination, embryogenesis, leaf senescence and stress responses to biotic and abiotic stressors.

ABA plays a role in seed dormancy and maturation, stomatal opening and closing, water management in the plant body and plant adaptability to external stressors. During the abiotic stress response, ABA production and redistribution are crucial in plant water relations as they close the stomata and lower transpiration rate, limiting cell development. Hormone-responsive TFs are upregulated by ABA, and phosphatases and kinases are involved in ABA signalling, which mediates quick reactions to a range of abiotic stressors. A wide range of biochemical defences, including the biosynthesis of proline, antioxidants, ROS detoxifying enzymes, heat shock proteins and unsaturated fatty acids, as well as the strengthening of cuticular waxes, are triggered and mobilised by ABA, allowing plants to partially combat with detrimental impacts of stress. Furthermore, chemically brassinosteroids (BRs) are polyhydroxysteroids in nature involved in the growth of epicotyls and hypocotyls, seed germination, rhizogenesis, xylem differentiation, flowering, cell division and elongation, abscission, photomorphogenesis, growth, reproduction, senescence and plant responses against stress conditions. IAA and BRs are connected because BRs regulate their s, coordinate the tropic responses of plants and encourage the beginning of lateral roots during the primordial stage of lateral root development. In addition to these, a phenolic plant hormone called salicylic acid (SA) controls a number of plant growth and development processes, such as seed germination, fruit ripening, flowering, senescence and defence against stresses (Pirasteh-Anosheh et al. 2023). It aids in the development of SAR to a variety of infections. However, ethylene (ET) is gaseous in nature and plays a significant role in numerous essential morpho-physiological processes such as triple response, fruit ripening, senescence, cell division and elongation, abscission, flowering, and in various stress responses. ET signalling controls genes linked to stress that are essential for plant development and survival. The fatty acid α-linolenic acid is the source of jasmonates (JAs), which

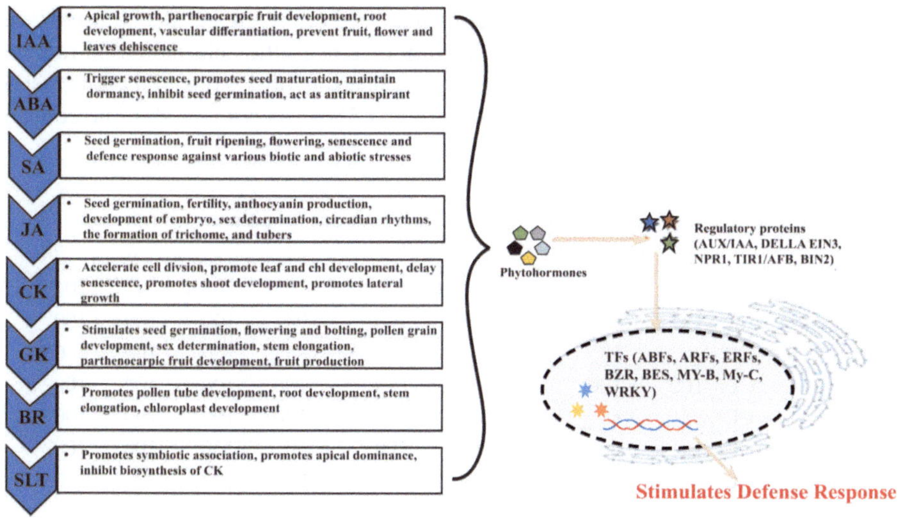

Fig. 3.1 Role of plant hormones in plant growth and stress responses

are oxylipins and control several aspects of plant growth and development, including senescence, anthocyanin synthesis, root growth, fertility, seed germination and plant defence. Jasmonates play a role in the stimulation of protection defences against wounding and herbivore attacks, both directly (by producing glucosinolate, methyl nicotinate, and cedrol) and indirectly (by activating defence gene signalling). They also have an important function that helps in abiotic stress tolerance; hence, research has concentrated on these chemicals due to their notable ability to defend plants from stress. JAs are responsible for germination and the development of the embryo as well as the reproductive organ development, sex determination, senescence, root growth, fruit ripening, trichome and tuber formation, leaf movement and gravitropism. Furthermore, the synthesis of secondary metabolites by JAs plays a role in plant steps that respond to seasonal and circadian rhythms (Jibran et al. 2013; El Sabagh et al. 2022) (Fig. 3.1).

3.1.2.1 Some Experimental Studies on the Germination and Role of the Hormones

An experimental study has been conducted to investigate the germination of culture forms of *Brassica juncea* cv. Obalakana and *Brassica campestris* cv. Shigatsu-shirona under dark and light conditions at various temperatures (10, 20 and 35 °C) using different concentrations of growth regulators.

An evaluation of the effects of growth regulators at given temperatures has revealed that the germination percentage of *B. juncea* seeds treated with gibberellic acid under continuous light was higher at 1 ppm and 10 °C than in the control group, whereas at other temperatures, it was similar to the control group. However, the radicle and hypocotyl lengths of the germinated seeds were greater than those of the control group at all temperatures. Application of gibberellic acid increased radicle and hypocotyl lengths. In *B. campestris*, application of 1 ppm gibberellic acid increased germination percentages at 10, 20, and 35 °C as compared to the control group (Tables 3.1 and 3.2).

Table 3.1 Percentages of germination at different temperatures for different growth regulators

			Brassica juncea				*Brassica campestris*			
		ppm	10 °C	20 °C	35 °C	10/35 °C	10 °C	20 OC	35 °C	10/35 °C
Gibberellin	CL	1	62	94	96	100	98	100	100	100
	CD		18	52	84	100	84	100	100	100
	CL	100	100	100	100	100	100	100	100	100
	CD		88	100	100	100	100	100	100	100
Kinetin	CL	1	60	88	98	100	96	100	100	100
	CD		22	58	64	100	88	100	100	100
	CL	50	84	100	100	100	98	98	100	98
	CD		34	68	92	100	80	100	100	100
Thiourea	CL	100	80	94	96	100	96	100	96	100
	CD		14	74	74	100	82	100	100	100
	CL	1000	84	100	100	100	100	100	94	90
	CD		36	78	66	100	92	100	100	100
KNO_3	CL	100	68	100	98	100	100	100	100	100
	CD		16	78	94	100	84	100	100	100
	CL	1000	94	100	100	100	100	100	100	96
	CD		8	60	68	100	94	94	84	100
IAA	CL	10	58	88	96	100	94	98	100	100
	CD		–	48	66	100	90	100	100	100
	CL	50	52	82	98	100	94	100	96	100
	CD		–	36	62	100	86	100	100	100
Ethrel	CL	10	38	78	96	100	96	100	90	98
	CD		4	6	–	40	76	90	66	90
	CL	50	54	86	66	100	86	96	98	98
	CD		10	44	–	54	80	86	74	92
	CL	100	40	98	74	98	94	98	96	94
	CD		14	44	24	66	62	80	90	84
	CL	200	18	80	20	50	80	92	72	80
	CD		4	44	14	36	60	80	60	84
	CL	400	10	60	6	18	52	82	52	78
	CD		–	46	8	18	46	70	32	76
ABA	CL	1	76	86	8	100	96	100	48	100
	CD		8	60	68	92	84	94	94	92
	CL	20	44	60	–	96	92	94	6	96
	CD		–	–	–	44	18	–	12	62
	CL	100	4	30	–	26	30	56	–	34
	CD		–	–	–	–	44	30	–	64

(continued)

Table 3.1 (continued)

			Brassica juncea				*Brassica campestris*			
		ppm	10 °C	20 °C	35 °C	10/35 °C	10 °C	20 OC	35 °C	10/35 °C
TIBA	CL	10	82	86	100	100	96	100	100	100
	CD		14	60	58	100	90	100	100	100
	CL	50	66	86	100	100	92	100	100	98
	CD		12	38	62	100	88	100	100	100
COM	CL	10	65	72	82	100	88	96	96	100
	CD		4	30	38	93	86	100	94	100
	CL	50	32	48	70	96	34	100	72	100
	CD		–	14	14	66	28	90	90	100
CCC	CL	1000	–	64	92	98	38	96	98	100
	CD		–	12	32	58	26	94	86	100
	CL	5000	–	56	70	98	14	98	84	94
	CD		–	6	8	50	5	74	42	100
MH	CL	100	62	96	94	100	96	100	96	100
	CD		62	37	28	95	84	96	30	100
	CL	1000	4	92	30	70	52	100	52	86
	CD		–	16	12	72	32	100	36	88
MNT	CL	1	64	96	90	100	88	100	96	100
	CD		36	42	38	100	92	100	80	100
	CL	10	60	88	100	100	94	100	90	100
	CD		16	46	20	100	94	100	34	100
PEG	CL	1	64	96	94	100	94	100	96	100
	CD		18	42	12	88	90	100	100	100
	CL	10	58	98	88	100	100	100	98	100
	CD		12	42	–	50	96	100	–	100
NaCl	CL	1000	68	94	94	100	92	100	100	98
	CD		14	28	2	94	82	100	80	100
	CL	20,000	–	–	–	–	–	–	–	–
	CD		–	–	–	–	–	2	–	–
Morphactin	CL	10	32	78	94	100	78	96	94	94
	CD		16	34	–	52	78	74	56	82
	CL	100	42	84	90	94	82	94	90	90
	CD		18	36	–	36	76	92	56	86
Control	CL		42	94	92	100	80	92	96	100
	CD		16	52	68	100	86	100	100	100

Looking at the effects of gibberellin in the dark, it was observed that there is a significant increase in *B. juncea* compared to the control group at all temperatures; however, no significant increase in radicle or hypocotyl lengths was observed. Similar results were obtained in *B. campestris* (Tables 3.1 and 3.2). In general, applying 1 ppm gibberellin was found to be more effective for both species when grown under light conditions, with increases observed in germination percentages as well as in radicle and hypocotyl lengths.

Table 3.2 Radicle and hypocotyl lengths (cm) under different growth regulators at different temperatures

			Brassica juncea								*Brassica campestris*							
			10 °C		20 °C		35 °C		10/35 °C		10 °C		20 °C		35 °C		10/35 °C	
		ppm	Rad	Hip	Rad	Hip	Rad	Hip	Rad	Hip	Rad	Hip	Rad	Hip	Rad	Hip	Rad	Hip
Gibberellin	**CL**	**1**	1.6	0.5	3.7	1.3	0.45	1.05	2	0.9	1.5	0.8	3.2	1.5	0.55	0.9	1.75	1.4
	CD		0.5	0.5	0.45	0.6	3.15	1.25	2.5	1	0.35	0.6	2.5	3.5	2.25	2.2	1.5	1.3
	CL	**100**	1.2	1.3	3.5	1.25	0.5	1.2	1.75	1.1	1.75	1.25	3.2	1.2	0.55	0.9	1.6	1.5
	CD		0.3	0.15	0.8	0.6	2.1	0.9	1.6	1	0.35	0.2	0.7	0.8	0.9	1.1	2.5	1.25
Kinetin	**CL**	**1**	1.25	0.5	2.7	1	0.5	0.8	1.1	0.75	1.4	0.75	3.7	1.3	0.3	1	1.5	1.2
	CD		0.2	0.15	0.55	0.7	2.3	1.8	0.8	0.6	0.35	0.2	0.4	0.55	2.5	2.2	1.2	1.3
	CL	**50**	0.75	0.5	0.9	0.9	0.45	0.9	0.4	0.55	1	0.7	1.6	1.5	0.25	0.9	0.5	1.3
	CD		0.2	0.15	0.5	0.6	0.35	0.85	0.3	0.8	0.3	0.2	0.25	0.6	1.4	2	0.5	0.7
Thiourea	**CL**	**100**	1.4	0.6	0.3	0.7	0.3	0.7	1.2	0.45	1.8	0.6	1.7	1.2	0.15	0.7	0.9	0.7
	CD		0.15	0.5	1.3	0.6	3.5	2.2	1.2	1	0.3	0.2	0.5	0.9	1.5	1.7	0.6	0.7
	CL	**1000**	1.1	0.45	2.5	1.1	0.35	0.45	1.1	0.6	1.3	0.6	2	1.4	0.15	0.4	0.7	0.75
	CD		0.3	0.2	1.2	0.7	3.7	3.25	1.5	0.7	0.3	0.25	1.5	1.2	2.5	3.2	1.3	1
KNO_3	**CL**	**100**	1.25	0.6	5.5	1.1	1.4	0.45	1.5	1	1.75	0.7	4.2	1.25	0.3	0.8	1.4	1
	CD		0.2	0.15	0.75	0.9	2.2	1.4	1.5	0.7	0.3	0.2	0.9	0.6	0.5	0.55	1.3	1
	CL	**1000**	1.3	0.6	2.5	1.7	0.5	0.9	1.2	0.7	0.9	0.8	3.2	1.5	0.35	0.9	0.7	1.2
	CD		0.6	0.6	0.55	0.9	1.6	1.9	0.9	0.6	0.3	0.2	0.6	1	1.5	2.2	1.2	1.5
IAA	**CL**	**10**	0.15	0.35	2	2	2.5	0.65	0.35	0.35	0.15	0.15	2	1.3	0.2	0.8	0.6	0.9
	CD		–	–	0.2	0.8	0.25	1.2	0.2	0.5	0.1	0.2	0.2	0.4	0.3	2	0.15	0.7
	CL	**50**	0.15	0.25	0.3	0.8	0.25	0.55	0.25	0.35	0.15	0.35	1.6	1.75	0.15	0.1	0.5	0.7
	CD		–	–	0.2	0.5	0.3	0.95	0.25	0.7	0.1	0.2	0.4	2.2	0.4	2	0.7	0.4

			Brassica juncea								*Brassica campestris*							
			10 °C		20 °C		35 °C		10/35 °C		10 °C		20 °C		35 °C		10/35 °C	
		ppm	Rad	Hip	Rad	Hip	Rad	Hip	Rad	Hip	Rad	Hip	Rad	Hip	Rad	Hip	Rad	Hip
Ethrel	**CL**	**10**	0.4	0.3	0.25	0.3	0.25	0.15	0.25	0.45	0.9	0.4	0.3	0.7	0.15	0.4	0.6	0.5
	CD		0.15	0.25	0.3	0.4	–	–	0.4	0.35	0.4	0.4	0.7	1	0.5	0.3	0.6	0.8
	CL	**50**	0.4	0.25	0.25	0.35	0.15	0.3	0.15	0.4	0.25	0.4	0.4	0.8	0.15	0.5	0.5	0.4
	CD		0.15	0.3	0.3	0.4	–	–	0.2	0.4	0.25	0.4	0.4	0.9	0.15	0.4	0.5	0.7
	CL	**100**	0.25	0.3	0.25	0.3	0.15	0.3	0.4	0.3	0.25	0.3	0.25	0.8	0.15	0.4	0.4	0.5
	CD		0.15	0.2	0.25	0.3	0.15	0.3	0.3	0.4	0.3	0.35	0.4	1	0.15	0.4	0.3	0.6
	CL	**200**	0.15	0.2	0.15	0.3	0.15	0.1	0.15	0.25	0.25	0.4	0.35	0.8	0.1	0.2	0.2	0.3
	CD		1	0.3	0.15	0.35	0.15	0.15	0.15	0.3	0.05	0.25	0.2	0.7	0.05	0.2	0.15	0.2
	CL	**400**	0.15	0.05	0.15	0.2	0.05	0.05	0.15	0.25	0.15	0.3	0.15	0.3	0.1	0.15	0.15	0.2
	CD		–	–	0.05	0.3	0.01	0.01	0.01	0.01	0.15	0.2	0.15	0.7	0.1	0.2	0.3	0.4
ABA	**CL**	**1**	1.1	0.4	3.2	1	0.5	0.7	1.3	0.6	1.2	0.6	3	1.5	0.15	0.6	1.1	0.9
	CD		0.15	0.05	1.2	0.5	0.3	0.6	0.5	0.8	0.2	0.15	0.6	0.5	0.4	0.9	0.9	1.1
	CL	**20**	0.15	0.15	0.01	0.01	–	–	0.15	0.15	0.15	0.15	0.1	0.1	0.15	0.2	0.1	0.05
	CD		–	–	–	–	–	–	0.01	0.01	–	–	–	–	0.15	1.5	–	
	CL	**100**	0.01	0.01	0.05	–	–	–	0.01	0.01	0.15	0.15	0.15	0.25	–	–	0.1	0.15
	CD		–	–	–	–	–	–	–	–	0.15	0.15	0.15	0.2	–	–	0.05	0.15
TIBA	**CL**	**10**	0.2	0.5	2.5	1.5	0.5	1	1.5	1	1.6	0.8	3	1.5	0.3	0.9	1.15	1.4
	CD		0.15	0.5	4.5	2.5	1.7	2.4	1.5	1	0.4	0.15	2.7	2.2	0.5	0.7	1.1	1
	CL	**50**	0.6	0.4	1.1	1.5	0.7	0.9	0.9	1.2	0.3	0.5	0.4	1.3	0.3	0.7	0.7	0.8
	CD		0.15	0.5	1.2	1.2	0.9	1.8	1.7	1	0.4	0.15	0.6	2.2	1.2	1.2	1	0.8

(continued)

Table 3.2 (continued)

			Brassica juncea								*Brassica campestris*							
			10 °C		20 °C		35 °C		10/35 °C		10 °C		20 °C		35 °C		10/35 °C	
		ppm	Rad	Hip	Rad	Hip	Rad	Hip	Rad	Hip	Rad	Hip	Rad	Hip	Rad	Hip	Rad	Hip
COM	**CL**	**10**	0.6	0.45	2.2	0.95	0.6	0.9	1.1	0.8	0.6	0.35	0.3	1.3	0.25	0.9	1	1
	CD		0.5	0.5	1.2	1.7	0.9	2.1	1.15	0.65	0.3	0.15	2.7	2.3	1.7	0.2	0.7	1
	CL	**50**	0.15	0.2	0.7	0.8	0.45	0.9	0.45	0.3	0.3	0.4	0.7	1.2	0.25	0.65	0.3	0.55
	CD		–	–	0.45	1.1	0.5	1.2	0.2	0.4	0.15	0.05	0.4	1.1	0.2	0.4	0.2	0.5
CCC	**CL**	**1000**	–	–	3.25	0.75	0.65	0.8	1.65	0.65	1.25	0.45	2.75	1.5	0.55	1.7	1.35	0.95
	CD		–	–	1	3	0.4	0.2	2.5	1.2	0.3	0.1	0.4	1.3	0.4	0.8	1	0.75
	CL	**5000**	–	–	1.5	0.55	0.6	0.65	1.1	0.55	0.75	0.25	0.95	0.6	1.5	0.35	0.6	0.5
	CD		–	–	–	–	0.4	0.5	1	0.5	–	–	0.8	0.3	–	–	0.5	0.5
MH	**CL**	**100**	0.55	0.4	0.55	1.1	0.6	0.95	0.65	0.55	1.15	0.65	1.15	1.1	0.35	0.55	0.8	0.9
	CD		0.2	0.2	0.5	1.5	–	–	0.5	0.5	0.3	0.2	0.9	1.9	–	–	1.7	1.2
	CL	**1000**	0.5	–	0.15	0.35	–	–	0.15	0.15	0.15	0.25	0.15	0.45	0.15	0.2	0.15	0.4
	CD		–	–	–	–	–	–	0.15	0.15	0.1	0.1	0.2	0.5	–	–	0.25	0.4
MNT	**CL**	**1**	1.5	0.45	3	0.75	0.55	0.9	1	0.6	2	0.75	3	1.25	0.4	0.75	0.7	0.7
	CD		0.1	0.1	4	2	1.2	1.7	1.5	0.7	0.3	0.4	0.9	1.5	–	–	1.5	0.9
	CL	**10**	1	0.45	3	0.9	0.55	0.85	1.2	0.6	2.5	0.65	2.25	1.3	0.45	1.25	1.25	0.9
	CD		0.1	0.1	3.7	2.7	0.5	0.9	1.5	0.7	0.2	0.2	0.5	0.6	–	–	0.8	1
PEG	**CL**	**1**	1.6	0.7	3.5	0.9	0.35	0.75	1.35	0.6	1.1	0.6	3.2	1.3	0.35	0.7	1.8	0.5
	CD		0.1	0.1	1.4	1.1	0.8	1.5	1.2	0.5	0.4	0.3	4	–	0.3	0.8	1.5	1
	CL	**10**	1.3	0.6	3.5	0.9	0.4	0.9	0.9	0.6	1.5	0.5	3.2	1.25	1.5	0.5	1.4	1.4
	CD		0.1	0.2	2	2.5	–	–	1.2	0.75	0.6	0.3	1.5	0.5	–	–	1.5	1.2
NaCl	**CL**	**1000**	1.2	0.45	2.5	1.25	0.55	0.8	1.5	0.9	2	0.8	2.5	2	0.3	0.6	1.1	1.25
	CD		0.1	0.1	2.5	3.5	0.2	0.3	1.5	1.2	0.2	0.2	0.3	0.6	–	–	0.9	0.7
	CL	**20,000**	**–**	–	–	–	–	–	–	–	–	–	–	–	–	–	–	–
	CD		**–**	–	–	–	–	–	–	–	–	–	–	–	–	–	–	–

			Brassica juncea								*Brassica campestris*							
			10 °C		20 °C		35 °C		10/35 °C		10 °C		20 °C		35 °C		10/35 °C	
		ppm	Rad	Hip	Rad	Hip	Rad	Hip	Rad	Hip	Rad	Hip	Rad	Hip	Rad	Hip	Rad	Hip
Morphactin	**CL**	**10**	0.35	0.35	0.3	0.35	0.15	0.3	0.55	0.35	0.3	0.4	0.35	0.9	0.15	0.3	0.6	0.5
	CD		0.7	0.3	0.8	0.4	–	–	0.35	0.45	0.7	0.6	1	0.7	0.15	0.4	0.5	0.7
	CL	**100**	0.25	0.3	0.3	0.3	0.1	0.3	0.45	0.25	0.4	0.3	0.3	0.8	0.15	0.45	0.4	0.6
	CD		0.4	0.3	0.65	0.5	–	–	0.45	0.35	0.6	0.5	0.6	11.2	0.15	0.25	0.4	0.65
Control	**CL**		0.4	0.25	0.8	0.4	0.25	0.3	0.55	0.95	1.25	0.45	1.75	0.85	0.2	0.4	1.75	0.6
	CD		0.4	0.3	0.4	2.5	2.8	1.7	1.5	0.75	0.3	0.15	0.6	0.55	3.25	2.7	1.25	0.75

The germination of all seeds at 3 temperatures was recorded by applying 100 ppm gibberellin under continuous light conditions, and radicle and hypocotyl lengths increased too. This is due to the germination-promoting effect of gibberellin, as the seeds treated with 100 ppm gibberellin and allowed to germinate in the dark achieved 100% germination in both species at all temperatures. However, the germination rate of *B. juncea* remained at 88 percent when seeds were treated with 100 ppm gibberellin and allowed to germinate at 10 °C. This may be because this species finds it more difficult to germinate than other species (Tables 3.1 and 3.2). In general, the seeds treated with 100 ppm gibberellin germinated under both light as well as dark conditions.

Examining different cultivars of *B. juncea* and *B. campestris* seeds treated with 1 and 500 ppm of kinetin at various temperatures has revealed that 50 ppm was effective at all temperatures, especially under dark conditions, compared to the control. An application of 1 ppm of kinetin under constant light conditions has increased the radicle and hypocotyl lengths, as compared to the control. The values obtained in the dark treatments were similar to those obtained in the continuous light (Tables 3.1 and 3.2).

A comparison of the germination percentages of *Brassica* seeds treated with thiourea at various temperatures has revealed that in *B. juncea*, it resulted in a significant increase at 10 °C under continuous light; in the case of *B. campestris*, the results were almost the same as the control. All increases at other temperatures were almost identical to control values. These results pointed out that thiourea application stimulates germination percentages in the *Brassica* cultivars used here. Thiourea application also stimulated radicle and hypocotyl growth under both light and dark conditions (Tables 3.1 and 3.2).

As far as KNO_3 application is concerned, it significantly increased germination percentages under continuous light compared to the control and reached up to 100% in *Brassica campestris* under all temperature treatments, but no such effect was observed under dark conditions. The results were similar to the control for both species (Tables 3.1 and 3.2).

Indole acetic acid (IAA) treatments were not significantly effective under continuous light conditions, producing results similar to those observed in the control group. No germination was observed in *B. juncea* under continuous darkness at 10 °C with either 10 or 50 ppm IAA; however, in *B. campestris* at 10 °C, the percentage exceeded control values. An inhibitory effect was also observed in *B. juncea* at other germination temperatures, except for the 10/35 °C treatment. In *B. campestris*, no inhibitory effects were observed at the used temperatures, and these were consistent with the control. Consequently, IAA had an inhibitory effect on *B. juncea* at low temperatures and under constant darkness. The results on radicle and hypocotyl length showed lower values than the control in all treatments (Tables 3.1 and 3.2).

Applying various concentrations of ethrel has revealed that there are inhibitory effects on both *B. juncea* and *B. campestris*, being parallel with increasing temperatures; this effect was more pronounced under continuous darkness (Tables 3.1 and 3.2).

Abscisic acid (ABA), a growth inhibitor, maintains its natural effect in both *B. juncea* and *B. campestris* in parallel with increases in temperature. In fact, 100 ppm ABA had no effect on germination in *B. juncea*, and only a minimal effect in *B. campestris* (Tables 3.1 and 3.2). These results suggest that *B. juncea* is more sensitive than *B. campestris*. TIBA, which has a germination-retarding effect, was also found to inhibit germination when percentages were examined under both light and dark conditions (Tables 3.1 and 3.2). Similar effects were observed in terms of radicle and hypocotyl length. The results from coumarin application also showed greater germination inhibition in *B. juncea* under constant darkness, in line with an increase in concentration (Tables 3.1 and 3.2).

Furthermore, the results obtained following application of CCC (a germination inhibitor) have demonstrated its effectiveness. Specifically, an application of 5000 ppm of chlormequat (CCC), a gibberellin antagonist, has proved more effective in both culture varieties, treated in the dark. However, it was ineffective in terms of germination percentages under dark or light conditions in the case of *B. campestris* at 10/35 °C (Tables 3.1 and 3.2). Moreover, the results of this experimental study indicate that CCC was the most effective inhibitor for both species at 10 °C.

The results regarding maleic hydrazide (MH) have shown that it inhibited germination compared to the control, and similar inhibitory effects were observed in terms of radicle and hypocotyl length throughout this series of applications (Tables 3.1 and 3.2).

As regards mannitol, the values obtained in the 1 ppm application series were similar to the control values. The germination percentages were unaffected at 10/35 °C temperature in the 10 ppm application series (Tables 3.1 and 3.2). Mannitol and polyethene glycol (PEG) have exhibited minimal inhibitory effects at some temperatures and negligible stimulatory effects at others.

Interestingly, *B. juncea* was more adversely affected by 1000 ppm NaCl treatment than the control group. As against this *B. campestris* exhibited greater salt tolerance. No germination was observed under dark or light treatments at 20,000 ppm NaCl; only 2% germination occurred in the dark at 20 °C (Tables 3.1 and 3.2).

When the values obtained from morphactin applications are compared with those from the control group, an inhibitory effect is observed in both species (Tables 3.1 and 3.2). An evaluation of Tables 3.1 and 3.2 reveals that CCC at 10 °C was the most effective inhibitor for both species under light conditions. At a maximum temperature of 35 °C, ABA was the most effective inhibitor for germination for both species under light conditions. No germination was observed in either species at 20,000 ppm NaCl treatment. At the optimal germination temperature of 10 °C, 5 ppm coumarin was a more effective inhibitor for *B. juncea* under light conditions; ABA was more effective for *B. campestris*. In the dark treatments, CCC was the most effective inhibitor for *B. juncea* at 10 °C, with almost all applied inhibitors, including maleic hydrazide, ABA and coumarin, proving effective. ABA was effective at both 20 and 35 °C. In light of the information presented above, inhibitors appear to be less effective at the optimal temperatures for these species. *B. campestris* appears to germinate more easily after 24 h in the dark, with an ideal temperature of 20 °C.

Table 3.3 The effects of applying different combinations of growth regulators with 1 ppm GB on the germination percentages of *Brassica juncea*

Applied plant growth regulators	Light	Germination percentages 10 °C	20 °C	30 °C
100 ppm IAA + 1 ppm GB	CD	–	8	–
100 ppm TIBA + 1 ppm GB	CD	–	6	2
100 ppm coumn + 1 ppm GB	CL	12	44	94
	CD	–	–	–
10.000 ppm MH + 1 ppm GB	CL	2	54	4
	CD	2	2	2
20 ppm ABA + 1 ppm GB	CL	30	100	6
	CD	6	10	–
100 ppm ABA + 1 ppm GB	CL	10	66	–
	CD	–	–	–
100 ppm MNI + 1 ppm GB	CD	6	26	4
100 ppm PEG + 1 ppm GB	CD	4	18	–
10 ppm ethrel + 1 ppm GB	CD	–	18	–
50 ppm ethrel + 1 ppm GB	CD	4	18	–
100 ppm ethrel + 1 ppm GB	CD	4	4	–
200 ppm ethrel + 1 ppm GB	CL	30	90	100
	CD	–	14	2
400 ppm ethrel + 1 ppm GB	CL	36	80	100
	CD	–	18	4
10 ppm morphactin + 1 ppm GB	CD	58	14	–
50 ppm morphactin + 1 ppm GB	CD	16	12	2
100 ppm morphactin + 1 ppm GB	CD	6	30	–
200 ppm morphactin + 1 ppm GB	CD	10	20	–
Control	CL	32	97	100
	CD	14	35	5

In another experimental series, two growth regulators with stimulatory properties (gibberellic acid and kinetin) were applied alongside growth regulators in the inhibitory or retarding group. Gibberellic acid and kinetin were applied at two different concentrations: a minimum of 1 ppm and a maximum of 100 ppm. Examining the germination percentages of *Brassica juncea* under varying combinations with 1 ppm GA_3 and other growth regulators (Table 3.3) revealed that no germination occurred in the presence of 100 ppm IAA and 1 ppm GB; the gibberellin eliminated ABA-inhibited germination under continuous light, and the germination approached the control level. In continuous darkness, however, this stimulating effect was halved. When the effect of morphantin, a known germination inhibitor, was compared to control values in continuous darkness, it was found to exceed the control level of 58 percent (Table 3.3).

An evaluation of the results of experimental applications of 100 ppm GA_3 instead of 1 ppm GA_3 at 10 °C (Table 3.4) revealed that an increase in the percentage was observed under continuous dark, when 100 ppm IAA and 100 ppm GA_3 were

Table 3.4 Germination percentages in *Brassica juncea* treated with different growth regulator combinations containing 100 ppm GB

Applied plant growth regulators	Light	Germination percentages		
		10 °C	20 °C	30 °C
100 ppm IAA + 100 ppm GB	CD	60	100	96
100 ppm TIBA + 100 ppm GB	CD	10	100	94
100 ppm coumn + 100 ppm GB	CL	10	100	100
	CD	–	44	16
10.000 ppm MH + 100 ppm GB	CL	2	70	10
	CD	2	12	22
20 ppm ABA + 100 ppm GB	CL	58	100	6
	CD	24	100	–
100 ppm ABA + 100 ppm GB	CL	4	7	–
	CD	12	28	–
100 ppm MNI + 100 ppm GB	CD	38	100	96
100 ppm PEG + 100 ppm GB	CD	28	100	74
10 ppm ethrel + 100 ppm GB	CD	58	100	92
50 ppm ethrel + 100 ppm GB	CD	88	100	88
100 ppm ethrel + 100 ppm GB	CD	34	100	98
200 ppm ethrel + 100 ppm GB	CL	90	100	100
	CD	42	100	92
400 ppm ethrel + 100 ppm GB	CL	88	100	100
	CD	62	100	84
10 ppm morphactin + 100 ppm GB	CD	14	100	56
50 ppm morphactin + 100 ppm GB	CD	66	96	40
100 ppm morphactin + 100 ppm GB	CD	26	94	32
200 ppm morphactin + 100 ppm GB	CD	36	100	20
Control	CL	32	97	100
	CD	14	35	5

applied together compared to the control group there was an increase of up to 90% under both continuous dark and continuous light, Ethrel gave the same result. However, the combined application of GA_3 with other inhibitors did not demonstrate a stimulatory effect. The results from the same application at 20 °C showed that GA_3 was highly effective in almost all cases, with germination values reaching 100%. At 35 °C, no germination was observed under light or dark conditions except in the ABA + GA_3 application. GA_3 exhibited a stimulatory effect in all applications except when combined with ABA.

An experimental series in which Kinetin was used instead of Gibberellin in *Brassica juncea* revealed that 1 ppm of Kinetin alongside other growth regulators resulted in a germination percentage that reached the control level at 10 °C under continuous darkness and in the presence of 20 ppm ABA and 1 ppm Kinetin. This treatment was also effective at 20 °C, but not at 35 °C. Similarly, applying 200 or 400 ppm ethrel and 1 ppm kinetin under continuous light resulted in a germination percentage that reached the control level at all three temperatures. However, the

Table 3.5 Germination percentages of *Brassica juncea* following the application of different growth regulator combinations at a concentration of 1 ppm kinetin

Applied plant growth regulators	Light	Germination percentages		
		10 °C	20 °C	30 °C
100 ppm IAA + 1 ppm kinetin	CD	4	8	2
100 ppm TIBA + 1 ppm kinetin	CD	2	16	6
100 ppm coumn + 1 ppm kinetin	CL	6	56	74
	CD	4	6	–
10.000 ppm MH + 1 ppm kinetin	CL	–	52	–
	CD	–	–	–
20 ppm ABA + 1 ppm kinetin	CL	34	88	6
	CD	4	6	2
100 ppm ABA + 1 ppm kinetin	CL	2	30	–
	CD	–	–	–
100 ppm MNI + 1 ppm kinetin	CD	4	22	2
100 ppm PEG + 1 ppm kinetin	CD	2	28	2
10 ppm ethrel + 1 ppm kinetin	CD	–	12	4
50 ppm ethrel + 1 ppm kinetin	CD	–	10	4
100 ppm ethrel + 1 ppm kinetin	CD	–	8	2
200 ppm ethrel + 1 ppm kinetin	CL	48	96	100
	CD	2	16	–
400 ppm ethrel + 1 ppm kinetin	CL	30	82	100
	CD	–	8	4
10 ppm morphactin + 1 ppm kinetin	CD	6	10	–
50 ppm morphactin + 1 ppm kinetin	CD	–	20	–
100 ppm morphactin + 1 ppm kinetin	CD	–	20	–
200 ppm morphactin + 1 ppm kinetin	CD	2	22	2
Control	CL	32	97	100
	CD	14	35	5

addition of kinetin was ineffective in continuous darkness (Table 3.5). No difference in germination percentages was observed when examining the results of the series in which kinetin was present at a higher concentration; these were similar to the results of the series in which 1 ppm kinetin was added (Table 3.6).

Examining the effects of applying 1 ppm GA_3 and other growth regulator combinations to *Brassica juncea* after germination (Table 3.7), it was found that the inhibitory effect of ABA was eliminated under continuous light, while the stimulating effect of GA_3 was observed. Under conditions containing 20 ppm ABA and 1 ppm GA_3 at 20 °C under continuous light, the hypocotyl length was 0.7 cm in the control group and 1.25 cm in the treatment group (Table 3.7). Whereas under combinations with 100 ppm gibberellin, the radicle and hypocotyl lengths increased accordingly (Table 3.8).

In this series of experiments involving *Brassica juncea*, no significant increase in radicle or hypocotyl lengths was observed after seed germination when combinations of kinetin and growth regulators at concentrations of 1 ppm and 100 ppm were

Table 3.6 Germination percentages of *Brassica juncea* following the application of different growth regulator combinations at a concentration of 100 ppm kinetin

Applied plant growth regulators	Light	Germination percentages		
		10 °C	20 °C	30 °C
100 ppm IAA + 100 ppm kinetin	CD	16	32	6
100 ppm TIBA + 100 ppm kinetin	CD	4	14	2
100 ppm coumn + 100 ppm kinetin	CL	12	80	88
	CD	6	6	–
10.000 ppm MH + 100 ppm kinetin	CL	4	78	8
	CD	–	–	–
20 ppm ABA + 100 ppm kinetin	CL	34	100	12
	CD	2	22	–
100 ppm ABA + 100 ppm kinetin	CL	6	78	–
	CD	–	8	–
100 ppm MNI + 100 ppm kinetin	CD	20	30	6
100 ppm PEG + 100 ppm kinetin	CD	8	22	4
10 ppm ethrel + 100 ppm kinetin	CD	14	16	6
50 ppm ethrel + 100 ppm kinetin	CD	2	10	6
100 ppm ethrel + 100 ppm kinetin	CD	2	12	6
200 ppm ethrel + 100 ppm kinetin	CL	54	86	100
	CD	–	18	10
400 ppm ethrel + 100 ppm kinetin	CL	46	88	96
	CD	4	30	24
10 ppm morphactin + 100 ppm kinetin	CD	6	44	–
50 ppm morphactin + 100 ppm kinetin	CD	4	40	–
100 ppm morphactin + 100 ppm kinetin	CD	10	32	–
200 ppm morphactin + 100 ppm kinetin	CD	4	26	–
Control	CL	32	97	100
	CD	14	35	5

applied, as compared to the control group (Tables 3.9 and 3.10). However, in another series of experiments it was noted that combinations containing gibberellins negated the effects of growth inhibitors and exhibited a potential synergistic effect (Table 3.7).

Table 3.11 shows the germination percentages of *Brassica juncea* seeds under experimental conditions, with various concentrations of ethrel (10, 50, 100, 200 and 400 ppm) applied in combination with growth-inhibitors like coumarin (100 ppm), MH (10,000 ppm), CCC (5000 ppm), NaCl (20,000 ppm) and ABA (20–100 ppm). The best germination percentage (88%) is observed when 100 ppm IAA and 50 ppm ethrel are used together at 20 °C under continuous light conditions; this value decreases in the dark (Table 3.11). At a temperature of 35 °C, the germination percentage increased at all Ethrel concentrations. The same results were observed in the control group (Table 3.11). When growth inhibitors and Ethrel were applied together, the inhibitory effect persisted (Table 3.11). However, when the effects of

Table 3.7 Radicle and hypocotyl lengths (cm) in *Brassica juncea* treated with different growth regulator combinations containing 1 ppm GB

Applied plant growth regulators	Light	Radicle and hypocotyl lengths					
		10 °C		20 °C		30 °C	
		Rad	Hip	Rad	Hip	Rad	Hip
100 ppm IAA + 1 ppm GB	CD	–	–	0.15	0.3	–	–
100 ppm TIBA + 1 ppm GB	CD	–	–	0.15	0.4	–	–
100 ppm coumn + 1 ppm GB	CL	0.05	0.1	0.4	0.45	0.5	0.55
	CD	–	–	–	–	–	–
10.000 ppm MH + 1 ppm GB	CL	–	–	0.05	0.15	–	–
	CD	–	–	–	–	0.1	0.2
20 ppm ABA + 1 ppm GB	CL	0.15	0.25	1.25	0.45	0.05	0.05
	CD	0.05	0.15	0.15	0.3	–	–
100 ppm ABA + 1 ppm GB	CL	–	–	0.15	0.2	–	–
	CD	–	–	–	–	–	–
100 ppm MNI + 1 ppm GB	CD	0.4	0.3	0.25	0.9	0.1	0.2
100 ppm PEG + 1 ppm GB	CD	0.25	0.25	0.4	0.95	–	–
10 ppm ethrel + 1 ppm GB	CD	–	–	0.15	0.7	–	–
50 ppm ethrel + 1 ppm GB	CD	0.1	0.3	0.15	0.5	–	–
100 ppm ethrel + 1 ppm GB	CD	0.1	0.3	0.1	0.3	–	–
200 ppm ethrel + 1 ppm GB	CL	0.25	0.3	0.75	0.5	0.25	0.35
	CD	–	–	0.15	0.35	0.1	0.2
400 ppm ethrel + 1 ppm GB	CL	0.25	0.3	0.55	0.35	0.3	0.4
	CD	–	–	0.1	0.35	0.1	0.3
10 ppm morphactin + 1 ppm GB	CD	0.65	0.3	0.25	0.45	–	–
50 ppm morphactin + 1 ppm GB	CD	0.35	0.35	0.35	0.55	–	–
100 ppm morpactin + 1 ppm GB	CD	0.3	0.15	0.4	0.6	–	–
200 ppm morphactin + 1 ppm GB	CD	0.25	0.3	0.15	0.5	–	–
Control	CL	0.45	0.27	0.7	0.95	0.9	0.8
	CD	0.05	0.15	0.45	1	0.2	0.2

the same experimental application on radicle and hypocotyl length were examined, the normal effects of the growth inhibitors were noted (Table 3.12).

Table 3.13 shows the results of the effects of various concentrations and combinations of Ethrel and growth inhibitors on the germination percentage of *Brassica campestris*. These values are almost identical to those obtained in *B. juncea*. Examining the effects of this series of experimental combinations on radicle and hypocotyl lengths (Table 3.14). It reveals an inhibitory effect, which is more pronounced in the application of 10 ppm Ethrel alongside inhibitors than in the control.

In another experimental setup, two concentrations of KNO_3 were added to the medium at different temperatures in *B. juncea* and *B. campestris*. The results of these experiments are outlined in Tables 3.15 and 3.16. The data obtained reveals that the germination percentage of *B. juncea* at 10 °C was 92% in the presence of 1000 ppm KNO_3 and 50 ppm TIBA under continuous light conditions compared to the control. The corresponding percentage in the control at the same temperature

Table 3.8 Radicle and hypocotyl lengths (cm) in *Brassica juncea* treated with different growth regulator combinations containing 100 ppm GB

		Radicle and hypocotyl lengths					
		10 °C		20 °C		30 °C	
Applied plant growth regulators	Light	Rad	Hip	Rad	Hip	Rad	Hip
100 ppm IAA + 100 ppm GB	CD	0.15	0.3	0.15	0.45	0.15	0.3
100 ppm TIBA + 100 ppm GB	CD	0.2	0.25	0.15	0.45	0.15	0.4
100 ppm coumn + 100 ppm GB	CL	0.15	0.2	0.4	0.5	0.15	0.2
	CD	–	–	0.35	0.65	0.1	0.25
10.000 ppm MH + 100 ppm GB	CL	–	–	0.05	0.2	–	–
	CD	–	–	–	–	–	–
20 ppm ABA + 100 ppm GB	CL	0.25	0.2	0.7	0.45	–	–
	CD	0.1	0.2	0.55	0.5	–	–
100 ppm ABA + 100 ppm GB	CL	–	–	0.15	0.2	–	–
	CD	–	–	–	–	–	–
100 ppm MNI + 100 ppm GB	CD	0.4	0.3	0.6	1.05	0.25	0.5
100 ppm PEG + 100 ppm GB	CD	0.8	0.4	0.5	0.65	0.3	0.45
10 ppm ethrel + 100 ppm GB	CD	0.2	0.4	0.4	0.75	0.15	0.35
50 ppm ethrel + 100 ppm GB	CD	0.15	0.35	0.25	0.65	0.15	0.25
100 ppm ethrel + 100 ppm GB	CD	0.15	0.3	0.25	0.65	0.15	0.3
200 ppm ethrel + 100 ppm GB	CL	0.3	0.3	0.45	0.5	0.15	0.4
	CD	0.15	0.3	0.3	0.65	0.15	0.3
400 ppm ethrel + 100 ppm GB	CL	0.3	0.4	0.25	0.45	0.15	0.45
	CD	0.1	0.3	0.25	0.55	0.15	0.4
10 ppm morphactin + 100 ppm GB	CD	0.7	0.5	0.45	0.5	0.1	0.3
50 ppm morphactin + 100 ppm GB	CD	0.75	0.6	0.35	0.65	0.1	0.35
100 ppm morphactin + 100 ppm GB	CD	0.3	0.3	0.35	0.65	0.1	0.2
200 ppm morphactin + 100 ppm GB	CD	0.35	0.25	0.35	0.6	0.1	0.3
Control	CL	0.45	0.27	0.7	0.95	0.9	0.8
	CD	0.05	0.15	0.45	1	0.2	0.2

was 60%. The most effective inhibitory treatment was a continuous light with 100 ppm KNO_3 and CCC added to 1000 ppm KNO_3, resulting in the percentage of germination at level 0. Furthermore, adding 20,000 ppm NaCl to the KNO_3 also resulted in zero germination. Another interesting result is that the inhibitory effect of ABA under constant light conditions is eliminated by KNO_3 (Table 3.15). When the results at 10 °C under constant darkness are compared with the control values in the same table, no significant stimulatory effect is observed, and inhibitory conditions appear to be more prevalent. Examining the results under constant light conditions reveals that 20 °C is more favourable than 35 °C; however, the 10/35 °C temperature application is optimal for all conditions. Furthermore, results close to the control values were obtained in both the 20,000 ppm NaCl and 20 ppm ABA applications (Table 3.15).

Therefore, it can be concluded that adding KNO_3 to the *B. juncea* medium counteracts the effects of inhibitors. This effect is particularly noticeable at temperatures

Table 3.9 Radicle and hypocotyl lengths (cm) in *Brassica juncea* treated with different growth regulator combinations containing 1 ppm kinetin

		Radicle and hypocotyl lengths					
		10 °C		20 °C		30 °C	
Applied plant growth regulators	Light	Rad	Hip	Rad	Hip	Rad	Hip
100 ppm IAA + 1 ppm kinetin	CD	–	–	0.3	0.6	–	–
100 ppm TIBA + 1 ppm kinetin	CD	–	–	0.4	0.5	0.2	0.4
100 ppm coumn + 1 ppm kinetin	CL	0.3	0.4	0.45	0.45	0.15	0.3
	CD	–	–	0.1	0.3	–	–
10.000 ppm MH + 1 ppm kinetin	CL	–	–	0.1	0.25	–	–
	CD	–	–	–	–	–	–
20 ppm ABA + 1 ppm kinetin	CL	0.2	0.2	0.3	0.35	0.05	0.05
	CD	–	–	0.2	0.3	–	–
100 ppm ABA + 1 ppm kinetin	CL	–	–	0.05	0.05	–	–
	CD	–	–	–	–	–	–
100 ppm MNI + 1 ppm kinetin	CD	0.1	0.2	0.4	0.95	0.4	0.6
100 ppm PEG + 1 ppm kinetin	CD	–	–	0.25	0.95	0.1	0.15
10 ppm ethrel + 1 ppm kinetin	CD	–	–	0.25	0.45	0.3	0.4
50 ppm ethrel + 1 ppm kinetin	CD	–	–	0.25	0.45	0.2	0.4
100 ppm ethrel + 1 ppm kinetin	CD	–	–	0.25	0.35	0.1	0.5
200 ppm ethrel + 1 ppm kinetin	CL	0.45	0.45	0.55	0.35	0.15	0.35
	CD	–	–	0.3	0.4	–	–
400 ppm ethrel + 1 ppm kinetin	CL	0.35	0.4	0.4	0.3	0.15	0.4
	CD	–	–	0.15	0.4	0.1	0.4
10 ppm morphactin + 1 ppm kinetin	CD	0.1	0.2	0.35	0.8	–	–
50 ppm morphactin + 1 ppm kinetin	CD	–	–	0.25	0.55	–	–
100 ppm morphactin + 1 ppm kinetin	CD	–	–	0.35	0.8	–	–
200 ppm morphactin + 1 ppm kinetin	CD	–	–	0.35	0.65	–	–
Control	CL	0.45	0.27	0.7	0.95	0.9	0.8
	CD	0.05	0.15	1.45	1	0.2	0.2

of 20 °C and 10/35 °C. A similar situation is observed for *B. campestris* as per the values in Table 3.16, with better germination at 10 °C than at 35 °C. Nevertheless, the optimal germination temperature appears to be 10/35 °C. Comparing the values in Tables 3.15 and 3.16, it is evident that *B. campestris* exhibits favourable germination under continuous darkness conditions. Notably, while the germination percentage was zero in the control group at 35 °C, it reached 92, 94 and 100 percent in the KNO_3 and TIBA-IAA treatments, respectively.

The results of the effects of potassium nitrate (KNO_3) and various combinations of growth regulators on the length of the radicle and hypocotyl in *B. juncea* and *B. campestris* are presented in Tables 3.17 and 3.18. The results presented in these tables reveal that adding KNO_3 to the medium and applying 1000 ppm NaCl at 10 °C produced a hypocotyl length of 1.4 cm in *B. juncea*, as opposed to 0.4 cm in the control. At 10/35 °C, the radicle and hypocotyl lengths were similar to the control values (Table 3.17). Similar results to the control group were obtained in

Table 3.10 Radicle and hypocotyl lengths (cm) in *Brassica juncea* treated with different growth regulator combinations containing 100 ppm kinetin

		Radicle and hypocotyl lengths					
		10 °C		20 °C		30 °C	
Applied plant growth regulators	Light	Rad	Hip	Rad	Hip	Rad	Hip
100 ppm IAA + 100 ppm kinetin	CD	0.1	0.15	0.15	0.6	0.1	0.4
100 ppm TIBA + 100 ppm kinetin	CD	0.1	0.2	0.15	0.35	–	–
100 ppm coumn + 100 ppm kinetin	CL	0.05	0.1	0.25	0.3	0.15	0.2
	CD	–	–	–	–	–	–
10.000 ppm MH + 100 ppm kinetin	CL	–	–	0.15	0.2	–	–
	CD	–	–	–	–	–	–
20 ppm ABA + 100 ppm kinetin	CL	0.3	0.2	0.35	0.4	0.1	0.2
	CD	–	–	0.45	0.55	–	–
100 ppm ABA + 100 ppm kinetin	CL	–	–	0.1	0.2	–	–
	CD	–	–	0.15	0.25	–	–
100 ppm MNI + 100 ppm kinetin	CD	0.15	0.3	0.3	0.55	0.2	0.3
100 ppm PEG + 100 ppm kinetin	CD	0.1	0.35	0.25	0.6	0.1	0.2
10 ppm ethrel + 100 ppm kinetin	CD	0.1	0.25	0.25	0.45	0.1	0.5
50 ppm ethrel + 100 ppm kinetin	CD	0.05	0.05	0.25	0.45	0.1	0.5
100 ppm ethrel + 100 ppm kinetin	CD	0.1	0.2	0.25	0.45	0.1	0.3
200 ppm ethrel + 100 ppm kinetin	CL	0.35	0.35	0.15	0.45	0.15	0.4
	CD	–	–	0.25	0.4	0.1	0.3
400 ppm ethrel + 100 ppm kinetin	CL	0.35	0.35	0.35	0.35	0.15	0.15
	CD	0.1	0.3	0.25	0.3	0.1	0.4
10 ppm morphactin + 100 ppm kinetin	CD	0.1	0.4	0.35	0.55	–	–
50 ppm morphactin + 100 ppm kinetin	CD	0.1	0.3	0.25	0.5	–	–
100 ppm morphactin + 100 ppm kinetin	CD	0.1	0.3	0.25	0.5	–	–
200 ppm morphactin + 100 ppm kinetin	CD	0.1	0.3	0.25	0.5	–	–
Control	CL	0.45	0.27	0.7	0.95	0.9	0.8
	CD	0.05	0.15	0.45	1	0.2	0.2

B. campestris, even with ABA application (Table 3.18). In *B. campestris* control, radicle and hypocotyl lengths were zero under continuous dark conditions at 35 °C; elongation was only observed in the presence of IAA and TIBA in the medium.

Tables 3.19 and 3.20 present data on the interactions between thiourea and other growth regulators in terms of germination percentages. According to these tables, under continuous light conditions in *B. juncea* at 10 °C, inhibitory effects were only observed when CCC and 20,000 ppm NaCl were applied alongside thiourea. No stimulatory effects were observed with any other applications. At 20 °C, the results generally mirrored those of the control, and at 35 °C, inhibitory effects were only observed with ABA. In the 10/35 °C application, inhibitory effects were observed with the 20,000 ppm NaCl + thiourea combinations, while stimulatory effects were observed with others (Tables 3.19 and 3.20). Under continuous darkness at 10 °C, an inhibitory effect was generally observed, while a stimulatory effect was only observed in the 1000 ppm thiourea and 1 ppm ABA treatment. While the control

Table 3.11 Germination percentages in *Brassica juncea* in the application of growth regulator combination with different ethrel concentrations

		10 ppm ethrel			50 ppm ethrel			100 ppm ethrel			200 ppm ethrel			400 ppm ethrel		
	Light	10–15 °C	20 °C	35 °C	10–15 °C	20 °C	35 °C	10–15 °C	20 °C	35 °C	10–15 °C	20 °C	35 °C	10–15 °C	20 °C	35 °C
100 ppm IAA	CL	46	82	92	60	88	90	26	59	92	60	66	90	28	80	92
	CD	22	32	–	26	22	–	16	20	–	–	30	–	2	2	–
100 ppm TIBA	CL	58	74	86	78	80	92	40	44	94	24	50	88	38	56	94
	CD	32	30	4	34	26	–	18	24	4	4	20	–	2	–	–
100 ppm couma	CL	8	34	34	–	38	32	6	18	78	6	28	78	6	6	90
	CD	2	4	–	–	–	–	2	–	–	–	–	–	–	–	–
10.000 ppm MH	CL	42	70	6	34	78	8	8	56	4	–	32	4	–	42	2
	CD	6	10	–	2	4	–	4	–	–	–	–	2	–	–	4
5.000 ppm CCC	CL	16	68	56	26	70	84	6	32	84	12	45	92	14	34	86
	CD	2	10	–	2	–	–	2	2	2	–	–	4	–	4	–
20.000 ppm NaCl	CL	24	92	56	42	90	56	–	90	42	–	58	78	14	78	100
	CD	4	14	–	–	–	–	4	–	–	–	–	–	–	2	–
20 ppm ABA	CL	58	76	–	62	84	4	14	72	6	12	50	–	20	62	–
	CD	28	26	–	28	–	–	24	–	–	2	–	–	–	–	–
100 ppm ABA	CL	28	42	–	22	20	–	–	6	–	–	8	–	–	6	–
	CD	10	–	–	6	–	–	10	–	–	–	–	2	–	–	–
Control	CL	54	78	92	48	92	98	40	74	98	60	78	94	48	72	96
	CD	38	48	4	32	30	8	40	48	6	38	36	6	40	42	–

Table 3.12 Radicle and hypocotyl lengths (cm) in *Brassica juncea* treated with combinations of growth regulators at different ethrel concentrations

		10 ppm ethrel						50 ppm ethrel						100 ppm ethrel						200 ppm ethrel						400 ppm ethrel					
	Light	10–15 °C		20 °C		35 °C		10–15 °C		20 °C		35 °C		10–15 °C		20 °C		35 °C		10–15 °C		20 °C		35 °C		10–15 °C		20 °C		35 °C	
		Rad	Hip	Rad	Hip	Rad	Hip	Rad	Hip	Rad	Hip	Rad	Hip	Rad	Hip	Rad	Hip	Rad	Hip	Rad	Hip	Rad	Hip	Rad	Hip	Rad	Hip	Rad	Hip	Rad	Hip
100 ppm IAA	CL	0.1	0.3	0.2	0.3	0.1	0.3	0.15	0.3	0.1	0.35	0.1	0.4	0.1	0.3	0.1	0.35	0.1	0.4	0.1	0.3	0.1	0.35	0.1	0.3	0.1	0.3	0.1	0.35	0.1	0.2
	CD	0.1	0.15	0.1	0.35	–	–	0.1	0.15	0.15	0.3	–	–	0.1	0.25	0.15	0.5	–	–	–	–	0.15	0.6	–	–	–	–	–	–	–	–
100 ppm TIBA	CL	0.15	0.3	0.15	0.4	0.1	0.3	0.15	0.3	0.15	0.35	0.15	0.35	0.1	0.35	0.1	0.35	0.1	0.3	0.1	0.35	0.1	0.35	0.1	0.3	0.1	0.3	0.1	0.35	0.1	0.3
	CD	0.1	0.3	0.2	0.3	0.1	0.3	0.1	0.2	0.15	0.3	–	–	0.15	0.35	0.15	0.3	0.1	0.3	0.1	0.2	0.1	0.3	–	–	–	–	–	–	–	–
100 ppm couma	CL	0.1	0.2	0.15	0.35	0.1	0.4	–	–	0.15	0.35	0.1	0.15	0.15	0.25	0.15	0.3	0.15	0.35	0.1	0.15	0.1	0.3	0.15	0.35	0.1	0.15	0.15	0.25	0.1	0.25
	CD	–	–	0.1	0.3	–	–	–	–	–	–	–	–	–	–	–	–	–	–	–	–	–	–	–	–	–	–	–	–	–	–
10.000 ppm MH	CL	0.1	0.15	0.1	0.15	–	–	0.1	0.15	0.1	0.25	–	–	–	–	0.1	0.15	–	–	–	–	0.05	0.15	–	–	–	–	0.05	0.1	–	–
	CD	–	–	0.1	0.2	–	–	–	–	–	–	–	–	0.1	0.1	–	–	–	–	–	–	–	–	–	–	–	–	–	–	–	–
5.000 ppm CCC	CL	0.2	0.3	0.15	0.35	0.15	0.4	0.25	0.3	0.25	0.3	0.15	0.4	0.1	0.2	0.35	0.35	0.15	0.35	0.1	0.2	0.3	0.35	0.15	0.4	0.15	0.3	0.35	0.3	0.1	0.15
	CD	–	–	0.1	0.3	–	–	–	–	–	–	–	–	–	–	–	–	–	–	–	–	–	–	0.1	0.4	–	–	0.1	0.3	–	–
20.000 ppm NaCl	CL	0.15	0.3	0.25	0.4	0.15	0.4	0.1	0.3	0.25	0.4	0.1	0.25	–	–	0.2	0.4	0.05	0.1	–	–	0.15	0.35	0.1	0.3	0.1	0.3	0.1	0.4	0.15	0.35
	CD	0.1	0.1	0.2	0.35	–	–	–	–	–	–	–	–	0.1	0.05	–	–	–	–	–	–	–	–	–	–	–	–	0.1	0.3	–	–
20 ppm ABA	CL	0.25	0.3	0.35	0.35	–	–	0.25	0.3	0.25	0.35	0.1	0.3	0.1	0.25	0.25	0.3	0.2	0.5	0.05	0.15	0.3	0.4	–	–	0.05	0.15	0.1	0.35	–	–
	CD	0.1	0.2	0.15	0.3	–	–	0.1	0.3	–	–	–	–	0.1	0.15	–	–	–	–	–	0.05	–	–	–	–	–	–	–	–	–	–
100 ppm ABA	CL	0.05	0.1	0.1	0.25	–	–	–	–	–	–	–	–	–	–	–	–	–	–	–	–	–	–	–	–	–	–	–	–	–	–
	CD	0.1	0.1	–	–	–	–	0.1	0.1	–	–	–	–	0.05	0.05	–	–	–	–	–	–	–	–	–	–	–	–	–	–	–	–
Control	CL	0.35	0.3	0.25	0.4	0.1	0.3	0.25	0.3	0.25	0.4	0.1	0.3	0.15	0.3	0.3	0.4	0.1	0.3	0.15	0.3	0.3	0.45	0.1	0.3	0.15	0.3	0.25	0.35	0.1	0.4
	CD	0.2	0.3	0.3	0.4	0.1	0.25	0.15	0.35	0.4	0.45	0.1	0.4	0.15	0.3	0.25	0.4	0.1	0.35	0.15	0.3	0.35	0.45	–	–	0.15	0.25	0.35	0.45	–	–

Table 3.13 Germination percentages in *Brassica campestris* in growth regulator combination applications with different ethrel concentrations

		10 ppm ethrel			50 ppm ethrel			100 ppm ethrel			200 ppm ethrel			400 ppm ethrel		
	Light	10–15 °C	20 °C	35 °C	10–15 °C	20 °C	35 °C	10–15 °C	20 °C	35 °C	10–15 °C	20 °C	35 °C	10–15 °C	20 °C	35 °C
100 ppm couma	CL	66	98	86	78	96	86	80	90	82	82	88	98	68	100	84
	CD	78	74	18	70	88	30	16	44	6	18	82	–	40	44	20
10.000 ppm MH	CL	94	100	82	90	98	90	94	98	82	98	96	62	80	100	70
	CD	78	6	48	82	96	46	34	54	4	20	94	10	28	46	50
5.000 ppm CCC	CL	86	96	90	92	100	98	96	98	98	92	96	78	90	98	98
	CD	23	82	28	74	86	20	68	48	6	32	90	6	34	48	16
20.000 ppm NaCl	CL	82	100	72	76	100	76	80	100	76	62	100	44	58	100	82
	CD	60	82	4	54	74	10	72	32	–	20	32	–	16	34	–
20 ppm ABA	CL	98	98	54	94	96	38	94	98	52	100	98	42	100	100	54
	CD	96	96	16	96	100	22	96	52	2	72	50	14	54	50	–
100 ppm ABA	CL	74	88	8	88	92	22	82	92	20	78	80	4	82	86	4
	CD	96	86	–	94	86	–	94	20	–	24	32	–	20	14	–
Control	CL	84	100	94	74	100	88	70	100	92	90	100	92	100	100	88
	CD	76	66	78	98	86	24	96	64	44	94	68	60	100	72	40

Table 3.14 Radicle and hypocotyl lengths (cm) in *Brassica campestris* in growth regulator combinations with varying ethrel concentrations

		10 ppm ethrel						50 ppm ethrel						100 ppm ethrel						200 ppm ethrel						400 ppm ethrel					
	Light	10–15 °C		20 °C		35 °C		10–15 °C		20 °C		35 °C		10–15 °C		20 °C		35 °C		10–15 °C		20 °C		35 °C		10–15 °C		20 °C		35 °C	
		Rad	Hip	Rad	Hip	Rad	Hip	Rad	Hip	Rad	Hip	Rad	Hip	Rad	Hip	Rad	Hip	Rad	Hip	Rad	Hip	Rad	Hip	Rad	Hip	Rad	Hip	Rad	Hip	Rad	Hip
100 ppm couma	CL	0.25	0.4	0.45	0.85	0.1	0.35	0.25	0.4	0.3	0.65	0.15	0.35	0.1	0.35	0.3	0.65	0.1	0.3	0.1	0.35	0.1	0.65	0.05	0.2	0.1	0.4	0.1	0.5	0.15	0.35
	CD	0.05	0.2	0.15	0.4	0.05	0.05	0.15	0.3	0.35	0.3	0.15	0.3	0.1	0.15	0.15	0.35	–	–	0.1	0.2	0.15	0.35	–	–	0.1	0.3	0.2	0.3	0.05	0.15
10.000 ppm MH	CL	0.1	0.2	0.1	0.3	0.1	0.2	0.1	0.15	0.1	0.2	0.05	0.1	0.1	0.15	0.15	0.2	0.05	0.1	0.1	0.15	0.1	0.25	0.05	0.1	0.1	0.15	0.1	0.3	0.05	0.1
	CD	0.1	0.2	0.1	0.25	0.05	0.1	0.1	0.15	0.1	0.2	0.05	0.05	0.1	0.15	0.05	0.15	–	–	0.1	0.15	0.1	0.15	0.1	–	0.1	0.25	0.1	0.15	0.05	0.1
5.000 ppm CCC	CL	0.3	0.4	0.3	0.45	0.1	0.2	0.15	0.8	0.15	0.45	0.05	0.2	0.15	0.35	0.3	0.45	0.15	0.4	0.15	0.4	0.2	0.45	0.15	0.35	0.15	0.4	0.25	0.45	0.05	0.3
	CD	0.15	0.3	0.3	0.45	0.05	0.1	0.15	0.4	0.25	0.4	0.05	0.05	0.25	0.3	0.15	0.3	–	–	0.15	0.3	0.25	0.45	0.1	0.2	0.25	0.4	0.35	0.45	0.1	0.25
20.000 ppm NaCl	CL	0.35	0.4	0.9	0.75	0.15	0.35	0.25	0.4	0.3	0.65	0.15	0.45	0.25	0.4	0.35	0.5	0.15	0.45	0.1	0.4	0.35	0.6	0.1	0.4	0.15	0.4	0.2	0.6	0.1	0.3
	CD	0.15	0.3	0.15	0.4	–	–	0.15	0.3	0.35	0.4	0.1	0.2	0.15	0.3	0.15	0.45	–	–	0.05	0.15	0.15	0.3	–	–	0.1	0.2	0.3	0.4	–	–
20 ppm ABA	CL	0.3	0.4	0.65	0.75	0.05	0.2	0.3	0.4	0.35	0.55	0.05	0.15	0.3	0.4	0.8	0.55	0.05	0.15	0.15	0.4	0.3	0.6	0.1	0.3	0.15	0.3	0.15	0.5	0.05	0.2
	CD	0.3	0.4	0.3	0.7	0.05	0.1	0.25	0.35	0.35	0.65	–	–	0.15	0.3	0.3	0.35	–	–	0.1	0.4	0.3	0.7	0.05	0.1	0.1	0.2	0.35	0.5	–	–
100 ppm ABA	CL	0.1	0.3	0.35	0.35	–	–	0.1	0.35	0.1	0.3	0.05	0.1	0.1	0.4	0.3	0.45	0.05	0.15	0.1	0.3	0.1	0.25	–	–	0.1	0.3	0.1	0.3	0.15	0.5
	CD	0.1	0.3	0.15	0.35	–	–	0.1	0.2	0.15	0.35	–	–	0.1	0.2	0.05	0.05	–	–	0.1	0.15	0.05	0.1	–	–	0.1	0.15	0.05	0.1	–	–
Control	CL	0.15	0.4	0.5	0.5	0.1	0.2	0.5	0.4	0.35	0.75	0.15	0.35	0.15	0.4	0.35	0.75	0.1	0.35	0.15	0.4	0.4	0.75	0.1	0.4	0.15	0.4	0.4	0.6	0.15	0.3
	CD	0.4	0.45	0.3	0.75	0.15	0.4	0.3	0.45	0.3	0.75	0.1	0.25	0.35	0.45	0.4	0.8	0.1	0.25	0.15	0.5	0.4	0.75	0.15	0.3	0.3	0.45	0.3	0.75	0.1	0.3

Table 3.15 Percentages of germination in *B. juncea* in two different combinations of KNO_3 concentrations

			Temperatures			
Applied plant growth regulators		Light	10 °C	20 °C	35 °C	10/35 °C
100 ppm KNO_3	10 ppm IAA	CL	78	94	90	100
		CD	8	62	92	100
	50 ppm IAA	CL	80	100	88	96
		CD	24	56	62	98
1000 ppm KNO_3	10 ppm IAA	CL	74	98	100	98
		CD	28	70	72	100
	50 ppm IAA	CL	62	98	98	100
		CD	30	72	90	98
100 ppm KNO_3	10 ppm TIBA	CL	86	98	98	100
		CD	8	72	86	100
	50 ppm TIBA	CL	74	98	96	98
		CD	4	72	34	98
1000 ppm KNO_3	10 ppm TIBA	CL	78	100	96	100
		CD	12	78	100	100
	50 ppm TIBA	CL	92	100	98	98
		CD	–	68	80	100
100 ppm KNO_3	10 ppm coumarin	CL	74	94	92	100
		CD	14	66	32	94
	50 ppm coumarin	CL	70	82	88	100
		CD	–	34	28	84
1000 ppm KNO_3	10 ppm coumarin	CL	74	96	96	100
		CD	20	62	60	100
	50 ppm coumarin	CL	72	88	90	100
		CD	4	30	22	82
100 ppm KNO_3	1 ppm ABA	CL	86	100	82	100
		CD	16	40	74	100
	20 ppm ABA	CL	74	90	–	96
		CD	4	20	–	94
1000 ppm KNO_3	1 ppm ABA	CL	80	100	84	100
		CD	24	86	90	100
	20 ppm ABA	CL	46	66	–	90
		CD	2	18	–	100
100 ppm KNO_3	1000 ppm CCC	CL	6	88	98	100
		CD	–	22	68	90
	5000 ppm CCC	CL	–	80	92	100
		CD	–	8	44	68
1000 ppm KNO_3	1000 ppm CCC	CL	2	90	100	100
		CD	–	32	70	94
	5000 ppm CCC	CL	–	86	90	100
		CD	–	6	64	90

(continued)

Table 3.15 (continued)

			Temperatures			
Applied plant growth regulators		Light	10 °C	20 °C	35 °C	10/35 °C
100 ppm KNO_3	100 ppm MH	CL	84	98	94	100
		CD	24	90	76	100
	1000 ppm MH	CL	80	94	72	96
		CD	4	80	74	100
1000 ppm KNO_3	100 ppm MH	CL	72	98	100	100
		CD	46	86	86	100
	1000 ppm MH	CL	78	98	68	100
		CD	2	68	60	100
100 ppm KNO_3	1 ppm manitol	CL	88	100	96	100
		CD	18	74	92	100
	10 ppm manitol	CL	88	100	98	100
		CD	32	90	100	100
1000 ppm KNO_3	1 ppm manitol	CL	74	100	94	98
		CD	16	68	96	100
	10 ppm manitol	CL	90	98	96	100
		CD	32	82	90	100
100 ppm KNO_3	1000 NaCl	CL	86	100	96	100
		CD	24	26	40	100
	20.000 NaCl	CL	–	82	–	72
		CD	–	24	–	58
1000 ppm KNO_3	1000 NaCl	CL	88	100	94	100
		CD	20	80	60	100
	20.000 NaCl	CL	–	88	6	94
		CD	–	8	–	80
Control		CL	60	96	86	100
		CD	26	84	80	100

value was 20%, it increased to 46% with the combination treatment. In the 20 °C test series, an inhibitory effect was generally observed, which decreased with the application of 1000 ppm thiourea and 1 ppm ABA alone. At 35 °C, almost all treatments exhibited inhibitory effects. In the 10/35 °C treatment, only CCC and 20,000 ppm NaCl combinations exhibited inhibitory effects, with other combination treatments showing results similar to the control (Table 3.19).

B. campestris generally germinates better at temperatures of 10 °C, 20 °C and 10/35 °C, with the exception of 35 °C. Under continuous light treatments, the combination of CCC and 20,000 ppm NaCl had an inhibitory effect at 10 °C; on other species, however, it had a stimulatory effect or was similar to the control. At 20 °C, the effects of CCC and 20,000 ppm NaCl combination treatment were similar to those of the control treatment; however, at 35 °C, the results were generally similar to the control, although the germination percentage dropped to 12% with the combination treatment of 100 ppm thiourea + 20 ppm ABA. Under 10/35 °C, the results

Table 3.16 Percentages of germination in *B. campestris* in two different combinations of KNO_3 concentrations

			Temperatures			
Applied plant growth regulators		Light	10 °C	20 °C	35 °C	10/35 °C
100 ppm KNO_3	10 ppm IAA	CL	100	100	96	100
		CD	92	100	100	100
	50 ppm IAA	CL	90	100	100	94
		CD	82	100	40	94
1000 ppm KNO_3	10 ppm IAA	CL	96	100	100	100
		CD	94	100	40	96
	50 ppm IAA	CL	96	100	98	96
		CD	84	98	50	98
100 ppm KNO_3	10 ppm TIBA	CL	96	100	100	100
		CD	84	100	–	100
	50 ppm TIBA	CL	98	100	98	100
		CD	82	98	92	100
1000 ppm KNO_3	10 ppm TIBA	CL	96	100	96	100
		CD	82	98	94	100
	50 ppm TIBA	CL	96	100	98	98
		CD	74	100	–	100
100 ppm KNO_3	10 ppm coumarin	CL	98	96	100	98
		CD	82	100	–	100
	50 ppm coumarin	CL	86	100	98	100
		CD	70	100	–	92
1000 ppm KNO_3	10 ppm coumarin	CL	100	100	96	98
		CD	78	100	–	100
	50 ppm coumarin	CL	96	100	92	98
		CD	54	100	–	100
100 ppm KNO_3	1 ppm ABA	CL	100	100	96	98
		CD	96	98	–	98
	20 ppm ABA	CL	96	96	30	98
		CD	52	74	–	90
1000 ppm KNO_3	1 ppm ABA	CL	100	98	98	100
		CD	100	100	–	100
	20 ppm ABA	CL	98	96	40	94
		CD	52	80	–	98
100 ppm KNO_3	1000 ppm CCC	CL	50	100	100	100
		CD	12	94	–	100
	5000 ppm CCC	CL	30	100	94	96
		CD	6	78	16	96
1000 ppm KNO_3	1000 ppm CCC	CL	78	100	98	96
		CD	12	96	16	98
	5000 ppm CCC	CL	52	100	100	100
		CD	10	96	–	96

(continued)

Table 3.16 (continued)

Applied plant growth regulators		Light	Temperatures 10 °C	20 °C	35 °C	10/35 °C
100 ppm KNO_3	100 ppm MH	CL	94	98	98	100
		CD	90	100	–	100
	1000 ppm MH	CL	94	100	98	100
		CD	68	100	–	96
1000 ppm KNO_3	100 ppm MH	CL	100	100	98	100
		CD	94	100	–	100
	1000 ppm MH	CL	92	100	98	98
		CD	68	98	–	96
100 ppm KNO_3	1 ppm manitol	CL	98	100	100	98
		CD	90	100	–	100
	10 ppm manitol	CL	94	100	100	100
		CD	92	100	–	100
1000 ppm KNO_3	1 ppm manitol	CL	96	100	100	100
		CD	96	100	22	100
	10 ppm manitol	CL	96	100	100	98
		CD	92	100	–	100
100 ppm KNO_3	1000 NaCl	CL	100	100	98	100
		CD	78	100	76	100
	20.000 NaCl	CL	26	96	72	100
		CD	6	94	–	82
1000 ppm KNO_3	1000 NaCl	CL	100	100	90	100
		CD	88	100	100	100
	20.000 NaCl	CL	78	94	84	90
		CD	8	86	–	90
Control		CL	92	100	98	98
		CD	84	100	–	100

were generally similar to the control, except for the 20,000 ppm NaCl treatment (Table 3.20).

In the experiments under continuous darkness, the combination of CCC and 20,000 ppm NaCl exhibited inhibitory effects at 10 °C, but at 20 °C, it had stimulatory or control-like effects. The combination of CCC and 20,000 ppm NaCl produced effects similar to the control group. When using 35 °C, it generally showed an inhibitory effect; only the application of 1000 ppm NaCl showed a stimulatory effect. Similar results comparable to the control were observed in all 10/35 °C application series, even at a 20,000-ppm NaCl application; the germination reached 92% (Table 3.20).

The data on interactions between thiourea and combinations of other growth regulators affecting the radicle and hypocotyl lengths of the species under question are presented in Tables 3.21 and 3.22. These tables reveal that at a temperature of 10 °C and under constant light conditions, the mannitol combination was the only

Table 3.17 Radicle and hypocotyl lengths (cm) in two different KNO_3 concentration combination applications in *B. juncea*

Applied plant growth regulators		Light	Temperatures							
			10 °C		20 °C		35 °C		10/35 °C	
			Rad	Hip	Rad	Hip	Rad	Hip	Rad	Hip
100 ppm KNO_3	10 ppm IAA	CL	0.6	0.35	3	1.5	0.25	0.4	1.1	0.4
		CD	0.15	0.1	0.35	1.5	0.7	0.75	0.25	0.45
	50 ppm IAA	CL	0.05	0.3	0.5	0.9	0.2	0.45	0.25	0.3
		CD	0.15	0.1	0.25	0.8	0.2	0.45	0.25	0.35
1000 ppm KNO_3	10 ppm IAA	CL	0.05	0.3	0.375	1.15	0.15	0.25	1.25	1.05
		CD	0.15	0.1	0.25	1	0.35	0.8	0.3	0.6
	50 ppm IAA	CL	0.05	0.2	0.4	0.85	0.15	0.3	0.25	0.45
		CD	0.15	0.1	0.25	0.95	0.15	0.4	0.25	0.35
100 ppm KNO_3	10 ppm TIBA	CL	1.05	0.45	2.25	1.1	0.25	0.65	1.5	0.6
		CD	0.15	0.1	2.75	2.75	0.85	0.7	1.75	1.5
	50 ppm TIBA	CL	0.5	0.3	1	1.5	0.2	0.4	1	0.7
		CD	0.15	0.1	1.5	2.25	0.6	0.9	1.35	1
1000 ppm KNO_3	10 ppm TIBA	CL	0.75	0.4	2.75	1.25	0.15	0.5	1.5	1.1
		CD	0.1	0.1	2.25	3.5	1.25	1.25	1.5	1.35
	50 ppm TIBA	CL	0.3	0.35	0.9	1.4	0.15	0.2	1.3	1.15
		CD	–	–	1.5	2.75	1.2	1.5	1.25	1.1
100 ppm KNO_3	10 ppm coumarin	CL	0.8	0.4	3.25	1.5	0.4	0.75	0.95	0.4
		CD	0.15	0.1	2.25	1.6	1.3	1.65	1.5	1
	50 ppm coumarin	CL	0.3	0.25	0.85	0.95	0.15	0.2	0.5	0.45
		CD	–	–	0.85	1.6	0.7	1.2	0.55	0.65
1000 ppm KNO_3	10 ppm coumarin	CL	0.55	0.45	1.75	1.15	0.45	0.55	1	0.75
		CD	0.15	0.1	2.25	3.5	1.15	1.75	1.1	1.5
	50 ppm coumarin	CL	0.3	0.35	1	0.9	0.3	0.5	0.6	0.75
		CD	0.1	0.1	0.55	2.25	0.65	1.25	0.35	0.55
100 ppm KNO_3	1 ppm ABA	CL	0.9	0.3	3.25	1.1	0.3	0.75	1.75	0.6
		CD	0.1	0.1	1.75	1.25	1.35	0.6	0.65	0.6
	20 ppm ABA	CL	0.3	0.25	0.25	0.85	–	–	0.15	0.2
		CD	0.1	0.1	–	–	–	–	0.15	0.15
1000 ppm KNO_3	1 ppm ABA	CL	0.85	0.3	2.75	1.25	0.3	0.55	1	0.55
		CD	0.15	0.1	1.3	3.25	2.75	3	0.65	0.75
	20 ppm ABA	CL	0.15	0.2	0.3	0.4	–	–	0.15	0.15
		CD	–	–	–	–	–	–	0.15	0.15
100 ppm KNO_3	1000 ppm CCC	CL	0.65	0.2	3.25	1.05	0.25	0.8	1	0.4
		CD	–	–	3.75	2.75	1.75	3.25	1.75	1.35
	5000 ppm CCC	CL	–	–	2.25	0.45	0.3	0.5	1	0.45
		CD	–	–	1.25	3.5	1.25	0.6	1.15	0.6

(continued)

Table 3.17 (continued)

Applied plant growth regulators		Light	Temperatures 10 °C		20 °C		35 °C		10/35 °C	
			Rad	Hip	Rad	Hip	Rad	Hip	Rad	Hip
1000 ppm KNO_3	1000 ppm CCC	CL	0.7	0.3	4	1.1	0.4	0.45	0.75	0.45
		CD	–	–	5	4	2.5	4	1.45	1.2
	5000 ppm CCC	CL	–	–	1.5	0.5	0.15	0.2	0.6	0.35
		CD	–	–	1.5	4	0.6	0.75	1.25	0.9
100 ppm KNO_3	100 ppm MH	CL	0.55	0.35	0.75	1.35	0.15	0.6	0.4	0.7
		CD	–	–	0.75	1.75	0.55	0.65	0.6	1.15
	1000 ppm MH	CL	0.15	0.3	0.25	0.4	0.15	0.2	0.15	0.25
		CD	–	–	0.25	0.55	0.25	0.45	0.25	0.55
1000 ppm KNO_3	100 ppm MH	CL	0.55	0.35	0.7	1.25	0.15	0.3	0.4	0.55
		CD	0.15	0.1	0.75	1	0.45	0.85	0.6	1.05
	1000 ppm MH	CL	0.15	0.35	0.15	0.35	0.15	0.2	0.15	0.3
		CD	–	–	2.5	0.55	0.25	0.5	0.25	0.45
100 ppm KNO_3	1 ppm manitol	CL	0.95	0.9	3	1.25	0.3	0.5	1.75	1
		CD	0.15	0.1	3.25	2	4	2	1.85	1.1
	10 ppm manitol	CL	1.15	0.35	3.75	1.1	0.4	0.4	2.5	0.75
		CD	0.15	0.1	3.5	2.25	1	0.8	2.5	1.3
1000 ppm KNO_3	1 ppm manitol	CL	0.8	0.35	2.75	0.75	0.15	0.35	1.5	0.8
		CD	0.15	0.1	3	3.5	1.75	3.5	2.35	1.75
	10 ppm manitol	CL	0.95	0.4	3.5	1.25	0.25	0.45	1.5	0.75
		CD	0.15	0.1	3.25	3.25	–	–	1.75	1.65
100 ppm KNO_3	1000 NaCl	CL	0.9	1.4	2.25	1.5	0.35	0.4	1.75	1.5
		CD	0.15	0.15	3.75	2.25	0.85	1.45	2.35	1.4
	20.000 NaCl	CL	–	–	1.1	0.5	–	–	0.35	0.45
		CD	–	–	2	2.5	–	–	0.4	0.35
1000 ppm KNO_3	1000 NaCl	CL	0.8	0.4	2.5	1.25	0.15	0.25	1.5	1
		CD	0.15	0.1	2.75	3.5	0.65	0.95	2.25	1.85
	20.000 NaCl	CL	–	–	1.35	0.4	0.15	0.15	0.75	0.4
		CD	–	–	1.5	3	–	–	0.4	0.3
Control		CL	1.05	0.4	5	0.7	0.4	0.5	2	0.65
		CD	0.15	0.15	6.5	1.75	1.6	1.25	2.75	0.9

treatment that stimulated radicle growth in *B. juncea*; radicle length was similar to that of the control group. Other treatments had an inhibitory effect. At 20 °C, radicle and hypocotyl lengths were similar to the control for both TIBA combinations and other treatments, but inhibitory for ABA, 5000 ppm CCC and 20,000 ppm NaCl; generally, the 35 °C treatment had an inhibitory effect. At 10/35 °C, only the TIBA treatment had a stimulatory effect; others had an inhibitory effect (Table 3.21). In the dark treatment of the *B. juncea* experimental series, IAA-TIBA-coumarin and ABA had an inhibitory effect at 10 °C, while mannitol had the same effect. At 20 °C, 1000 ppm CCC had a stimulatory effect, but others had an inhibitory effect.

Table 3.18 Radicle and hypocotyl lengths (cm) in two different KNO_3 concentration combination applications in *B. campestris*

Applied plant growth regulators		Light	Temperatures							
			10 °C		20 °C		35 °C		10/35 °C	
			Rad	Hip	Rad	Hip	Rad	Hip	Rad	Hip
100 ppm KNO_3	10 ppm IAA	CL	0.3	0.35	0.37	1.25	0.35	1.15	1	0.85
		CD	0.2	0.1	0.35	1	0.45	0.7	0.6	0.8
	50 ppm IAA	CL	0.3	0.35	2.5	1.25	0.15	0.2	0.3	0.6
		CD	0.2	0.1	0.25	0.6	0.8	0.5	0.25	0.45
1000 ppm KNO_3	10 ppm IAA	CL	0.3	0.4	4.5	1.75	0.15	0.35	0.25	0.95
		CD	0.2	0.1	0.6	2	0.15	0.15	0.55	1
	50 ppm IAA	CL	0.15	0.25	0.65	1.25	0.3	1.25	0.2	0.7
		CD	0.2	0.1	0.65	3.5	0.35	0.2	0.6	0.9
100 ppm KNO_3	10 ppm TIBA	CL	1.35	0.6	4	1.5	0.2	0.45	1.4	1
		CD	0.2	0.1	0.55	0.9	–	–	0.5	1.6
	50 ppm TIBA	CL	0.4	0.6	0.75	1.5	0.25	1.75	0.7	1.05
		CD	0.2	0.1	0.4	0.65	0.2	0.25	1.4	1.1
1000 ppm KNO_3	10 ppm TIBA	CL	0.5	0.75	4	1.75	0.8	1.5	1.5	1.05
		CD	0.2	0.1	0.6	0.75	1.1	1.35	0.85	0.5
	50 ppm TIBA	CL	0.45	0.5	1.25	2	0.4	1.05	0.9	1.5
		CD	0.2	0.1	2	2.25	–	–	2.2	1.5
100 ppm KNO_3	10 ppm coumarin	CL	0.9	0.45	2.75	1.25	0.3	0.7	1.35	1.15
		CD	0.15	0.1	0.6	0.75	–	–	1	1.25
	50 ppm coumarin	CL	0.4	0.35	1.25	0.95	0.15	0.35	0.7	1
		CD	0.15	0.1	0.4	0.65	–	–	0.35	0.7
1000 ppm KNO_3	10 ppm coumarin	CL	0.85	0.5	2	1.75	0.3	0.6	1.15	1.1
		CD	0.15	0.1	1.5	2.5	–	–	1.15	1.25
	50 ppm coumarin	CL	0.3	0.65	1.75	1.5	0.35	0.5	0.4	0.95
		CD	0.15	0.1	0.9	2.25	–	–	0.55	1.2
100 ppm KNO_3	1 ppm ABA	CL	1.15	0.2	4.25	1.25	0.3	0.85	2.5	1.25
		CD	0.2	0.1	1.25	1.1	–	–	1.35	1
	20 ppm ABA	CL	1.05	0.4	3.5	1	0.3	0.45	0.6	0.85
		CD	0.1	0.1	–	–	–	–	0.2	0.1
1000 ppm KNO_3	1 ppm ABA	CL	0.75	0.65	2	2.25	0.45	0.75	1.15	1.5
		CD	0.2	0.15	1.1	1.25	–	–	0.75	1.2
	20 ppm ABA	CL	0.5	0.4	1.75	1.75	0.35	0.45	0.3	0.5
		CD	–	–	–	–	–	–	0.15	0.1
100 ppm KNO_3	1000 ppm CCC	CL	1	0.4	4.5	1.05	0.35	0.85	1.25	0.9
		CD	0.15	0.1	3.75	2.75	–	–	2.1	1.5
	5000 ppm CCC	CL	0.7	0.3	2	0.45	0.35	1.25	0.4	0.55
		CD	0.1	0.1	2	1.5	–	–	0.9	0.7

(continued)

Table 3.18 (continued)

			Temperatures							
Applied plant growth regulators		Light	10 °C		20 °C		35 °C		10/35 °C	
			Rad	Hip	Rad	Hip	Rad	Hip	Rad	Hip
1000 ppm KNO_3	1000 ppm CCC	CL	0.85	0.5	3.5	1.25	0.3	1.05	1.1	1.25
		CD	0.15	0.1	3	3.5	–	–	1.75	1.4
	5000 ppm CCC	CL	0.75	0.35	1.5	0.65	0.3	0.8	0.8	0.75
		CD	0.1	0.1	1.5	1.5	–	–	1	1.25
100 ppm KNO_3	100 ppm MH	CL	0.95	0.45	0.75	1.25	0.35	0.55	0.95	1
		CD	0.15	0.1	0.75	1.25	–	–	1	1.2
	1000 ppm MH	CL	0.2	0.35	0.25	0.35	0.4	0.6	0.35	0.65
		CD	0.15	0.1	0.25	0.65	–	–	0.25	0.55
1000 ppm KNO_3	100 ppm MH	CL	0.95	0.65	1.35	1.75	0.45	1.25	0.6	1.2
		CD	0.2	0.15	0.75	2.25	–	–	1	1.35
	1000 ppm MH	CL	0.2	0.45	0.3	0.55	0.3	0.6	0.35	0.6
		CD	0.15	0.1	0.4	0.1	–	–	0.35	0.65
100 ppm KNO_3	1 ppm manitol	CL	1.4	0.6	5.5	1.25	0.35	1	2.25	1.05
		CD	0.25	0.15	1.1	1.2	–	–	1.25	1.15
	10 ppm manitol	CL	1.05	0.65	4.25	1.25	0.4	1.25	3	1.5
		CD	0.3	0.15	3	1.75	–	–	1.7	1.25
1000 ppm KNO_3	1 ppm manitol	CL	0.95	0.65	3.75	1.75	0.35	0.7	1.75	1.35
		CD	0.4	0.2	3.25	3.5	–	–	1.75	1.6
	10 ppm manitol	CL	1.15	0.8	0.75	2.25	0.3	0.55	2.5	1.5
		CD	0.3	0.15	0.65	1.1	–	–	2	1.6
100 ppm KNO_3	1000 NaCl	CL	2	0.6	3.25	1.75	0.3	0.85	2.75	1.75
		CD	0.25	0.15	1.75	2.5	0.5	0.6	2	1.75
	20.000 NaCl	CL	0.45	0.25	0.5	0.6	0.55	0.65	0.35	0.65
		CD	0.15	0.1	1.25	0.9	–	–	0.4	0.55
1000 ppm KNO_3	1000 NaCl	CL	1.15	0.8	3.25	2.25	0.3	0.45	1.5	1.5
		CD	0.25	0.15	4.25	5.05	0.3	0.45	1.25	1.5
	20.000 NaCl	CL	0.3	0.35	1	0.9	0.2	0.35	0.6	1.35
		CD	0.15	0.1	1.5	1.45	–	–	0.8	0.6
Control		CL	1	0.5	3.25	1.25	0.5	1	2.5	1.25
		CD	0.5	0.2	6.25	2.25	–	–	2.25	1.25

An inhibitory effect was observed in all 35 °C treatments, and it was also generally the case in the 10/35 °C (Table 3.21).

Considering the lengths of radicle and hypocotyl data for *B. campestris*, an inhibitory effect is observed under constant light conditions at 10 °C, but at 20 °C, radicle length is almost entirely stimulatory, while hypocotyl length generally follows control levels (Table 3.22). At 35 °C, an inhibitory effect is generally observed in terms of both radicle and hypocotyl lengths; at 10/35 °C, a parallel or inhibitory effect is observed.

Table 3.19 Germination percentages in *B. juncea* in the application of thiourea and other growth regulator combinations

			Temperatures			
Applied plant growth regulators		Light	10 °C	20 °C	35 °C	10/35 °C
100 ppm thiourea	10 ppm IAA	CL	74	96	100	96
		CD	20	60	8	88
	50 ppm IAA	CL	74	96	98	100
		CD	12	35	12	86
1000 ppm thiourea	10 ppm IAA	CL	86	100	96	100
		CD	26	68	0	98
	50 ppm IAA	CL	94	96	98	100
		CD	24	77	2	98
100 ppm thiourea	10 ppm TIBA	CL	80	100	98	100
		CD	34	52	62	82
	50 ppm TIBA	CL	80	98	96	98
		CD	20	58	18	88
1000 ppm thiourea	10 ppm TIBA	CL	92	100	98	100
		CD	38	60	50	98
	50 ppm TIBA	CL	86	100	100	100
		CD	22	78	0	98
100 ppm thiourea	10 ppm coumarin	CL	56	98	84	100
		CD	18	34	20	84
	50 ppm coumarin	CL	76	90	84	96
		CD	2	10	8	38
1000 ppm thiourea	10 ppm coumarin	CL	82	98	90	100
		CD	20	60	26	74
	50 ppm coumarin	CL	94	94	88	100
		CD	10	18	0	46
100 ppm thiourea	1 ppm ABA	CL	84	100	98	100
		CD	16	54	60	92
	20 ppm ABA	CL	70	98	0	94
		CD	6	0	0	80
1000 ppm thiourea	1 ppm ABA	CL	96	100	98	100
		CD	46	60	8	96
	20 ppm ABA	CL	94	100	6	96
		CD	8	30	0	84
100 ppm thiourea	100 ppm CCC	CL	4	92	94	100
		CD	0	2	16	48
	1000 ppm CCC	CL	0	76	88	94
		CD	0	4	32	40
1000 ppm thiourea	100 ppm CCC	CL	10	90	98	98
		CD	0	18	40	50
	1000 ppm CCC	CL	4	90	96	96
		CD	0	4	60	50

(continued)

Table 3.19 (continued)

			Temperatures			
Applied plant growth regulators		Light	10 °C	20 °C	35 °C	10/35 °C
100 ppm thiourea	100 ppm MH	CL	74	100	100	100
		CD	16	46	28	82
	1000 ppm MH	CL	78	100	94	100
		CD	4	44	32	72
1000 ppm thiourea	100 ppm MH	CL	94	100	94	98
		CD	40	94	6	92
	1000 ppm MH	CL	86	100	96	100
		CD	6	75	12	96
100 ppm thiourea	1 ppm manitol	CL	92	100	96	98
		CD	30	42	46	82
	10 ppm manitol	CL	86	100	94	98
		CD	24	80	96	100
1000 ppm thiourea	1 ppm manitol	CL	98	100	100	100
		CD	50	96	92	100
	10 ppm manitol	CL	90	98	100	100
		CD	40	68	70	90
100 ppm thiourea	1000 NaCl	CL	90	100	100	100
		CD	30	34	32	100
	20.000 NaCl	CL	0	98	90	92
		CD	0	10	0	70
1000 ppm thiourea	1000 NaCl	CL	92	100	96	100
		CD	44	74	34	100
	20.000 NaCl	CL	0	76	82	64
		CD	0	4	–	44
Control		CL	68	92	96	100
		CD	20	70	80	84

In *B. campestris* applications at 10 °C under continuous dark conditions were found to stimulate radicle growth and inhibit hypocotyl growth, but the results of applications at 20 °C showed an inhibitory effect on radicle length at all concentrations except at 1000 ppm NaCl. In the 35 °C and 10/35 °C application series, reductions in radicle and hypocotyl length have been recorded when compared to the control (Table 3.22).

An experimental study was undertaken to determine the effects of red (R), far-red (FR) and blue (B) light of different durations on *Brassica juncea* and *Brassica campestris* in different combinations. The plants were exposed to different light conditions at three temperatures: 10 °C, 20 °C and 35 °C. A comparison of the germination percentages under red light with the control group revealed that *B. juncea* germinated more successfully when exposed to red light for 5 or 15 min. Overall, R, B and FR light produced better results at 10 °C and 20 °C (Table 3.23).

Table 3.20 Germination percentages in *B. campestris* in the application of Thiourea and other growth regulator combinations

			Temperatures			
Applied plant growth regulators		Light	10 °C	20 °C	35 °C	10/35 °C
100 ppm thiourea	10 ppm IAA	CL	96	100	98	98
		CD	94	100	48	100
	50 ppm IAA	CL	96	100	96	98
		CD	92	100	–	100
1000 ppm thiourea	10 ppm IAA	CL	98	100	98	94
		CD	94	100	–	100
	50 ppm IAA	CL	100	100	96	96
		CD	96	100	–	98
100 ppm thiourea	10 ppm TIBA	CL	98	100	94	98
		CD	88	100	–	100
	50 ppm TIBA	CL	100	100	96	98
		CD	92	100	–	100
1000 ppm thiourea	10 ppm TIBA	CL	98	96	96	92
		CD	92	100	–	96
	50 ppm TIBA	CL	100	98	100	96
		CD	92	100	–	98
100 ppm thiourea	10 ppm coumarin	CL	90	100	96	100
		CD	86	100	–	100
	50 ppm coumarin	CL	80	100	88	100
		CD	64	98	–	98
1000 ppm thiourea	10 ppm coumarin	CL	100	100	92	100
		CD	94	100	–	100
	50 ppm coumarin	CL	88	100	96	96
		CD	60	100	–	100
100 ppm thiourea	1 ppm ABA	CL	100	100	90	100
		CD	98	100	–	98
	20 ppm ABA	CL	88	100	12	96
		CD	86	96	–	96
1000 ppm thiourea	1 ppm ABA	CL	100	98	86	96
		CD	96	100	–	100
	20 ppm ABA	CL	100	100	60	92
		CD	84	86	–	94
100 ppm thiourea	100 ppm CCC	CL	48	100	92	100
		CD	34	94	–	100
	1000 ppm CCC	CL	28	100	68	100
		CD	22	72	–	92
1000 ppm thiourea	100 ppm CCC	CL	84	100	90	96
		CD	18	98	–	100
	1000 ppm CCC	CL	60	96	98	100
		CD	12	82	6	98

(continued)

Table 3.20 (continued)

			Temperatures			
Applied plant growth regulators		Light	10 °C	20 °C	35 °C	10/35 °C
100 ppm thiourea	100 ppm MH	CL	98	100	88	98
		CD	92	100	–	100
	1000 ppm MH	CL	92	100	86	94
		CD	76	100	–	98
1000 ppm thiourea	100 ppm MH	CL	100	100	92	100
		CD	96	100	–	98
	1000 ppm MH	CL	100	100	76	96
		CD	90	100	–	100
100 ppm thiourea	1 ppm manitol	CL	96	98	98	96
		CD	92	100	–	100
	10 ppm manitol	CL	100	100	98	100
		CD	94	100	–	100
1000 ppm thiourea	1 ppm manitol	CL	100	100	96	100
		CD	90	100	–	100
	10 ppm manitol	CL	100	100	100	100
		CD	98	100	–	100
100 ppm thiourea	1000 NaCl	CL	96	100	98	98
		CD	88	100	96	98
	20.000 NaCl	CL	24	94	44	16
		CD	8	90	–	92
1000 ppm thiourea	1000 NaCl	CL	100	100	100	100
		CD	94	100	70	100
	20.000 NaCl	CL	26	100	44	60
		CD	4	77	–	94
Control		CL	94	100	98	98
		CD	98	100	14	100

A decrease in the germination percentages was recorded in the experimental series in which R-FR were applied simultaneously. However, an increase in germination percentage was observed when the plants were first exposed to FR, followed by R (Table 3.23). In the series where R and B treatments were given together, the germination percentage decreased compared to the control. A significant decrease was observed, particularly at 35 °C. An increase in germination percentage was observed when the plants were exposed to B followed by R, compared to exposure to the other two lights. In the experimental series in which R, FR and B were applied at equal intensities, an increase in germination percentage was observed at all three temperatures compared with the FR-R and FR series.

In the series where R and B were applied for equal periods of time, the results were similar to those of the control group. However, an increase was observed in the trial series in which all three lights were given for 30 min compared to the control (Table 3.23).

Table 3.21 Radicle and hypocotyl lengths (cm) in *B. juncea* in combination with the application of thiourea and other growth regulators

			Temperatures							
Applied plant growth regulators		Light	10 °C		20 °C		35 °C		10/35 °C	
			Rad	Hip	Rad	Hip	Rad	Hip	Rad	Hip
100 ppm thiourea	10 ppm IAA	CL	0.1	0.1	1.5	1.5	0.15	1.1	0.3	0.3
		CD	0.15	0.25	2	1.25	0.25	0.5	1.25	0.65
	50 ppm IAA	CL	0.1	0.1	0.2	0.3	–	–	0.4	0.4
		CD	0.05	0.25	0.15	0.5	0.35	0.35	0.3	0.4
1000 ppm thiourea	10 ppm IAA	CL	0.1	0.1	0.2	1.25	–	–	0.35	0.45
		CD	0.15	0.35	1.6	1	0.3	0.4	0.7	0.55
	50 ppm IAA	CL	0.1	0.1	0.2	1.25	–	–	0.2	0.9
		CD	0.05	0.35	1.25	0.9	0.35	0.55	1.1	0.5
100 ppm thiourea	10 ppm TIBA	CL	0.2	0.15	3	2.25	1.35	0.6	2.25	1.5
		CD	1.3	0.45	2.75	1.4	0.3	0.45	2.25	0.75
	50 ppm TIBA	CL	0.2	0.15	1.75	2.5	1	1.8	1.75	1.1
		CD	0.6	0.35	1.75	1	0.3	0.5	1.75	0.9
1000 ppm thiourea	10 ppm TIBA	CL	0.2	0.15	3.5	2	0.8	1	2	1.1
		CD	1.15	0.35	3	1.35	0.35	0.45	1.65	0.5
	50 ppm TIBA	CL	0.15	0.15	2.75	2.25	–	–	2.25	1.35
		CD	0.9	0.4	2.75	1.35	0.4	0.5	2	0.8
100 ppm thiourea	10 ppm coumarin	CL	0.1	0.15	2	2.25	2.9	2.5	1.3	0.75
		CD	0.45	0.25	2.75	0.8	0.4	0.7	1.25	0.8
	50 ppm coumarin	CL	–	–	1	2	2	2	0.5	0.3
		CD	0.75	0.4	1.25	0.75	0.2	0.35	1.25	0.7
1000 ppm thiourea	10 ppm coumarin	CL	0.15	0.15	2.5	2	0.6	1.1	1.25	0.85
		CD	0.45	0.35	2.5	1	0.35	0.55	0.9	0.5
	50 ppm coumarin	CL	0.1	0.1	0.5	1.4	–	–	0.5	0.65
		CD	0.65	0.4	0.75	0.75	0.35	0.55	0.6	0.45
100 ppm thiourea	1 ppm ABA	CL	0.15	0.1	1.5	3.2	1.3	1	1.25	0.4
		CD	0.5	0.4	4.25	1.1	0.35	0.5	1.15	0.8
	20 ppm ABA	CL	–	–	–	–	–	–	0.05	0.05
		CD	0.6	0.35	4	0.85	–	–	0.2	0.25
1000 ppm thiourea	1 ppm ABA	CL	0.15	0.1	2.6	1.5	1.4	1.6	1.6	0.7
		CD	1.15	0.35	2.25	1.25	0.35	0.4	1.5	0.4
	20 ppm ABA	CL	–	–	–	–	–	–	–	–
		CD	0.35	0.2	3.75	0.75	0.15	0.35	0.2	0.2
100 ppm thiourea	1000 ppm CCC	CL	–	–	5	4	1.4	1.8	1.2	0.75
		CD	0.25	0.15	4	0.6	0.5	0.55	1.35	0.6
	5000 ppm CCC	CL	–	–	1.6	2	1.75	2	1	0.5
		CD	–	–	0.9	0.4	0.4	0.5	1.25	0.55

(continued)

Table 3.21 (continued)

			Temperatures							
Applied plant growth regulators		Light	10 °C		20 °C		35 °C		10/35 °C	
			Rad	Hip	Rad	Hip	Rad	Hip	Rad	Hip
1000 ppm thiourea	1000 ppm CCC	CL	–	–	2.5	2.5	1	1	0.2	0.2
		CD	0.4	0.15	3.25	0.75	0.25	0.4	1.25	0.55
	5000 ppm CCC	CL	–	–	1.5	3	0.2	0.4	0.5	0.8
		CD	0.15	0.15	2	0.55	0.15	0.45	1.25	0.55
100 ppm thiourea	100 ppm MH	CL	0.2	0.15	1.1	1.7	0.75	1.5	0.65	0.8
		CD	0.9	0.35	0.9	0.85	0.4	0.6	0.55	0.45
	1000 ppm MH	CL	0.1	0.1	0.2	0.5	0.1	0.5	0.15	0.45
		CD	0.15	0.35	0.25	0.6	0.05	0.35	0.15	0.45
1000 ppm thiourea	100 ppm MH	CL	0.3	0.2	0.9	1.5	0.3	0.6	1	0.8
		CD	0.7	0.4	1.15	1.1	0.35	0.4	0.55	0.55
	1000 ppm MH	CL	0.15	0.15	0.35	0.6	0.3	0.7	0.15	0.4
		CD	0.15	0.35	0.15	0.45	0.15	0.35	0.15	0.55
100 ppm thiourea	1 ppm manitol	CL	0.15	0.15	2.7	2	2	1.75	0.5	0.75
		CD	0.6	0.4	2.25	0.9	0.4	0.5	1.4	0.75
	10 ppm manitol	CL	0.15	0.15	3	1.8	0.2	0.4	2.5	1.25
		CD	1.4	0.35	3.5	1	0.45	0.55	2.25	1.1
1000 ppm thiourea	1 ppm manitol	CL	0.2	0.3	2.25	1.5	0.6	0.5	1.25	0.6
		CD	1.5	0.7	3.25	1.15	0.35	0.45	1.75	0.65
	10 ppm manitol	CL	0.15	0.15	3.25	1.6	1.8	1	2.3	1
		CD	1.4	0.4	3	1.25	0.5	0.4	2	0.75
100 ppm thiourea	1000 NaCl	CL	0.2	0.15	3	3.8	0.8	0.7	1.5	1
		CD	1.25	0.45	3.25	1.25	0.4	0.7	1.6	0.65
	20.000 NaCl	CL	–	–	1.1	2.1	–	–	0.45	0.3
		CD	–	–	1.25	0.4	0.15	0.15	0.4	0.3
1000 ppm thiourea	1000 NaCl	CL	0.15	0.15	2	2.25	1.6	1.75	1.4	0.8
		CD	1.25	0.4	3.25	1.75	0.3	0.45	2	0.7
	20.000 NaCl	CL	–	–	1.5	1	–	–	0.3	0.2
		CD	–	–	1.25	0.65	0.15	0.45	0.65	0.4
Control		CL	0.2	0.15	3.25	1.75	0.7	0.3	1.75	0.9
		CD	0.9	0.4	3.25	1.25	0.5	0.55	2.25	0.75

We can therefore say that *B. juncea* exhibits better germination rates at 10 °C and 20 °C compared to 35 °C, under various lighting conditions. When the same treatments are applied to *B. campestris*, increases are observed in all three temperature groups. As emphasised above in the series in which R and B were applied for 30 min each resulted in increased germination percentages under all three temperature groups (Table 3.23).

A comparison and evaluation of the values obtained from this experimental series with the control values in terms of their effect on radicle and hypocotyl length, it is evident that both radicle and hypocotyl are longer in *B. campestris* at 20 °C than

Table 3.22 Radicle and hypocotyl lengths (cm) in *B. campestris* in combination with the application of thiourea and other growth regulators

Applied plant growth regulators		Light	Temperatures							
			10 °C		20 °C		35 °C		10/35 °C	
			Rad	Hip	Rad	Hip	Rad	Hip	Rad	Hip
100 ppm thiourea	10 ppm IAA	CL	0.3	0.55	2.5	1.25	0.2	0.8	0.6	1.1
		CD	0.2	0.15	0.3	1.15	0.5	0.1	0.15	0.9
	50 ppm IAA	CL	0.15	0.35	0.4	1.1	0.2	0.35	0.3	0.55
		CD	0.2	0.15	0.4	1	–	–	0.1	0.55
1000 ppm thiourea	10 ppm IAA	CL	0.45	0.7	1.5	1.25	0.2	0.4	0.5	0.6
		CD	0.4	0.15	0.85	2.25	–	–	0.6	0.75
	50 ppm IAA	CL	0.9	0.5	1.75	1.25	0.15	0.3	0.15	0.4
		CD	0.2	0.15	0.25	0.7	–	–	0.4	0.4
100 ppm thiourea	10 ppm TIBA	CL	1	0.55	2.75	1.25	0.4	0.4	1.5	0.8
		CD	0.2	0.15	1.25	0.75	–	–	1.1	0.9
	50 ppm TIBA	CL	0.55	0.7	0.9	1.4	0.25	0.65	0.65	0.75
		CD	0.2	0.15	0.7	3.25	–	–	1.5	1.75
1000 ppm thiourea	10 ppm TIBA	CL	1.15	1.15	2.25	1.25	0.15	0.45	0.8	0.65
		CD	0.3	0.2	1.6	0.9	–	–	2.25	1.25
	50 ppm TIBA	CL	1.25	0.8	2.75	1.25	0.3	0.65	1.25	0.7
		CD	0.45	0.2	1	0.85	–	–	2	1.25
100 ppm thiourea	10 ppm coumarin	CL	0.9	0.5	2.25	1.35	0.3	0.65	0.8	1.15
		CD	0.35	0.15	2.25	2.25	–	–	1.75	1.3
	50 ppm coumarin	CL	0.6	0.35	0.35	0.9	0.3	0.55	0.5	0.65
		CD	0.2	0.15	0.3	0.9	0.1	0.1	0.6	1
1000 ppm thiourea	10 ppm coumarin	CL	1	0.6	3.25	1.35	0.3	0.75	1	0.75
		CD	0.3	0.15	2.5	2.25	–	–	2.2	1.25
	50 ppm coumarin	CL	0.5	0.35	0.9	1.25	0.15	0.15	0.4	0.8
		CD	0.2	0.15	0.3	0.7	–	–	0.25	0.6
100 ppm thiourea	1 ppm ABA	CL	1.4	0.6	2.25	1.5	0.15	0.2	1.25	1.15
		CD	0.2	0.15	2.75	2	–	–	1.5	0.9
	20 ppm ABA	CL	0.6	0.4	1.5	1.15	0.15	0.05	0.9	0.6
		CD	0.15	0.1	0.6	1.25	–	–	0.1	0.2
1000 ppm thiourea	1 ppm ABA	CL	1.35	0.5	2.75	1.5	0.15	0.35	1.5	0.8
		CD	0.35	0.2	2.25	2.25	–	–	1.25	1.35
	20 ppm ABA	CL	0.75	0.55	2.5	0.75	0.15	0.75	0.15	0.45
		CD	0.15	0.1	0.1	1.1	–	–	0.1	0.15
100 ppm thiourea	1000 ppm CCC	CL	1.25	0.35	4	1.05	2.5	0.75	0.9	0.75
		CD	0.2	0.15	0.95	1.75	–	–	2.25	1.2
	5000 ppm CCC	CL	0.75	0.4	1.25	0.6	0.2	0.6	0.7	0.4
		CD	0.1	0.1	1.5	1.2	0.2	1.35	1.4	0.9

(continued)

Table 3.22 (continued)

			Temperatures							
Applied plant growth regulators		Light	10 °C		20 °C		35 °C		10/35 °C	
			Rad	Hip	Rad	Hip	Rad	Hip	Rad	Hip
1000 ppm thiourea	1000 ppm CCC	CL	0.85	0.4	2.5	1.1	0.15	0.45	0.8	0.6
		CD	0.15	0.1	3	2.25	–	–	1.75	1.1
	5000 ppm CCC	CL	0.65	0.5	1.25	0.6	0.15	0.45	0.4	0.5
		CD	0.15	0.1	1.25	1.1	–	–	1.35	0.75
100 ppm thiourea	100 ppm MH	CL	1	0.65	0.9	0.75	0.25	0.55	0.45	0.5
		CD	0.2	0.3	0.9	0.85	–	–	0.95	1
	1000 ppm MH	CL	0.2	0.4	0.25	0.65	0.15	0.2	0.15	0.4
		CD	0.2	0.15	0.55	0.9	–	–	0.3	0.6
1000 ppm thiourea	100 ppm MH	CL	2.1	0.5	1.25	1.4	0.25	0.6	0.9	0.85
		CD	0.25	0.15	1.35	2.35	–	–	1.5	1.5
	1000 ppm MH	CL	0.15	0.35	0.4	0.7	0.15	0.2	0.25	0.45
		CD	0.2	0.15	0.4	0.5	–	–	0.15	0.65
100 ppm thiourea	1 ppm manitol	CL	1.5	0.55	1.25	0.75	0.3	0.45	1.25	0.9
		CD	0.55	0.3	1	1.5	–	–	2.2	1.3
	10 ppm manitol	CL	1.35	0.6	4	1.15	0.6	1.1	1.15	0.85
		CD	0.3	0.25	1.5	1.7	–	–	2.25	1.2
1000 ppm thiourea	1 ppm manitol	CL	1.75	0.6	3	1.8	0.3	0.5	1.25	1
		CD	0.45	0.4	1.5	1.75	–	–	2.75	1.25
	10 ppm manitol	CL	1.15	0.45	2.4	1.25	0.4	0.9	1.25	1
		CD	0.6	0.25	3.25	2.25	–	–	1.75	1.25
100 ppm thiourea	1000 NaCl	CL	1.5	0.6	2	1.15	0.25	0.6	1	0.85
		CD	0.25	0.3	3.5	2.75	0.5	0.5	2.35	1.75
	20.000 NaCl	CL	0.15	0.3	0.75	0.65	0.15	0.15	0.15	0.15
		CD	0.1	0.15	0.75	2.5	–	–	1.35	0.75
1000 ppm thiourea	1000 NaCl	CL	0.9	0.6	2.5	1.5	0.2	0.55	1.35	0.85
		CD	0.35	0.15	3.25	2.75	–	–	1.75	1.05
	20.000 NaCl	CL	0.25	0.25	1.5	0.9	0.2	0.25	0.4	0.55
		CD	0.1	0.1	1.1	1.55	–	–	1.75	1.25
Control		CL	2	0.65	0.35	1.15	0.45	1.25	1.5	1.25
		CD	0.15	0.2	3	1.75	–	–	2.5	1.5

in the control. However, results similar to those of the control are generally observed (Table 3.24).

Table 3.25 shows the numerical data on the germination percentages of *Brassica juncea* and *B. campestris* when exposed to various growth regulators and R-FR-B lights at three temperatures: 10 °C, 20 °C and 35 °C. The table enlightens the fact that, application of GA_3 increases the germination percentages of both species under all light conditions compared to the control.

The germination percentages in *B. juncea* increase when kinetin is applied, depending on the concentration. At 10 °C, an increase in the R-FR has the same

Table 3.23 Percentage of germination in R, FR and B light at different durations and combinations in *Brassica* species

		Brassica juncea			*Brassica campestris*		
		Temperatures			Temperatures		
	Duration–minutes	10 °C	20 °C	35 °C	10 °C	20 °C	35 °C
Red (R)	5	43	50	32	95	78	84
	15	42	60	26	95	82	84
	30	38	66	48	92	90	82
Far-red (FR)	5	36	62	22	91	66	78
	15	32	30	10	90	80	68
	30	42	44	28	92	94	78
Blue (B)	5	41	36	24	95	80	68
	15	35	34	10	93	84	72
	30	31	34	8	96	86	80
R-FR	5 + 5	24	50	22	90	76	72
	15 + 15	25	24	22	85	88	72
	30 + 30	23	28	30	92	82	84
FR-R	5 + 5	39	50	32	95	80	72
	15 + 15	40	42	34	95	74	84
	30 + 30	41	56	32	93	80	64
R-B	5 + 5	32	48	4	90	90	37
	15 + 15	30	32	6	84	70	62
	30 + 30	24	34	–	93	90	32
B-R	5 + 5	40	42	4	93	94	58
	15 + 15	46	68	12	90	90	60
	30 + 30	51	72	8	98	86	62
R-FR-R	5 + 5 + 5	55	64	28	96	86	64
	15 + 15 + 15	57	44	28	97	82	72
	30 + 30 + 30	61	80	38	98	88	70
FR-R-FR	5 + 5 + 5	37	38	16	91	74	64
	15 + 15 + 15	41	36	4	95	60	58
	30 + 30 + 30	32	50	10	91	76	74
R-B-R	5 + 5 + 5	47.5	64	14	97	90	58
	15 + 15 + 15	50	76	28	97	86	72
	30 + 30 + 30	69	72	26	93	96	80
B-RB	5 + 5 + 5	47	52	4	91	86	74
	15 + 15 + 15	40	42	6	95	82	66
	30 + 30 + 30	34	68	6	92	82	76
Control		36	48	34	91	86	72

effect as B, but at 20 °C, increases are observed following R-FR-B application. However, at 35 °C, an increase is only observed with R, but no effect is observed with FR, B. The data on *B. campestris* stresses the fact that an increase in the germination percentages is noted under all three types of lights at 10 and 20 °C, but a decrease is observed at 35 °C compared to the control (Table 3.25).

Table 3.24 Radicle and hypocotyl lengths (cm) in R-FR-B light conditions following their application for different durations and varying combinations in *Brassica* species

		Brassica juncea						*Brassica campestris*					
		Temperatures						Temperatures					
	Duration–minutes	10 °C		20 °C		35 °C		10 °C		20 °C		35 °C	
		Rad	Hip	Rad	Hip	Rad	Hip	Rad	Hip	Rad	Hip	Rad	Hip
Red	5	0.65	0.4	3	2.25	–	–	0.87	0.55	3.75	1.75	0.5	1.25
	15	0.67	1.45	3.5	1.8	0.75	1.5	0.87	0.57	3.75	2	0.45	0.85
	30	0.75	0.37	4	2	0.75	1.5	1.12	0.58	3.25	1.75	0.35	1.2
Far-red	5	0.6	0.375	4	2.25	0.65	1.5	0.73	0.55	2	1.5	0.4	0.7
	15	0.72	0.45	3.75	2.5	0.75	1.15	0.83	0.45	3	2	0.4	1.3
	30	0.55	0.5	3.5	2.5	0.85	1.4	0.95	0.52	3.15	2	0.4	1.55
Blue	5	0.57	0.42	4	2.25	0.6	1.5	0.95	0.6	3.25	1.9	0.6	1.5
	15	0.62	0.32	3.25	2.25	1	1.35	0.8	0.6	3	2.25	0.4	0.8
	30	0.45	0.35	3.75	2.25	0.9	1.1	0.89	0.55	3	2.25	0.4	0.7
R-FR	5 + 5	0.55	0.3	3.5	2.25	0.75	1.55	0.68	0.53	4.5	2	0.45	1.05
	15 + 15	0.57	0.37	3.5	4.5	0.5	1.25	0.42	0.47	3	2.25	0.45	1.05
	30 + 30	0.53	0.32	3.75	2	0.45	1.1	0.9	0.45	2.75	2.25	0.5	1
FR-R	5 + 5	0.72	0.6	3.5	1.75	2	1.1	0.8	0.45	3.5	2.25	0.3	0.65
	15 + 15	0.65	0.3	2.75	2.25	5.5	1.15	0.68	0.43	3	2	0.45	1
	30 + 30	0.57	0.37	3.5	1.6	3	0.9	0.85	0.35	3.5	2.5	0.4	0.75
R-B	5 + 5	1.07	0.65	3.5	2.5	3	0.8	1.18	0.8	3	2.5	0.35	0.55
	15 + 15	0.73	0.55	3.75	2.25	3	1.25	0.77	0.78	3.25	1.75	0.25	0.75
	30 + 30	0.83	0.65	5.25	2.5	4	0.75	1.43	0.55	3.5	2.25	0.2	0.3

(continued)

Table 3.24 (continued)

			Brassica juncea						*Brassica campestris*				
			Temperatures						Temperatures				
	Duration–minutes	10 °C		20 °C		35 °C		10 °C		20 °C		35 °C	
		Rad	Hip	Rad	Hip	Rad	Hip	Rad	Hip	Rad	Hip	Rad	Hip
B-R	5 + 5	0.88	0.45	4	2	–	–	1.1	1.18	3.25	2	0.25	0.7
	15 + 15	0.78	0.53	3	1.75	0.35	1.15	1.2	0.68	4.25	2	0.4	1
	30 + 30	0.8	0.5	3.75	1.75	0.4	0.9	0.73	0.78	3.25	2.1	0.25	0.45
R-FR-R	5 + 5 + 5	0.65	0.6	3.5	1.75	0.2	0.25	0.93	0.45	3.75	2.1	0.4	1
	15 + 15 + 15	0.8	0.6	4.5	2.15	0.4	1	1.05	0.63	3.75	2	0.3	0.75
	30 + 30 + 30	0.65	0.35	4	2	0.3	0.5	1.02	0.5	3.25	2	0.6	1.1
FR-R-FR	5 + 5 + 5	0.83	0.5	3.5	1.75	0.75	1.2	0.93	0.6	3.75	2.25	0.45	0.9
	15 + 15 + 15	0.8	0.43	4	2.1	0.25	0.35	0.88	0.6	3.5	1.75	0.4	0.95
	30 + 30 + 30	0.6	0.53	3.75	1	1.5	2	0.9	0.47	3.75	2	0.35	1.25
R-B-R	5 + 5 + 5	0.6	0.35	3.5	1.75	0.55	1.75	1.18	0.8	5.25	2.5	0.4	0.75
	15 + 15 + 15	0.6	0.43	3.5	2	0.5	1.1	1.18	0.6	2.25	1.75	0.4	1.1
	30 + 30 + 30	0.85	0.4	2.75	2	0.6	1.45	1.02	0.45	1.5	1.5	0.55	1.3
B-R-B	5 + 5 + 5	0.65	0.4	3.25	1.8	0.65	0.2	1.07	0.5	2.25	1.5	0.2	0.25
	15 + 15 + 15	0.55	0.3	3.5	2.5	0.15	0.55	0.9	0.55	3	1.6	0.3	0.75
	30 + 30 + 30	0.58	0.38	2.25	1.75	0.15	0.7	0.75	0.5	2.75	2.25	0.25	0.7
Control		0.5	0.4	4.5	2.25	1.75	0.8	1.15	0.58	1.1	2	0.7	0.75

Table 3.25 Percentages of germination in *Brassica* species when exposed to different light conditions and growth regulators

		Brassica juncea									*Brassica campestris*								
		10 °C			20 °C			35 °C			10 °C			20 °C			35 °C		
		Red	Far red	Blue	Red	Far red	Blue	Red	Far red	Blue	Red	Far red	Blue	Red	Far red	Blue	Red	Far red	Blue
Applied plant growth regulators	ppm																		
GA_3	1	60	2	12	92	82	46	4	–	–	70	38	52	100	92	98	–	–	–
	100	80	34	10	100	100	100	100	–	50	94	88	66	100	100	94	–	–	–
Kin	1	66	–	4	90	44	60	70	–	–	60	36	62	100	90	98	–	–	–
	50	70	2	2	100	64	80	100	–	–	86	76	64	98	100	92	6	–	–
	100	64	–	2	100	80	100	100	–	40	88	88	74	100	96	100	–	–	50
IAA	10	36	4	2	90	32	48	–	–	–	78	20	50	100	96	98	–	–	–
	50	18	2	–	80	22	18	–	–	–	68	18	38	100	100	94	30	–	–
	100	8	–	–	70	14	36	–	–	–	52	20	22	98	98	96	80	–	–
TIBA	10	46	2	–	74	14	54	46	–	14	72	44	38	100	92	100	–	–	32
	50	42	2	–	90	30	50	100	–	100	70	26	40	98	98	100	–	2	100
	100	18	–	–	96	10	26	100	–	56	64	14	26	100	92	100	–	–	–
Coumarin	10	36	–	–	88	4	20	100	–	32	46	14	26	100	70	70	–	–	–
	50	8	–	–	52	–	–	100	–	24	12	–	–	100	60	88	–	–	–
	100	–	–	–	34	–	4	52	–	–	2	–	–	92	32	82	–	–	–
ABA	1	34	–	–	98	8	30	40	–	12	74	42	52	96	94	80	72	–	–
	20	4	–	–	62	2	6	–	–	–	54	18	22	100	72	80	10	–	–
	100	–	–	–	12	–	–	–	–	–	10	–	6	20	4	45	–	–	–
CCC	1000	–	–	–	88	12	12	–	–	–	12	4	–	98	74	80	–	–	86
	5000	–	–	–	80	2	–	–	–	–	4	2	–	100	68	78	6	–	–

(continued)

Table 3.25 (continued)

		Brassica juncea									*Brassica campestris*								
		10 °C			20 °C			35 °C			10 °C			20 °C			35 °C		
		Red	Far red	Blue	Red	Far red	Blue	Red	Far red	Blue	Red	Far red	Blue	Red	Far red	Blue	Red	Far red	Blue
Applied plant growth regulators	ppm																		
MH	100	58	2	2	94	26	72	92	–	–	78	34	45	98	92	98	–	–	–
	1000	–	–	–	54	4	–	–	–	–	16	4	4	90	60	22	–	–	–
NaCl	1000	54	–	–	100	10	58	–	–	–	74	26	42	98	88	84	100	–	6
	20.000	–	–	–	–	–	–	–	–	–	–	–	–	–	–	–	–	–	–
T.U	100	72	–	–	100	40	40	100	–	80	76	30	50	100	84	100	40	12	100
	1000	66	2	8	98	50	38	100	–	100	90	66	66	96	98	100	14	24	100
KNO_3	100	62	4	2	100	40	40	–	–	–	94	28	66	100	100	100	90	–	90
	1000	64	–	4	100	66	30	–	44	–	84	60	60	100	100	100	–	–	100
Ethrel	50	26	4	–	98	60	100	–	–	20	92	60	72	100	96	100	–	–	–
	100	28	2	–	100	80	100	–	–	–	96	54	90	100	96	100	–	–	–
	200	18	–	–	100	84	80	–	–	30	76	64	72	96	98	100	–	–	–
	400	12	2	4	84	60	56	–	–	40	66	54	60	86	68	80	–	–	–
PEG	1	42	–	–	100	16	12	–	14	86	74	48	57	100	88	100	100	36	100
	10	46	–	4	94	24	22	12	–	76	76	34	50	100	94	98	100	88	100
	100	54	–	4	92	28	14	80	–	80	72	44	52	100	100	100	100	58	100
Man.	1	40	–	8	90	24	80	–	–	40	80	38	60	100	98	88	–	–	100
	10	50	6	–	98	14	50	100	–	56	62	50	24	100	96	100	30	–	44
	100	40	2	–	96	26	32	–	–	–	72	50	58	100	100	100	–	–	34
Morf.	50	6	4	–	90	8	80	–	–	–	44	28	38	100	100	94	–	–	–
	100	–	–	–	84	30	50	–	–	60	56	14	40	100	96	100	100	–	48
	200	–	–	–	94	18	34	96	–	56	66	20	28	100	98	100	–	–	30

		Brassica juncea									*Brassica campestris*								
		10 °C			20 °C			35 °C			10 °C			20 °C			35 °C		
		Red	Far red	Blue	Red	Far red	Blue	Red	Far red	Blue	Red	Far red	Blue	Red	Far red	Blue	Red	Far red	Blue
Applied plant growth regulators	ppm																		
Control		40	–	2	98	18	78	22	–	88	78	32	50	94	96	100	90	22	92

The results on the IAA application depict that there is an increase in germination percentage in *B. juncea* at 10 °C under R; under other conditions, however, the percentage was similar to or lower than that observed in the control group. In *B. campestris*, a decrease in germination percentage occurred as the concentration increased, but a significant decrease was recorded at 35 °C compared to the control group (Table 3.25).

Coming to the TIBA application, an increase in *B. juncea* was observed after the application at 35 °C and R, compared to the control. A decrease was observed in all other applications. In *B. campestris*, a decrease was observed under all applications (Table 3.25).

The results of coumarin applications were similar to those of TIBA; an increase from 22 to 100% was observed in *B. juncea* alone at 35 °C in R. In *B. campestris*, the results were similar to the control at 20 °C; decreases were observed in other applications. The decreases were observed in both species under different light conditions with ABA applications; however, *B. campestris* was unaffected at 1 and 20 ppm at 20 °C, with a decrease observed at 100 ppm. The decreases were also observed in other applications as well (Table 3.25).

Similar results to those observed in ABA were recorded in CCC and MH applications. However, only the MH application showed an increase, from 22 to 100% at 100 ppm, at a temperature of 35 °C, and under R. In NaCl applications, both *Brassica* species were less affected naturally at 1000 ppm than at 20,000 ppm; however, a significant decrease was observed compared to the control. Similar results to the control were observed in both species in Thiourea + KNO_3 applications. However, only Thiourea showed an increase from 22 to 100% at 35 °C under R (Table 3.25).

In ethrel applications, an increase in germination percentages was observed in both species at low concentrations and at 20 °C, while a decrease was noted at 35 °C. The results resembling those of control were observed in both species in PEG applications, but, unlike other growth regulators, increases were observed in *B. campestris* at 35 °C compared to the control (Table 3.25). The results for mannitol and morphactin applications in both species were the same. However, increases of 22 and 96%, respectively, were observed under applications of 10 ppm mannitol and 200 ppm morphactin at 35 °C under R (Table 3.25).

Generally, the germination percentages of *B. juncea* at 20 °C and of *B. campestris* at 10 °C and 20 °C are comparable to those of the control group; most negative results have been recorded at 35 °C (Table 3.25). The effects of similar applications on radicle and hypocotyl lengths are presented at length in Tables 3.26 and 3.27. As with the germination percentages in Table 3.25, all combinations applied to *B. juncea* and *B. campestris* produced similar results to the control group (Tables 3.26 and 3.27).

Table 3.26 Radicle and hypocotyl lengths (cm) in *Brassica juncea* under different growth regulators and lighting conditions

		Brassica juncea																	
		10 °C						20 °C						35 °C					
		Red		Far Red		Blue		Red		Far Red		Blue		Red		Far Red		Blue	
Applied plant growth regulators	ppm	Rad	Hip	Rad	Hip	Rad	Hip	Rad	Hip	Rad	Hip	Rad	Hip	Rad	Hip	Rad	Hip	Rad	Hip
GA$_3$	1	0.35	0.25	0.2	0.3	0.05	0.05	0.26	0.4	2.25	0.4	0.35	0.55	–	–	–	–	–	–
	100	0.4	0.25	0.35	0.25	0.05	0.05	0.2	0.85	0.5	1.3	0.35	0.55	0.55	0.45	–	–	0.15	0.15
Kin	1	0.2	0.15	–	–	0.15	0.15	4.25	1.1	2.25	0.35	0.75	3.25	0.35	0.25	–	–	–	–
	50	0.15	0.25	–	–	0.05	0.05	0.55	0.9	1	0.5	0.6	0.7	0.3	0.5	–	–	–	–
	100	0.15	0.25	–	–	0.05	0.05	0.35	0.85	0.5	0.7	0.7	0.4	0.25	0.5	–	–	0.2	0.2
IAA	10	0.05	0.05	–	–	–	–	1.25	1.25	2	0.45	0.45	0.55	–	–	–	–	–	–
	50	–	–	–	–	–	–	0.15	0.25	0.15	0.45	0.15	0.3	–	–	–	–	–	–
	100	0.1	0.1	–	–	–	–	0.35	0.45	0.1	0.15	0.25	0.25	–	–	–	–	–	–
TIBA	10	0.15	0.15	–	–	–	–	1.1	1.15	2	0.6	2.25	1.25	0.3	0.2	–	–	0.2	0.3
	50	0.15	0.15	–	–	–	–	1.25	0.65	0.9	0.8	0.4	0.3	1	0.5	–	–	0.3	0.3
	100	0.15	0.15	–	–	–	–	0.75	0.55	0.7	0.5	1.6	1.35	1	0.5	–	–	0.3	0.35
Coumarin	10	0.15	0.25	–	–	–	–	1.25	1	1	0.5	0.3	0.2	3.5	1.25	–	–	0.3	0.4
	50	0.05	0.05	–	–	–	–	0.4	0.6	–	–	–	–	1	0.5	–	–	0.2	0.2
	100	–	–	–	–	–	–	0.15	0.35	–	–	0.1	0.1	0.3	0.2	–	–	–	–
ABA	1	0.3	0.15	–	–	–	–	3.75	0.5	1.5	0.4	1.25	0.4	0.3	0.25	–	–	–	–
	20	0.05	0.05	–	–	–	–	0.2	0.35	0.15	0.15	0.3	0.15	–	–	–	–	–	–
	100	–	–	–	–	–	–	–	–	–	–	–	–	–	–	–	–	–	–
CCC	1000	–	–	–	–	–	–	0.2	0.7	1.5	0.4	0.25	0.3	–	–	–	–	–	–
	5000	–	–	–	–	–	–	0.45	0.35	–	–	–	–	–	–	–	–	–	–
MH	100	0.3	0.15	–	–	–	–	0.55	0.35	0.55	0.6	0.35	0.45	0.3	0.2	–	–	–	–
	1000	–	–	–	–	–	–	0.15	0.05	0.15	0.15	–	–	–	–	–	–	–	–

(continued)

Table 3.26 (continued)

		Brassica juncea																	
		10 °C						20 °C						35 °C					
		Red		Far Red		Blue		Red		Far Red		Blue		Red		Far Red		Blue	
Applied plant growth regulators	ppm	Rad	Hip	Rad	Hip	Rad	Hip	Rad	Hip	Rad	Hip	Rad	Hip	Rad	Hip	Rad	Hip	Rad	Hip
NaCl	1000	0.2	0.15	–	–	–	–	3.5	1.2	5	0.4	1	0.5	–	–	–	–	–	–
	20.000	–	–	–	–	–	–	–	–	–	–	–	–	–	–	–	–	–	–
T.U	100	0.3	0.15	–	–	–	–	3.25	0.9	2.5	0.25	1.25	1.1	1	0.35	–	–	0.45	0.4
	1000	0.45	0.2	0.4	0.3	0.2	0.15	0.8	0.5	0.9	0.4	2.25	0.75	0.8	0.35	–	–	0.45	0.45
KNO_3	100	0.25	0.15	0.25	0.15	0.1	0.1	2.75	0.75	3.5	0.5	2.25	0.95	–	–	–	–	–	–
	1000	0.3	0.15	–	–	–	–	3	1.75	1	0.5	0.6	0.5	–	–	0.4	0.3	–	–
Ethrel	50	0.15	0.15	–	–	–	–	0.15	0.3	0.15	0.3	0.3	0.4	–	–	–	–	0.15	0.15
	100	0.15	0.15	–	–	–	–	0.15	0.35	0.15	0.25	0.2	0.2	–	–	–	–	–	–
	200	0.05	0.05	–	–	–	–	0.15	0.35	0.1	0.2	0.2	0.4	–	–	–	–	0.2	0.2
	400	0.05	0.05	–	–	–	–	0.15	0.3	0.1	0.3	0.2	0.4	–	–	–	–	0.15	0.3
PEG	1	0.2	0.2	–	–	–	–	3.75	0.6	3.25	0.5	3	1	–	–	0.3	0.45	0.8	0.3
	10	0.2	0.15	–	–	0.05	0.05	3.25	0.6	3.75	0.55	4	0.9	0.2	0.2	–	–	0.4	0.35
	100	0.2	0.15	–	–	0.15	0.15	3.25	0.7	3	0.5	2.75	0.75	0.45	0.25	–	–	0.6	0.25
Man.	1	0.3	0.2	–	–	0.3	0.15	3.75	0.9	3.25	0.6	2	1	–	–	–	–	0.2	0.3
	10	0.3	0.15	0.4	0.15	–	–	3.5	0.55	5	1	0.6	0.4	0.3	0.35	–	–	0.35	0.35
	100	0.2	0.15	–	–	–	–	2.5	0.45	4.5	0.6	2	1	–	–	–	–	–	–
Morf.	50	–	–	0.2	0.2	–	–	0.8	0.75	1.5	0.5	0.25	0.2	–	–	–	–	–	–
	100	–	–	–	–	–	–	2.25	0.75	0.85	0.45	1.25	0.9	–	–	–	–	0.45	0.35
	200	–	–	–	–	–	–	1.1	0.85	1.3	0.5	1.5	1.25	0.35	0.3	–	–	0.3	0.35
Control		0.15	0.15	–	–	–	–	2.75	0.85	3.25	0.55	1.25	0.75	0.2	0.15	–	–	0.45	0.35

Table 3.27 Radicle and hypocotyl lengths (cm) in *Brassica campestris* under different growth regulators and lighting conditions

		Brassica campestris																	
		10 °C						20 °C						35 °C					
		Red		Far red		Blue		Red		Far red		Blue		Red		Far red		Blue	
Applied plant growth regulators	ppm	Rad	Hip	Rad	Hip	Rad	Hip	Rad	Hip	Rad	Hip	Rad	Hip	Rad	Hip	Rad	Hip	Rad	Hip
GA_3	1	0.3	0.2	0.3	0.2	0.25	0.15	1.25	1.35	3.5	1	0.4	0.5	–	–	–	–	–	–
	100	0.4	0.3	0.4	0.35	0.3	0.15	2.25	1.75	0.7	1.1	0.6	1	–	–	–	–	–	–
Kin	1	0.3	0.25	0.35	0.3	0.35	0.2	0.6	1	2	0.5	0.55	0.45	–	–	–	–	–	–
	50	0.2	0.3	0.25	0.2	0.25	0.15	0.4	1.15	1.5	0.75	0.5	0.4	0.15	0.15	–	–	–	–
	100	0.2	0.35	0.15	0.25	0.25	0.15	0.75	1.6	0.75	0.85	0.6	0.4	–	–	–	–	0.3	0.3
IAA	10	0.15	0.05	0.15	0.05	0.05	0.05	0.55	0.6	1.25	0.4	1.5	1	–	–	–	–	–	–
	50	0.15	0.05	0.15	0.05	0.15	0.15	0.55	1	0.4	0.3	0.15	0.25	0.15	0.15	–	–	–	–
	100	0.05	0.05	–	–	0.05	0.05	0.3	0.55	0.3	0.3	1	0.5	0.45	0.5	–	–	–	–
TIBA	10	0.35	0.3	0.3	0.15	0.3	0.15	2.75	1.5	2	1.1	1.25	0.75	–	–	–	–	0.4	0.4
	50	0.15	0.4	0.25	0.15	0.2	0.15	0.75	0.55	1.4	0.9	0.15	0.2	–	–	–	–	0.45	0.35
	100	0.15	0.35	0.15	0.15	0.15	0.15	0.85	1.25	0.45	0.5	0.3	0.2	–	–	–	–	–	–
Coumarin	10	0.15	0.3	0.15	0.15	0.15	0.15	2	1.5	1.25	0.7	0.25	0.2	–	–	–	–	–	–
	50	0.15	0.15	–	–	–	–	0.4	1	0.45	0.5	0.6	1	–	–	–	–	–	–
	100	0.2	0.1	–	–	–	–	0.2	0.3	0.15	0.35	0.15	0.3	–	–	–	–	–	–
ABA	1	0.3	0.25	0.35	0.2	0.2	0.05	2.5	1.25	1.5	0.6	0.5	0.4	0.4	0.3	–	–	–	–
	20	0.25	0.15	0.15	0.05	0.05	0.05	2	1.5	1.5	0.5	0.4	0.3	0.25	0.15	–	–	–	–
	100	0.05	0.05	–	–	–	–	0.15	0.05	0.15	0.35	0.2	0.15	–	–	–	–	–	–
CCC	1000	0.2	0.05	–	–	–	–	2.75	1.35	0.55	0.35	0.25	0.2	–	–	–	–	0.3	0.85
	5000	0.15	0.05	–	–	–	–	0.5	0.45	0.4	0.3	0.15	0.15	–	–	–	–	–	–
MH	100	0.25	0.15	0.4	0.15	0.2	0.15	0.4	0.75	0.55	0.45	0.6	0.5	–	–	–	–	–	–
	1000	0.05	0.05	–	–	–	–	0.25	0.35	0.15	0.25	0.15	0.15	–	–	–	–	–	–

(continued)

Table 3.27 (continued)

		Brassica campestris																	
		10 °C						20 °C						35 °C					
		Red		Far red		Blue		Red		Far red		Blue		Red		Far red		Blue	
Applied plant growth regulators	ppm	Rad	Hip	Rad	Hip	Rad	Hip	Rad	Hip	Rad	Hip	Rad	Hip	Rad	Hip	Rad	Hip	Rad	Hip
NaCl	1000	0.35	0.15	0.4	0.2	0.3	0.15	2.5	2	0.5	0.3	0.3	0.2	0.75	0.5	–	–	0.15	0.15
	20.000	–	–	–	–	–	–	–	–	–	–	–	–	–	–	–	–	–	–
T.U	100	0.5	0.3	0.5	0.25	0.3	0.15	1.7	0.9	3.5	0.75	0.5	0.35	0.45	0.35	0.15	0.15	0.35	0.15
	1000	0.4	0.3	0.4	0.2	0.35	0.15	1.25	0.65	1.25	0.95	0.3	0.35	0.15	0.15	0.3	0.35	0.45	0.35
KNO_3	100	0.45	0.3	0.3	0.15	0.3	0.2	2.75	1.1	2.5	0.6	0.25	0.15	0.35	0.2	–	–	0.3	0.35
	1000	0.35	0.3	0.3	0.15	0.35	0.2	1.75	1.6	1.25	1.25	0.25	1.25	–	–	–	–	0.5	0.45
Ethrel	50	0.2	0.15	0.15	0.15	0.2	0.15	0.25	0.45	0.3	0.45	0.15	0.65	–	–	–	–	–	–
	100	0.2	0.15	0.15	0.15	0.35	0.2	0.2	0.9	0.15	0.45	0.3	0.2	–	–	–	–	–	–
	200	0.15	0.15	0.15	0.15	0.2	0.15	0.2	0.5	0.15	0.35	0.25	0.85	–	–	–	–	–	–
	400	0.15	0.15	0.15	0.15	0.15	0.05	0.15	0.3	0.15	0.25	0.25	0.2	–	–	–	–	–	–
PEG	1	0.3	0.2	0.4	0.25	0.2	0.15	2.25	1.25	2.5	0.75	1.1	0.35	0.45	2	0.35	0.4	0.35	0.75
	10	0.3	0.2	0.45	0.15	0.45	0.2	1.8	1.25	0.6	0.7	0.8	0.2	0.35	0.3	0.3	0.4	0.5	0.35
	100	0.4	0.25	0.55	0.25	0.4	0.2	3.25	1.5	1.5	0.6	0.3	0.35	0.4	0.5	0.3	0.2	0.3	0.35
Man.	1	0.35	0.35	0.25	0.15	0.3	0.15	1.4	0.65	3.25	0.75	0.75	0.6	–	–	–	–	0.3	0.2
	10	0.25	0.2	0.3	0.2	0.2	0.15	2.25	0.8	1.2	0.5	0.15	0.35	0.15	0.15	–	–	0.3	0.3
	100	0.3	0.2	0.5	0.3	0.35	0.15	3.25	1.25	1.25	0.55	0.15	0.3	–	–	–	–	0.3	0.3
Morf.	50	0.25	0.15	0.15	0.15	0.3	0.2	0.65	1.25	0.5	0.3	0.3	0.65	–	—	–	–	–	–
	100	0.35	0.25	0.15	0.15	0.3	0.15	0.8	1.25	1	0.35	0.25	0.3	0.25	0.2	–	–	0.2	0.2
	200	0.25	0.2	0.15	0.15	0.3	0.15	1	1.25	0.9	0.4	0.8	0.5	–	–	–	–	0.2	0.2
Control		0.2	0.15	0.25	0.15	0.3	0.15	1.85	1.5	1.5	0.65	0.6	0.4	0.45	1	0.3	0.65	0.25	0.2

3.2 Potential Abiotic Stresses in the *Brassicas*

Abiotic stresses such as drought, heat and salinity, resulting from climate change driven by anthropogenic activities, adversely affect *Brassica* species. The *Brassica* genus, comprising key oilseed and vegetable crops, is particularly vulnerable to these stresses. These disruptions lead to reduced photosynthesis, nutrient uptake, ionic balance, biomass, altered leaf morphology, impaired reproductive development and, ultimately, significant yield losses. The strategies to lessen or alter the abiotic stress effects on *Brassica* must study the impact of different stress factors on the crop plants. This chapter provides an overview of the impacts of stress on various *Brassica* species. The effects of different stresses on *Brassica* are summarised in Tables 3.28, 3.29, 3.30, and 3.31.

3.2.1 Drought Stress

Drought is a significant abiotic stress arising from rapid global warming and climate change, posing a severe challenge to global crop production (Batool et al. 2022; Ozturk et al. 2025a, b). According to UNCCD COP-15, drought stress continues to escalate, affecting approximately 2.3 billion people and causing socio-economic losses exceeding $120 billion (Rosendahl et al. 2022). Drought stress affects almost 45% of the world's agricultural land due to water scarcity. It affects many people around the globe and is an expensive disaster, particularly in developing countries of Africa and Asia (Kogan et al. 2019). According to the the "Water Resources of India" report by the Central Water Commission states that around 65% of India's agricultural land depends on rainfall. Drought stress impacts irrigation availability, leading to water shortages and inadequate irrigation, reducing crop yields. Drought can be described as a prolonged insufficient water supply, which disrupts transpiration rates and disturbs cellular water levels due to the roots' inability to absorb adequate water from the root zone. It dehydrates plants, disrupts metabolism, leads to stomatal closure, restricts gas exchange, severely damages the photosynthetic machinery and accumulates reactive oxygen species. This stress leads to leaf rolling, wilting, senescence and eventually leaf death, resulting in stunted plant growth and development (Ramzan et al. 2023). The effects on different parameters are described here.

3.2.1.1 Germination

The germination of seeds is a crucial stage in any plant's life cycle, and it determines the crop yield. Successful seedling establishment and faster germination are essential for agricultural production (Batool et al. 2022). The germination process starts with the absorption of water and swelling of the seed, culminating in cell

Table 3.28 Effect of drought stress on different plants of *Brassica*

Plant species with cultivars	Concentration and duration	Effect	References
Brassica napus cvs. P287 and T88	Soil relative water content (15–20%) for 7 days	Reduction in RWC and chl content, Soluble protein and MDA increased, enhanced antioxidant (POD, SOD, CAT)	Zhu et al. (2021)
Brassica napus cvs. (Hyola308, Hyola420, Olga, Sarigol, Amica, and RGS003)	PEG(6000), with field capacity 70% (mild), 50% severe for two weeks	Reduction in dry weight, height and leaf area, Chl a and b decreased, soluble sugar, proline increases as stress increases as well as antioxidants and ROS (CAT, POD, PPO)	Kazemi et al. (2024)
Brassica napus cvs. Ks7 and Nima	PEG (150, 300 g/l)	Plant height, stem height, root length, fresh and dry weight all decrease at 300 g/l, photosynthetic pigments (chl a, chlb and total chl) all decrease at 300 g/l, antioxidants (CAT, POD, SOD, MDA) increase with proline accumulation, enhance sugars (glucose, galactose, rhamnose, xylose) in leaf and root	Khodabin et al. (2020)
Brassica rapa ssp. *chinensis*	Field capacity of 30–40% (water deficit condition) for 15 days	Stomatal density increases on the upper and lower epidermis, but stomatal width and aperture decrease. Chl a, b and total chl decrease, whereas chl a/b increases. Pn, Gs, Tr and Ci all decrease, whereas O_2^- production as well as SOD, POD, CAT activity and MDA enhance	Li et al. (2022)
Brassica oleracea var. cross	Not irrigated for two weeks	Root length and weight increase, high plant weight, ten glucosinolates (six aliphatic, two indoles, and two aromatics) were detected, which are higher in leaves compared to roots.	Ben Ammar et al. (2023)
Brassica campestris	Field capacity is 50% after 21 days of sowing.	Decrease in root length, root fresh weight, and root dry weight; minimised the contents of chlorophylls *a* and *b*; electrolyte leakage increase; more oxidants (MDA and H_2O_2) enhanced the enzymatic antioxidant (CAT, SOD, and POD) production; reduced the content of anthocyanin, carotenoids, and flavonoids; ascorbic acid, total soluble proteins (TSP) and total soluble sugars (TSS) contents were significantly increased; NO_3^-, SO_4^{2-}, Ca^{2+} and K^+ contents were reduced and reduced all yield attributes (1000-grain weight, number of silique/plants, number of seeds/siliqua, and biological yield)	Ramzan et al. (2023)

(continued)

Table 3.28 (continued)

Plant species with cultivars	Concentration and duration	Effect	References
Brassica rapa	To create drought conditions, irrigation is done at 2-week intervals	Decrease in shoot and root length with a decrease in fresh and dry weight, with a reduction in Chl *a*, Chl *b* and total chlorophyll, decrease in total phenolic levels and total soluble protein with increased Na^+ uptake in root and leaves and a reduction in K^+ and Zn^+ uptake	Hasnain et al. (2023)
Brassica juncea cvs. Weiliang and Shuidong	−0.5 Mpa (moderate stress) and −1.0 Mpa (severe stress)	Germination percentage reduces, antioxidants (CAT, POD and SOD) and proline are enhanced	Wei et al. (2024)
Brassica rapa cv. Chunkwang	10% soil water content for 3 weeks	Total leaf number, total leaf area, fresh and dry weights decrease, reduction in flavonoids	Shawon et al. (2020)
Brassica napus	10%PEG-6000 solution for 4 and 8 days	Fresh and dry biomass, along with RWC, reduced; electron transport rate with enhanced NPQ, reduction in PSII efficiency, electron transport rate and non-photochemical quenching increase, as well as BnSOD, BnPOD and BnCAT, and BnAPX enhances	Chen et al. (2022)
Brassica rapa	14 days	Shoot dry weight decreased with a reduction in chlorophyll and carotenoid contents ROS (O_2^- and H_2O_2), content of MDA the activities of antioxidant enzymes, such as SOD, CAT, GPOD and APOD, osmotic stress decreased, and proline content was significantly increased, and reduced GSH content	La et al. (2019)

division and elongation, protein biosynthesis and stored reserve food mobilisation and metabolism, which result in the emergence of the radicle and the growth of the stem (Torabi and Ardestani 2013; Han et al. 2017). Water scarcity hinders germination and seed establishment during drought stress by decreasing water intake, restricting energy availability and interfering with enzyme activity, or it also reduces osmotic potential, limiting water absorption, resulting in a decrease in turgor pressure and inhibiting cell division, ultimately impairing radical growth (Batool et al. 2022; Torabi and Ardestani 2013). As a result of all these actions, the seed coat becomes impervious to water, which causes a delay in germination (Batool et al. 2022). In addition, during drought stress in the endosperm, there is slower hydrolysis of storage compounds or a decrease in the transportation of hydrolysed material to the developing embryonal axis (Torabi and Ardestani 2013). Furthermore, alteration in enzymatic activity and hormone levels in seeds leads to a delay in germination (Channaoui et al. 2017). In a study of *Brassica napus*, two cultivars demonstrated

Table 3.29 Effect of salt stress on different plants of *Brassica*

Plant species with varieties or cultivars	Concentration and duration	Effects	References
Brassica juncea cv.CS-52 Ashirwad	25 mM, 50 mM, 75 mM, 100 mM and 125 mM (in a stepwise manner for 2 days) and the next 5 days for 150 mM	Total dry weights, stem, root, leaf dry weight; leaf area reduced, transpiration rates, rates of stomatal conductance are reduced, leaf temperatures, reduction in RWC, increase in WUE, electrolytic leakage, multifold increase in Na concentrations, but K, Ca, Mg, Fe, Zn and Co concentrations reduced and B, Mn and Cu increase	Singh et al. (2019)
Brassica juncea	180 mM for 7 days	Reduction in shoot length and root length; increased MDA and H_2O_2, rate of electrolyte leakage, reduced viability of cells, reduced soluble protein	Kaur et al. (2018)
Brassica oleracea var. ***italica***	20, 40, 60 and 80 mmol L^{-1} for 6 weeks	Leaf area, root length, shoot length, root and shoot dry weights decrease; *reduced Fv/Fm*, LEF, qP, activity of PS-II efficiency, *K*$^+$, Ca^{2+}, Mg^{2+}, PO_4^{3-} and NO_3^-; leaf chlorophyll content, qL, Na^+ and Cl^- increase	Ali et al. (2022)
Brassica oleracea cvs. Marathon and GreenPort	40, 80 and 120 mM after 14 days of transplant of plants	Reduction in morphological characters (plant height, head fresh weight, head dry weight, head diameter, number of florets, shoot fresh weight, shoot dry weight, root length, root fresh and dry weight, number of leaves, and relative water content), with reduction in Chl a, Chl b and carotenoids; enhanced SOD, POD and CAT activity with production of MDA and H_2O_2 along with Na^+ and Cl^-ions which reduces K^+ content	Sardar et al. (2023)
Brassica oleracea cv. Olimpia	4 dS m^{-1} for the entire crop cycle	Lower water (Ψ_w) and osmotic (Ψ_π) potential; reduction of the root dry biomass and a reduction of the leaf area and of the total plant dry weight; lower stem, root, head and total plant dry weights, root shoot ratio; lower K^+/Na^+, Mg^{2+}/Na^+ and Ca^{2+}/Na^+ ratios, as well as inflorescence appearance decreases	Di Gioia et al. (2018)
Brassica juncea	10, 30, 50,70 and 100 mM for 4 days	After 50 mM, reduction of fresh weight, dry weight, root and shoot length, and proline level. Highest MDA content and H_2O_2 found at 100 mM	Khator and Shekhawat (2020)
Brassica rapa	100 mM	Decreased shoot and root biomass with anthocyanin content reduction, enhanced MDA and H_2O_2 content, reduced total soluble protein	Bano et al. (2022)

(continued)

Table 3.29 (continued)

Plant species with varieties or cultivars	Concentration and duration	Effects	References
Brassica juncea cv. RH0-749	100 mM after 10 days of sowing	Leaf area and plant dry mass were reduced; accumulation of Na^+ and Cl^- ions, whereas N was reduced; H_2O_2 and TBARS increased; inhibited photosynthetic characteristics (P_N, g_s, C_i and chlorophyll content) and ubisco activity, enhanced proline; activity of APX, GR and SOD increased with contents of Cys and GSH, whereas N content and NR activity are reduced	Majid et al. (2021)
Brassica napus	3 g NaCl Kg^{-1} soil	Decrease in Pn, Gs, Tr and CE; increase in SOD, CAT and AsA content as well as soluble sugar, protein, and proline; enhance lipid peroxidation	Wang et al. (2022)
Brassica rapa	80 mM, 160 mM for 2 weeks	Plant height, no. of leaves, shoot length, root length, leaf length, and leaf area all get reduced, as well as a reduction in weights of roots and shoots, both fresh and dry. Yield weight and diameter decreased, as well as the levels of carotenoids, total chlorophyll, Chl. b and Chl. a. Yield weight and diameter decreased, and increase expression of SOD, POD, and CAT and Proline, total soluble sugar, and total soluble protein levels decreased, and MDA and H_2O_2 increased	Fatima et al. (2024)

strong germination under normal conditions. However, under stress, germination rates dropped significantly, with one cultivar showing a steep decline of 93.01% while the other experienced a modest decrease of 7.02% (Batool et al. 2022). While Toosi et al. (2014) found in *B. juncea* that the length and weight of hypocotyl decreased as water stress increased, as well as the radical growth was reduced at 4 MPa and was completely inhibited at 10 and 12 MPa of PEG solution. In a study on cauliflower (*B. oleracea*) and canola (*B. napus*), the germination rate in both plants was reduced when the drought stress was induced by PEG (12%) and mannitol (2.5%) compared to the control (Hadi et al. 2014). These studies show that insufficient water creates an osmotic imbalance, causing dormancy or impairing seed germination, negatively impacting plant health.

Table 3.30 Effect of heat stress on different plants of *Brassica*

Plant species with cultivars	Temperature and duration	Effects	References
Brassica oleracea var. *capitata* f. *alba*	At 42 °C for 3 h and 5 h (HS) in an incubator	Closing stomata at high temperatures reduces the chlorophyll content and reduces photosynthesis, affecting the intercellular CO_2	Moradpour et al. (2021)
Brassica campestris (WS-6 (heat-sensitive) and WS-1 (heat-tolerant)	30 °C/26 °C (d/n) 35 °C/28 °C (d/n) 40 °C/30 °C (d/n) treatment for 5 days	WS-6 showed more morphological alterations than WS-1, demonstrating that high temperatures affected Wucai plant height, stem diameter, and single plant weight based on cultivar. At 40 °C, the Chl concentrations of both Wucai cultivars significantly decreased (only WS-6 exhibited a reduction at 35 °C), and their cellular ultrastructure changed in distinct ways	Zou et al. (2017)
Canola - *Brassica napus*	2 treatments (T1 = 18 C and T2 = 28 C)	Reduced photosynthetic rate, stomatal conductance, and inter CO_2 concentration, leading to the yield under heat stress conditions	Waraich et al. (2022)
B. rapa. ssp. chinensis cv. Wu-11	Heat shock at 46 C for 1 h	Elevated temperatures impede plant development, resulting in leaf etiolation and potential mortality	Yu et al. (2012)
Brassica napus cv. Aviso	Repeated stress sequence (RSS) composed of 5 early full days under mild warming (26 oC ± 2.2) followed by 5 days under natural thermoperiod 7 days under mild warming with daily heat peaks on 4 days (35.5 oC ± 1.6 from 10 am till 3 pm)	Different intensities of heat stress impact seed growth. Yield components experience mild heat stress during the seed set and after seed filling. Heatwave timing, frequency, length, and intensity impact seed development with pod age. Stresses may delay crop performance-related factors, altering enzymes, hormones, and precursors later in seed development	de Almeida et al. (2023)

3.2.1.2 Shoot Growth

Drought significantly impacts plant height, which is strongly correlated with cell enlargement and leaf senescence. Plant height is reduced under drought stress due to factors such as decreased cell expansion, increased leaf shedding, or disruption of the mitosis of the cell cycle (Yang et al. 2021). Cell elongation is affected by the

Table 3.31 Effect of heavy metal stress on different plants of *Brassica*

Plant species with cultivars	Concentration and duration	Effects	References
Two species of *Brassica* (*B. carinata* and *B. juncea*)	-Cd (10, 20, 40) mg kg^{-1} -Cr (50, 100, 150) mg kg^{-1} -Pb (100, 150, 200) mg kg^{-1} for 30 days	The lowest shoot length was observed in plants treated with 40 mg kg^{-1} Cd. The shoot length of *B. carinata* was greater than that of *B. juncea*	Iqbal et al. (2012)
Brassica juncea	The soil was intentionally poisoned with ten concentrations of lead (Pb)—0, 25, 50, 100, 250, 500, 750, 1000, 1250 and 1500 mg kg^{-1} of soil	Increased the content and uptake of Pb by the plant. The elevated concentration of lead (Pb) led to reduced content and absorption of micronutrients by plants, indicating that Pb is not only directly hazardous to plants but also induces nutritional imbalances in crop species	Sikka et al. (2010)
B. napus cvs. Cadeli, Viking, and Navajo	Cd (1, 5 and 10 μM of $CdCl_2$) and Zn (1, 10 and 20 μM of $ZnSO_4$) for 14 days	Cadmium treatment adversely affected plant growth parameters, including root length and the weight of both roots and shoots. Concerning the translocation of Cd and Zn, none of the three evaluated cultivars achieved the requisite threshold to exhibit any phytoextraction capacity, as shown by the transfer factors (TFs)	Benáková et al. (2017)
Brassica juncea	Five levels of chromium, namely, 0, 20, 40, 60 and 80 ppm	It was discovered that chromium had a detrimental effect on the growth characteristics. As the amount of chromium rose, there was a considerable reduction in the growth of mustard	Kumar et al. (2018)
Brassica species (*B. juncea*, *B. ampestris* and *B. napus*)	Pb (0, 50, 100, 150) mM	In all three species, shoot length was reduced to 150 mM. At a dosage of 100 mM, shoot length was similarly reduced in *B. napus* and *B. campestris*. Plant biomass in *B. napus* and *B. campestris* was reduced when Pb concentrations were 100 and 150 mM	Shehzad et al. (2023)
B. juncea L. Czern	Pb ((0, 125, 250, 500, 700 and 1000 mg Pb/kg soil)	The height and dry weight of the shoots and roots increased	Naaz and Chauhan (2019)
B. juncea	Cd and Pb (150 and 300 mg/kg soil)	Both cultivars showed increased lipid peroxidation due to Cd and Pb treatments. Proline accumulation increased in both cultivars when Cd and Pb concentrations increased	Ahmad et al. (2012)

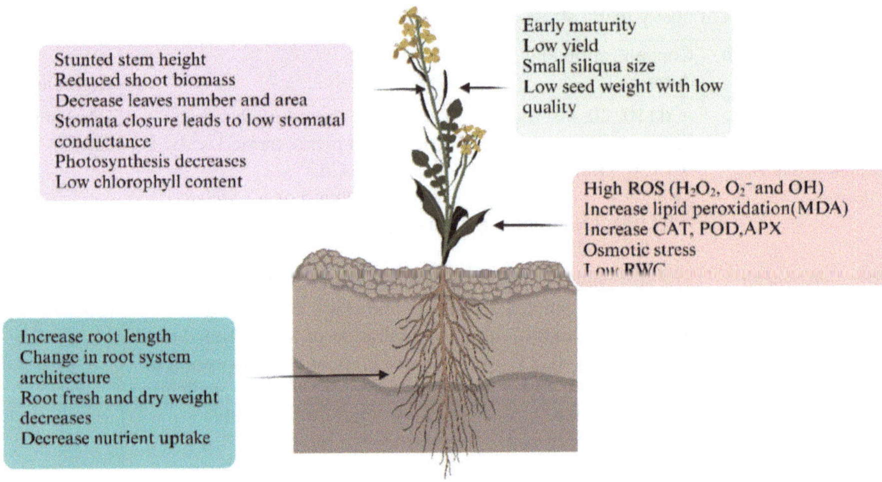

Fig. 3.2 Impact of drought stress on different parts of *Brassica* spp.

reduction in water potential and turgor pressure, which hampers cell division and expansion of cells or affects cell wall biosynthesis, leading to a reduction in plant growth and development, ultimately leading to yield loss (Sheoran et al. 2024; Pakdel et al. 2020). A study on *B. rapa* suggests a reduction of about 23% in shoot length under drought stress (Hasnain et al. 2023). This could be the result of reduced nutrient uptake or a drop in transpiration rate brought on by ABA-induced stomata closure, which effectively reduces water loss and intensifies the stress response (Sánchez-Rodríguez et al. 2010). Achard et al. (2006) suggested that under drought stress, elevated ABA levels cause stunted plant growth, as shown in Fig. 3.2. The assessment of drought stress can be indicated by the fresh and dry weight of the shoot, which are indicative of growth characteristics that are substantially diminished compared to the control in *B. rapa* (Hasnain et al. 2023). In *B. napus*, these factors were reduced by 78% by the fourth day and 85% by the eighth day, which resulted in a considerable loss in plant weight (Chen et al. 2022). The duration of stress has an impact on these factors as well; the longer the stress lasts, the more severe it becomes. This is proven by the study of *B. rapa,* in which shoot dry weight decreased by 14% after 14 days of drought stress, which suggests that species-to-species variations may exist (La et al. 2019). The complex physiological impacts of drought stress can be shown by the fact that decreases in turgor pressure, enzymatic activity, and photosynthetic rate are important contributors to the morphological factors, such as decline in total biomass, fresh weight, dry weight, and height of shoot (Pakdel et al. 2020).

3.2.1.3 Root Growth

The plant's anchoring organ, the roots, is responsible for absorbing nutrients and water from the surrounding soil, which is crucial for the survival and productivity of the plant. Drought stress affects the roots of the plant more than the aboveground components; it restricts the amount of water available, which harms the growth and development of the entire plant. Deepness, branching, angle of lateral roots, and other aspects of roots that regulate soil water intake are all impacted by drought stress (Kalra et al. 2024; Kage et al. 2004). Brassica's roots grow longer under drought stress because they are buried deeper to draw out water (Raza et al. 2017). Another reason for the increase in root length and dry mass is that they can enhance the allocation of assimilates to the roots, which promotes the elongation of roots under drought stress (Kage et al. 2004). SUC/SWEET sugar transporters help in sugar unloading in roots, and their overexpression enhances water uptake from deeper soil layers, promoting root elongation by increasing carbon supply during drought stress (Kalra et al. 2024). According to Raza et al. (2017), there is an increment in root length under 50% of the field capacity under drought stress. But recently, a contradictory study found that *B. rapa* has the length of the root reduced under drought stress (Hasnain et al. 2023). Additionally, root length decreases under drought stress caused by PEG and mannitol as these compounds extract water from the plant (Hadi et al. 2014). In *B. napus*, root dry biomass decreased by 46% on the fourth day and 61% on the eighth day under drought stress, while root fresh weight declined by 41% and 57%, respectively (Chen et al. 2022).

3.2.1.4 Leaves

The main organ of any plant for assimilation and transpiration is the leaves, which are a direct indicator of the degree of water shortage. When drought stress occurs, water loss lowers turgor and potential pressure, which causes wilting, rolling, and a reduction in leaf area. This, in turn, lowers the photosynthetic rate and crop output (Rucker et al. 1995; Yang et al. 2021). In *B. rapa*, drought stress at 10% soil water for 3 weeks reduced total leaf area and fresh weight by 26% and 34%, respectively (Shawon et al. 2020). Under water-deficit conditions, leaf osmotic potential decreased in rapeseed, likely due to the accumulation of solutes in leaf cells. Additionally, under drought conditions, the number of leaves in Chinese kale (*B. oleracea*) dropped by 38% (Issarakraisila et al. 2007). This decrease was probably caused by rapid leaf abscission and senescence triggered by elevated ABA levels during drought stress (Raza et al. 2017).

3.2.1.5 Relative Water Content

Drought stress primarily affects the water balance in plants, with reduced water availability leading to a decline in tissue water content (Sheoran et al. 2024). Relative water content (RWC) is a key indicator of leaf water status and is an important measure of drought stress (Dogra et al. 2018). RWC reduction results in lower water potential and a subsequent loss of turgor pressure. The loss of turgor pressure impedes critical physiological processes such as cell elongation, ultimately affecting plant growth and development (Sheoran et al. 2024). In addition to this, RWC is also associated with leaf water potential, and cell potential represents the level of hydration, which is important for plant metabolism, or it can influence other processes like stomatal conductance, transpiration, etc. (Raza et al. 2017; Soni et al. 2021). In a study by Zhu et al. (2021), in *B. napus*, there is a decrease in RWC when the drought stress is increased. Hadi et al. (2014) observed that drought stress induced by PEG and mannitol led to a decrease in seedling water content as the concentration of PEG and mannitol increased in *B. napus* and *B. oleracea*. This indicates a negative correlation between water content and the concentration of drought-inducing solutions. A study on *B. juncea* involving different genotypes-Pusa Bold, Kranti, RSPR-03 and RSPR-01-revealed a significant reduction in RWC under drought stress compared to the control. The RWC decreased progressively with the duration of drought stress (Dogra et al. 2018).

3.2.1.6 Oxidative Damage

The oxidative stress is produced when the production of reactive oxygen species (ROS), such as superoxide radicals, hydroxyl radicals, hydrogen peroxide, alkoxy radicals and singlet oxygen, exceeds the plant's critical concentration threshold (Ahmad et al. 2012; Sheoran et al. 2024). ROS form in cell organelles, like chloroplasts via Mehler reaction, mitochondrial dysfunction, peroxisomes via photorespiration, etc., in normal conditions (Salehi-Lisar and Bakhshayeshan-Agdam 2016; Saibi 2021), but under stress conditions, excessive ROS production can damage DNA by causing nicks, oxidise amino acids or break disulphide bonds, result in denaturation of proteins, pigments, and lipids and lead to lipid peroxidation. These processes can disrupt the structure and function of macromolecules (Salehi-Lisar and Bakhshayeshan-Agdam 2016). ROS oxidises chlorophyll, impairing photosynthesis and reducing photosynthate production, ultimately leading to cell death responsible for the reduction in crop yield (Sheoran et al. 2024; Saibi 2021). In addition to this, ROS adversely affects the membrane permeability and is unstable (Latif et al. 2016). ROS damage to proteins primarily occurs through carbonylation and glycosylation. Oxygen-free radicals can interact with the sulfhydryl groups at the enzyme's active site, oxidising them into disulfide bonds, which leads to enzyme inactivation. These enzymes can be photosynthesis-related and growth-related. Also, breakage of DNA interrupts cell division, resulting in the inhibition of growth and development (Yang et al. 2021). A study on *B. rapa* revealed that 14 days of drought stress accelerates O_2^-, H_2O_2 and MDA levels by 2.3-fold, 68.5% and 27%, respectively. To counteract these oxidations, the plants have the defence of antioxidant enzymes such as SOD, CAT, GPOD and APOD enhanced by 160.3%, 71.9%, 87.5% and 64.4%, respectively, compared to the control (La et al. 2019).

3.2.1.7 Photosynthesis

Photosynthesis is a vital physiological process essential for plant and human survival. Plants protect their photosynthetic mechanisms from stressors that impact photosynthesis and its apparatus. During drought stress, reduced stomatal conductance and relative water content decrease CO_2 availability due to stomatal closure, which activates Rubisco's oxygenase activity, resulting in additional ATP loss (Antunović Dunić et al. 2023). Drought disrupts the oxygen-evolving complex (OEC), alters PSI redox states, impairs electron transfer in PSII, reduces PSI and PSII efficiency, and inhibits overall electron transfer (Zhou et al. 2019). These effects, combined with ROS generation, cause photoinhibition, potentially leading to permanent PSII damage under severe conditions. Rubisco concentration declines due to its instability, reduced small subunits, or inhibitor binding during drought (Fahad et al. 2017). Additionally, drought stress lowers PSII's quantum yield (QY), indicating photo-inhibitory damage and disintegration of the LHCII complex of PSII (Chen et al. 2022). A study on *B. napus* found that drought stress reduced PSII efficiency by increasing the electron transport rate through PSII and enhancing non-photochemical quenching (La et al. 2019). A study on *B. napus* reported a significant reduction in Rubisco protein levels under drought stress, with decreases of 75.6% and 51.1% observed in two genotypes after 72 h of drought stress. Similarly, the net photosynthesis rate declined by 74.6% and 58.5% in these genotypes compared to the control, highlighting the impact of drought on key photosynthetic components (Lee et al. 2016). The expression of important photosynthetic and metabolic genes encoding enzymes (RBCS, TPI, FBP, FBPA and TKL) was markedly down-regulated in *B. napus* during drought stress, causing detrimental effects on photosynthesis and associated processes (Liu et al. 2016).

Chlorophyll (Chl) is also an essential component of photosynthesis, which is impacted by unfavourable environmental conditions as well as the deformed shape and structure of thylakoids. Furthermore, drought stress decreases the activity of the enzymes by ROS that produce chlorophyll or increases the activity of the enzymes that break down chlorophyll, such as chlorophyllase (Sheoran et al. 2024) or distorts the thylakoid membrane (Shen et al. 2020). Additionally, the ratio of chlorophyll a/b gets reduced under drought-stressed conditions. In drought-stressed leaves, water potential (Ψw) decreased, along with a reduction in chlorophyll content. Under drought stress, chlorophylls a and b are reduced by about 41.9% and 40.6% *B. campestris* (Ramzan et al. 2023). Similarly, under stress conditions, chlorophyll levels decreased in proportion to the severity of the stress, as shown in Fig. 3.2. In *B. napus*, chlorophyll content declined by 24% and 46% on the fourth and eighth days, respectively, resulting in a reduction in net photosynthesis by 75% and 86% over the same period (Chen et al. 2022). The decline of chlorophyll may be caused by changes in the leaf area, or it has been shown that when plants are under drought stress, the colour of the chlorophyll turns yellowish brown (Yang et al. 2021).

3.2.1.8 Glucosinolates

In the Brassicaceae family, drought stress influences the levels of secondary metabolites. *Brassica* crops are renowned for their glucosinolates (GL), sulphur-containing compounds that contribute to their distinctive flavours and spiciness. Moreover, these compounds are valued for their potential health benefits (Sheoran et al. 2024). Their concentration in the plant varies with different stages of growth and development. A recent study by Huang et al. (2023) revealed that the total GL content in mustard roots initially decreased but later increased during development, peaking at the pod-setting stage. In contrast, the leaves showed an opposite pattern, with GL content increasing initially, reaching its highest level at the bolting stage, and then declining as the plant matured. However, abiotic stress affects its concentration in the plant. Del Carmen Martínez-Ballesta et al. (2013) found that the accumulation of glucosinolates in plants is influenced by the duration and severity of drought stress. Under 2 weeks of drought stress, *B. rapa* showed elevated GL levels (Shawon et al. 2020). In *B. oleracea*, GL concentration was higher in leaves compared to roots, potentially as a plant tolerance strategy through osmotic adjustment (Schreiner et al. 2009), which also enhances leaf antioxidant status or promotes root growth by lowering GL levels in the roots because downregulation of GL metabolism in roots during drought indicates a shift in sulphur metabolism towards processes that promote root growth (Ben Ammar et al. 2023). Another study by Eom et al. (2018) found that GL accumulation, directly or indirectly linked to stomatal closure, helps minimise water loss under drought stress. Enhancing GL levels, crucial for stress tolerance, can be achieved by upregulating genes like gluconapin and 4-hydroxyglucobrassicin involved in the GL biosynthesis pathway. Furthermore, in *B. napus*, drought stress led to a sixfold increase in the expression of the big-26 gene, which is likely involved in the synthesis of osmoprotective compounds and may contribute to the enhanced production of glucosinolates (López-Berenguer et al. 2008).

3.2.2 Salinity Stress

Salt stress is a significant global challenge, particularly in arid and semi-arid regions, severely impacting agricultural productivity, food security, and sustainability (Ozturk et al. 1992, 1997, 2006, 2023b, 2025a, b; Raza et al. 2019a; Shahrasbi et al. 2021; Ilyas et al. 2024; Pirasteh-Anosheh et al. 2025). Approximately 397 million hectares of land are affected by salinity, with 230 million hectares of irrigated fields and 45 million hectares of salt-affected land. In India alone, an estimated 6.73 million hectares are affected (Singh et al. 2021). Salinisation reduces crop yields by over 50%, primarily due to the disruption of the plant's water and nutrient balance (Ilyas et al. 2024). This stress arises from rock weathering, low precipitation, irrigation with saline water and poor cultural practices (Aggarwal et al. 2024). Salt accumulation in the root zone, particularly Na^+, K^+, Cl^- and SO_4^{2-}, causes osmotic changes that impair water uptake (Balasubramaniam et al. 2023). Soil salinity is measured by electrical conductivity (EC) and sodicity by the exchangeable sodium percentage (ESP) or sodium adsorption ratio (SAR) (Andrade Foronda 2022). When EC exceeds 4 dS m^{-1}, and ESP is less than 15, the soil becomes saline (Uçarlı 2020). Salt stress disrupts crop yields by causing osmotic stress, nutrient deficiencies, and oxidative damage, leading to the generation of reactive oxygen species (ROS) that damage cellular structures (Ondrasek et al. 2022). Salt ions accumulate in plant tissues, causing ionic and osmotic stress and triggering secondary oxidative stress (Feng et al. 2023). This oxidative stress damages photosynthetic machinery, membranes and enzymes (Wu et al. 2010). Severe salinity can lead to plant death due to high sodium and osmotic potential in the soil (Ozturk et al. 2023b).

3.2.2.1 Germination

Seed germination is a crucial stage for seedling establishment and crop production, yet it is highly vulnerable to adverse environmental conditions (Ren et al. 2020). Salinity disrupts metabolic processes, suppressing germination, growth and yield. Early growth stages are more sensitive to salinity, negatively impacting germination rates and percentages due to ion toxicity and nutrient imbalances (Kataria et al. 2017). Soil salinity can affect canola (*B. napus*) seed germination by creating an osmotic potential that prevents water uptake or through the toxic effects of Na^+ and Cl^- ions (Khajeh-Hosseini et al. 2003). It may also reduce seed germination phytohormones like GAs, increase ABA levels and alter the seed's membrane permeability and water movement (Uçarlı 2020). The delay in water uptake and reduced α-amylase activity with increasing NaCl concentration are key factors in delayed germination. This decreases α-amylase activity, limits sugar translocation, is crucial for embryo development and lowers sugar concentrations, altering the osmotic potential and reducing water uptake (El-Hendawy et al. 2019). A study on canola (*B. napus*) revealed that seedling emergence and its rate declined as salinity levels increased from 6 to 11 dS m^{-1}, with germination rates dropping significantly from 70 to 20% (Ahmadi and Ardekani 2006). In *B. napus*, even if germination occurs under salt stress, the length of the radicle and plumule decreases as salinity intensifies, adversely impacting root and shoot development (Bybordi 2010). Under extreme salinity levels, *B. insularis* seeds were observed to lose viability at a concentration of 200 mM (Santo et al. 2017).

3.2.2.2 Root Architecture

The primary root length is typically measured manually as it represents the first organ of the root system and is easily identifiable (Lema et al. 2019). Under salt stress, Ca^{2+} levels in roots remain stable, but K^+ levels decrease, and Na^+ accumulates, causing notable root damage. This stress impacts root morphology, reduces activity and water content, and disrupts ion homeostasis. Consequently, cell membrane integrity is compromised, leading to oxidative stress with elevated MDA and H_2O_2 levels, ultimately resulting in cellular damage (Li et al. 2022). Pavlović et al. (2019) revealed that salt stress significantly alters the root morphology in Chinese cabbage (*Brassica rapa*), primarily due to a marked reduction in essential nutrients like potassium (K), magnesium (Mg) and trace elements such as iron (Fe) and zinc (Zn), alongside the accumulation of toxic elements. Furthermore, a study by Ranjit et al. (2016) showed that that when roots of *B. juncea* were subjected to severe salt stress (250 mM) exhibited a rapid increase in MDA levels within 30 min, along with the accumulation of H_2O_2 and superoxide radicals, leading to an oxidative burst. Membrane damage rate intensified, with a 2.8-fold increase after 30 min, rising to 4.2-fold after 2 h of stress, resulting in cell death.

3.2.2.3 Photosynthesis

Plants' growth and productivity rely on efficient photosynthesis, which must adjust under environmental stress. Salinity severely disrupts photosynthesis, hindering fundamental processes and posing a significant threat to plant survival. Initially, the accumulation of Na^+ and Cl^- ions in leaf cells leads to water loss from guard cells, altering stomatal morphology and density, thereby reducing CO_2 intake and restricting photosynthesis (Lu et al. 2023). Moreover, the water relation is also disturbed by salt stress, which affects the stomatal conductance, which negatively influences the local ABA synthesis (Farooq et al. 2015). A study by Nazar et al. (2015) suggests that under 100 mM of NaCl applied for 15 days reduces stomatal conductance (Gs) by 26% and intercellular CO_2 concentration (Ci) by 41% in *B. juncea* cv. Pusa Jai Kisan. Furthermore, it was found that salt stress disrupts the electron transport chain (ETC) between photosystem II (PSII) and photosystem I (PSI), leading to excessive ROS generation and causing oxidative stress, which damages nucleic acids, proteins, lipids, membranes and key photosynthetic enzymes, thereby impairing CO_2 fixation and diminishing crop yields (Lu et al. 2023). The factors discussed above may also be a reason for the drastic decline in leaf area, which may be the primary reason for low photosynthetic capacity (Singh et al. 2019). Some studies also prove that the membrane of thylakoids is distorted, due to which the energy conversion efficiency of the plant gets reduced (Alamri et al. 2020). The study by Pavlović et al. (2019) shows that salt stress leads to the accumulation of reduced Q_A and Q_B, halting electron transfer along the electron transport chain. Furthermore, chlorophyll fluorescence is significantly impacted, as evidenced by a decline in photosystem II quantum yield and a reduction in non-photochemical quenching (Ma et al. 2017). The D1 protein exhibits sensitivity to salinity at concentrations of 100 mM, 150 mM and 200 mM NaCl, leading to PSII inhibition due to its degradation (Mittal et al. 2012). Moreover, the key enzyme responsible for photosynthesis, i.e. Rubisco, also gets reduced due to osmotic stress induced by salinity, due to which the Calvin cycle gets affected (Yousuf et al. 2016).

Chlorophyll also gets reduced under salt stress due to the overproduction of ROS generated by the cell organelles. Premature leaf senescence is also a sign of stress conditions due to an imbalance between chlorophyll biosynthesis and degradation under stress conditions, as chlorophyll degradation and the activity of chlorophyll-degrading enzymes (Chlase, Chl-POX, and PPH) were elevated compared to the activity of the chlorophyll biosynthesis pathway (Alamri et al. 2020). A study by Ahmad et al. (2015) found that in *B. juncea*, Chl a, Chl b and carotenoids decrease by 52%, 25.26% and 28.57%, respectively, under 200 mM NaCl concentration.

3.2.2.4 Oxidative Stress

Salt stress induces ionic toxicity by causing excessive accumulation of Na^+ and Cl^- ions and a reduction in the potassium ion in plant cells. This disruption in ion homeostasis impairs guard cell function, leading to stomatal closure, reduced carbon fixation due to limited CO_2 availability and accelerated production of ROS like H_2O_2, $^1O_2$, $O_2{\bullet}-$, $HO_2^{\bullet}$, $RO^{\bullet}$ and $^{\bullet}OH$ (Hasanuzzaman et al. 2021; Ozturk et al. 2023b). This is called oxidative stress. Salinity induces drought-like conditions in plants by reducing water potential, simultaneously increasing ROS production through the disruption of photosynthetic processes (Jalil and Ansari 2020). In addition, high Na^+ concentrations in the cytoplasm impair enzymes and metabolic activities, while salinity-induced oxidative stress activates antioxidant defences to neutralise ROS. Failure to efficiently eliminate ROS causes damage to proteins, lipids and nucleotides, as well as high H_2O_2 content leads to cellular dehydration and loss of turgor pressure due to lipid peroxidation, ultimately leading to cell death (Pavlović et al. 2019; Sardar et al. 2023) as shown in Fig. 3.3. Ilyas et al. (2024) reported that exposure to 200 mM NaCl led to an increase in H_2O_2 and TBARS content in *B. napus* varieties Faisal (V1) and Super (V2), an increase by 48% and 49% in V1 and 43% and 55% in V2, respectively. Similarly, *B. oleracea* exhibited increased MDA and H_2O_2 levels under 100 mM NaCl, with 2.31- and 1.71-fold elevations observed in the Marathon cultivar and 2.22 and 1.95-fold increase in the Greenport cultivar, respectively, which hampers its physiological and biochemical processes (Sardar et al. 2023).

3.2.2.5 Nutrient Imbalance

Salinity stress induces osmotic stress by dehydrating cells through excessive Na^+ uptake, thereby disrupting essential biochemical processes in plants. Elevated Na^+ and Cl^- levels reduce intracellular K^+ concentrations, causing metabolic imbalances

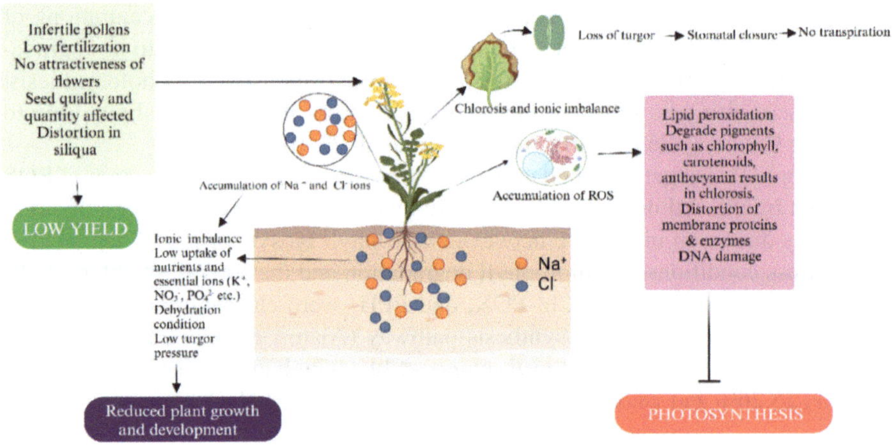

Fig. 3.3 Impact of salt stress on different parts of *Brassica* spp.

(Sardar et al. 2023). Salt stress in the root zone reduces osmotic and water potential, limiting water availability to the plant. Furthermore, the presence of salt stress near the root zone disrupts the nutrient balance and restricts the uptake of essential nutrients (Acosta-Motos et al. 2017). Na^+ accumulation under salt stress interferes with the absorption of key macronutrients (N, P, Ca, K) and micronutrients (Zn, Fe, Mn) as well as restricts the uptake of other essential cations and anions (NO_3^-) (Raihan et al. 2023; Ali et al. 2022). As salinity increases, the levels of major elements (Mg, P, S, K, Ca) and minor elements like Fe, Zn and Co decrease. However, B, Mn and Cu, which are needed in minimal amounts, increase, leading to a nutrient imbalance in *Brassica juncea*, impacting overall plant growth (Singh et al. 2019). A statistically significant negative correlation between Na^+ and Cl^- contents and concentrations of NO_3^-, Ca^{2+}, K^+ and P in leaves indicates a disruption in mineral uptake and ion homeostasis, likely occurring due to nutrient diffusion limitations, mobility constraints and altered nutrient uptake kinetics by roots (Ali et al. 2022). Additionally, according to Acosta-Motos et al. (2017) reported reduced K^+ content and increased Cl^- ions were reported in *B. rapa* ecotype PTWG-HL under salt stress. In *B. juncea*, nitrate reductase (NR) activity decreased by 14.92% due to low NR gene expression in roots, while NO levels increased by 39.10% under 50 mM NaCl but significantly declined at 100 mM NaCl, indicating NO concentration depends on stress intensity, which may result in stunted plant growth (Khator and Shekhawat 2020). Due to the downregulation of 26 genes responsible for nitrogen uptake transporters (NRT, AMT) and assimilation enzymes (NR, NiR, GS, GOGAT, GDH, ASN) under salt stress, depending on the duration of exposure to salt stress (Goel and Singh 2015). Also, in *B. napus* treated with 150 mM NaCl, NR activity declined by 27%, 58% and 52% in the cultivars 'SLM046,' 'Licord,' and 'Okapi,' respectively. This reduction in NR activity due to elevated NaCl levels was associated with a decrease in total nitrogen content and nitrate uptake (Bybordi 2010).

3.2.2.6 Reproductive Phases

Salinity significantly affects various reproductive stage processes in Brassicaceae plants, including crucial processes such as flower initiation, anther and pollen grain development, fertilisation, siliqua formation and development, and seed filling. In *B. napus*, seed production potential is reflected by activities before and after flowering. The reproductive stage is considered the most sensitive to stress (Raza 2021). The flowering and seed-filling stages are the most susceptible phases for oilseed crops. Arif et al. (2019) observed that 100 mM salt treatment caused noticeable changes in two *B. napus* genotypes (Binasarisha-5 and BARI Sarisha-8). Within 2 days, leaves and flowers became discoloured and rolled, and by 7 days, leaves were nearly dead, young siliquae were severely damaged, and mature siliquae showed discolouration, wrinkling, and premature maturity. The first genotype (Binasarisha-5) was more affected than the second (BARI Sarisha-8). As a result, seed formation is completely inhibited, leading to significant crop yield losses. Salt stress has a direct impact on the male gametophyte by reducing pollen grain production, causing anthers to become abnormal, colourless, and smaller and leading to the destruction of the anther wall, which is unable to fertilise the female gametophyte in *B. napus* (Mahmoodzadeh and Bemani 2008). The assimilation of carbon and nitrogen is disrupted when enzymes are inactivated under salt stress, which lowers photosynthetic rates and leaf area, resulting in shorter phases of pod maturity and reduced seed formation. For instance, research on two *B. napus* cultivars, Yanyouza 3 and Ningza 1818, found that high salt conditions resulted in yield losses of 44.92% and 46.11%, respectively (Wang et al. 2022). Furthermore, salt stress hinders filament growth in the male reproductive structure, leading to a 14- to 24-h delay in pollination and preventing fertilisation. Additionally, the meiosis of microspore mother cells is highly sensitive to water deficiency caused by salt stress, disrupting pollen grain development and resulting in male sterility (Mahmoodzadeh and Bemani 2008).

3.2.3 Heat Stress

Thermotolerance is regulated developmentally and tissue-specific; thus, it is essential to examine the effects of high temperatures on various developmental stages to develop heat-tolerant cultivars that can withstand heat stress (Kourani et al. 2022).

3.2.3.1 Seedling and Vegetative Growth

The morphological traits of crops may be significantly impacted by high temperatures, especially for cool-season crops like Indian mustard (Raza et al. 2019b). Heat stress generally occurs when soil and ambient temperatures are too high for a plant's growth and photosynthetic temperature range, causing irreversible damage (Frenck et al. 2011). The simultaneous presence of other stressors, such as drought, may adversely affect a plant's performance under heat stress (Nawaz et al. 2023). When experiencing heatwave conditions during the early vegetative stage, adequately irrigated *B. napus* displays increased saturating light levels, transpiration, and the ratio of intercellular to ambient CO_2 concentration (Dikšaitytė et al. 2019). Plants mitigate overheating by opening their stomata and enhancing transpiration, thereby preventing leaf damage (Elferjani and Soolanayakanahally 2018). A notable morphological alteration observed under elevated temperature conditions is a significant reduction in plant height, which may diminish by as much as 22.8% when seeded later (Pillai and Walia 2024). This decrease in height is attributed to limited soil moisture, constraining the plant's ability to reach its genetic potential.

During the flowering stage, *B. napus* displays a rapid closure of stomata in response to the combined effects of heat and drought stress. This mechanism prevents further water loss through transpiration, thereby preserving evaporative cooling. As a result, leaf temperature significantly increases, leading to enhanced photosynthetic damage (Elferjani and Soolanayakanahally 2018). This contradiction necessitates that plants maintain a balance between preventing overheating and reducing water loss (Zandalinas et al. 2018). In the study of Hasanuzzaman et al. (2014), the impairment of chlorophyll production and the disruption of biochemical events in photosystems were observed in 10-day-old *B. napus* seedlings exposed to 38 °C. Heat stress triggers physiological changes in photosynthesis and transpiration, impacting the synthesis and utilisation of assimilates and consequently leading to morphological abnormalities in floral organs (Echer et al. 2014) as shown in Fig. 3.4. Furthermore, at the molecular level, heat stress can structurally damage membranes and disrupt the stability of proteins embedded within them. Additionally, it can metabolically disrupt enzyme processes and create toxic byproducts. Both structural and metabolic damage can occur due to heat stress (Mittler et al. 2012). Applying heat stress to *A. thaliana* generated reactive oxygen species (ROS), which increased lipid peroxidation and membrane integrity (Rocco et al. 2013). The breakdown of proteins, the bleaching of pigments, and the breakage of DNA strands are all consequences of heat stress, respectively.

3.2.3.2 Reproduction

The developmental progression of plants from seed to seed depends on the efficacy of sexual reproduction; consequently, vulnerability to heat stress during this phase may adversely impact yield potential. Plants exhibit heightened sensitivity to substantial temperature fluctuations during the reproductive phase compared to the vegetative growth stage. This often results in stunted growth and diminished seed yield (Zhang et al. 2017). During their reproductive stage, plants exhibit greater vulnerability to significant temperature fluctuations than in their vegetative phase (Baron et al. 2012). This increased sensitivity often leads to impaired growth and reduced seed production, particularly during meiosis and fertilisation. Among plant

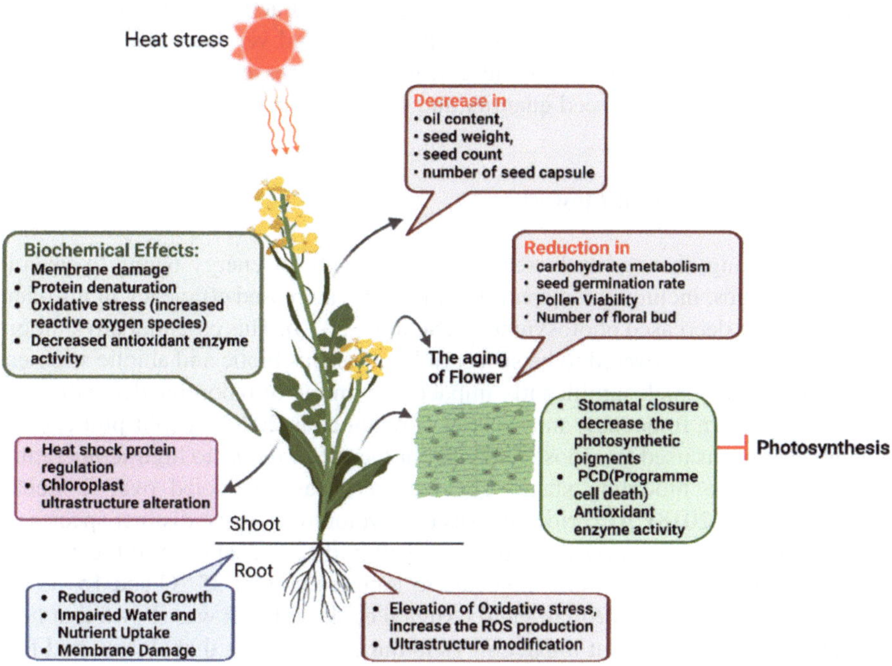

Fig. 3.4 Impact of heat stress on different parts of *Brassica* spp.

structures, male reproductive organs are most susceptible to these temperature variations (Giorno et al. 2013).

Heat stress during reproduction impacts plant physiology in three primary ways: it diminishes the number of flowers before anthesis, decreases flower fertility and impairs the growth of pods and seeds (Morrison and Stewart 2002). However, this topic has not been extensively researched, and the specific responses to each process remain unclear. Likewise, only a few studies have examined the impact of heat stress on the reproductive development of *A. thaliana* (Bac-Molenaar et al. 2015; Chao et al. 2017). A genome-wide association (GWA) mapping study in *A. thaliana* identified distinct quantitative trait loci (QTLs) linked to heat response before and after flowering when subjected to brief periods of high temperature. This finding indicates that the genetic factors controlling heat response are specific to different developmental stages (Bac-Molenaar et al. 2015). Structurally, exposure to high temperatures resulted in deformities of the reproductive structures, including the formation of vacuoles, disintegration within microspores, and misshapen pollen sacs that could not store carbohydrates (Mondal et al. 2023). Additionally, when *A. thaliana* was subjected to 37 °C for 24 h, it had a significant development impairment of stamen filaments, which eliminated the possibility of pollination (Baron et al. 2012). Heat stress generated anomalies in the reproductive organs, such as vacuolation, collapse inside microspores and irregularly formed pollen sacs that could not accumulate carbohydrates. These abnormalities were caused at the structural level (Baron et al. 2012). Pre-anthesis-linked activities in spring wheat, such as gametogenesis and gamete development, proved very susceptible to high-temperature treatment. This was because heat stress decreased floret fertility, which led to a significant drop in seed quantity and weight (Bheemanahalli et al. 2019).

3.2.3.3 Development of Flowers

During blooming, there is an increase in the demand for energy owing to several distinct activities, including the creation of flowers, increased efficiency in nitrogen utilisation, and decreased photosynthesis (Schiessl 2020). This essential physiological time has been discovered to be susceptible to various biotic and abiotic stresses, particularly heat, which significantly impacts the timing of blooming depending on the stress's length, frequency, and severity (Lohani et al. 2020). A split-plot experimental design was used to expose many genotypes of *B. napus* to high-temperature stress during the blooming stage. The research was conducted over 2 weeks (Koscielny et al. 2018). At various phases of development, three distinct species of Brassica (*B. juncea*, *B. napus*, and *B. rapa*) suffered severe damage to their reproductive organs when they were subjected to temperatures of 35 °C and 15 °C for 7 days. This resulted in a drop in yield (Angadi et al. 2000). It was shown that heat stress had a more significant impact on yield during blooming than during pod formation. This suggests pods had crossed a crucial developmental barrier, making them more heat-resistant. Among the three species of *Brassica* that were investigated, it is interesting to note that only *B. napus* could not recover entirely once the heat stress was eliminated, generating abnormally fat and short pods (Kourani et al. 2022).

Air temperature is critical throughout the blooming phase in rice, influencing reproductive outcomes (Bheemanahalli et al. 2019), with premature or postponed flowering adversely affecting crop growth and quality. Recent research conducted by Pokharel et al. (2020) found that high night temperature stress caused a change in the peak flower opening time of *B. napus* into early and more incredible morning hours. This shift suggests that the plant has adapted to the heat escape response; however, this adaptation was accompanied by a considerable drop in yield. Although *B. napus* is not entirely reliant on pollinators, research conducted by Zou et al. (2017) has shown that pollinators improve both the quantity and quality of seeds produced by the plant.

3.2.4 Pollution Stress

Air pollutants are primarily chemical substances of human origin that impact biological systems due to their accumulation in regions of human activity (Ozturk et al. 2025b). A significant portion of air pollutants originates from natural sources, such as sulphur compounds emitted during volcanic eruptions, carbon dioxide, nitrogen oxides and methane produced via microbial activity, and they undergo standard atmospheric processes. Higher concentrations of these molecules often result from human activity; however, they may also be distributed to distant places or undergo atmospheric interactions to generate breakdown products or secondary pollutants. One example is the photochemical generation of ozone from nitrogen oxides (Sitch et al. 2007; Blande et al. 2014). According to Calatayud and Barreno (2004), air pollution significantly disrupts plants' physiological and biochemical functions, eventually reducing yield. Reduced photosynthetic pigments, suppression of specific physiological processes, changes in metabolic processes and altered enzyme activity are all signs of air pollution's effects on plants (Heath 2008).

Plants can only perceive and be affected by air pollution. Because plants are subjected to gases and particles, air pollution is a multistressor. Other elements, including water, light, relative humidity, ambient temperature and nutrient availability, may affect plant responses to air pollution. Plants function as dust interceptors and may signal pollution load by changing physiological indicators such as pigment degradation, membrane damage, antioxidant metabolite synthesis and antioxidant enzyme activity.

Researchers from various regions of the globe have researched the influence of primary air pollutants on numerous crops. Prolonged, low-concentration exposure to air pollution adversely affects plant leaves without causing noticeable damage (Bytnerowitz et al. 1987). Previous studies have shown that nitrogen dioxide (NO_2) adversely affects the concentrations of chlorophyll a, chlorophyll b and total chlorophyll in soybeans. Conversely, sulphur dioxide (SO_2) has a detrimental impact, diminishing chlorophyll a and chlorophyll b levels and the seed weight in *Brassica campestris* (Tomer and Parkash 1989; Joshi et al. 2009).

3.2.5 Heavy Metal Stress

Toxic metals and metalloids with a longer persistence and rapid escalation have garnered significant attention in the industrialised world. Various human activities are contributing to its rise, including the discharge of hazardous wastes and agricultural and industrial activity (Ozturk et al. 2012; Hasanuzzaman et al. 2020). The presence of heavy metals in high concentrations disrupts the biological processes that occur in the soil. Lead (Pb) toxicity in *Brassica* varieties caused changes in the transport of ions such as nitrogen (N), calcium (Ca), magnesium (Mg), phosphorus (P), potassium (K), and zinc (Zn) in the roots and shoots of the various *Brassica* varieties (Gopal and Rizvi 2008).

3.2.5.1 Seed Germination and Vegetative Growth

Heavy metals are recognised to impact seed germination and seedling growth adversely. The reduction in germination derives from suppressing the seed's physiological and metabolic functions. Cadmium (Cd) diminishes the water stress tolerance of plants (Barcelo et al. 1986), resulting in decreased turgor pressure at elevated relative water content compared to untreated plants. The influence of cadmium on the water state of germinating rice seeds has been investigated (Barcelo and Poschenrieder 1990), demonstrating that cadmium significantly impacts the water status of these seeds. The first observable signs of heavy metal poisoning often include slowed growth, diminished root development and changes in root shape (Feigl et al. 2013). Typically, roots are more affected than leaves since they are in close contact with the harmful material. Numerous studies on plants from the genus *Brassica* indicate reduced root development in the presence of certain heavy metals. The work by Ebbs and Kochian (1997) revealed that different metals can produce diverse effects. Notably, copper was found to have a more pronounced impact on lateral root development compared to the influence of zinc (Zn). Nouairi et al. (2006) reported that exposing *B. juncea* and *B. napus* to cadmium stress reduced the development of both the roots and the shoots (Fig. 3.5). They hypothesised this may have been due to limited iron and manganese accumulation in the shoots, though they did not provide evidence to support this hypothesis. In the presence of toxic metals and metalloids, rapeseed could not absorb key components necessary for plant development; thus, yield decreased (Ali et al. 2018). In addition, a reduction

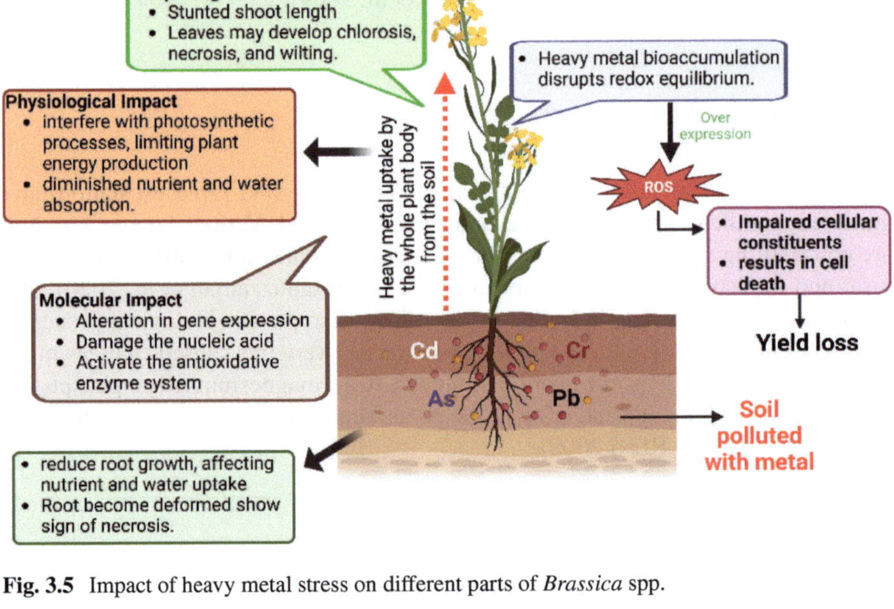

Fig. 3.5 Impact of heavy metal stress on different parts of *Brassica* spp.

in the number of chloroplasts led to a decline in photosynthetic activity that occurred when heavy metals were present (Kamran et al. 2020; Raza 2021).

Although Cd is not required for plants or other living things, it may be found in high concentrations in several soil types. Exposure to increasing concentrations of Cd (0.25 and 0.5 mM $CdCl_2$) resulted in decreased biomass production, leaf water content and light-harvesting pigments in *Brassica* species, including rapeseed. Conversely, this exposure elevated these plants' MDA, proline, H_2O_2, lipoxygenase activity and methylglyoxal levels. The observed effects were proportional to the concentration of Cd applied (Al Mahmud et al. 2017).

3.2.5.2 Plant Leaf Gas Exchange and Photosynthetic Metabolism

According to Amari et al. (2014), the levels of total chlorophyll content and stomatal conductance that are reduced due to hazardous HM ion concentrations directly impact the amount of carbon dioxide plants can absorb. According to the published research, the presence of hazardous heavy metal (HMs) ions leads to alterations in the activity of Rubisco, which is accountable for carbon fixation in light-dependent processes (the first stage of photosynthesis) (Mancilla-Leyton et al. 2016). Prior studies indicated that *B. juncea* seedlings subjected to Ni substantially reduced net photosynthetic rate, stomatal conductance, CO_2 levels, transpiration and concentration inside the substomatal chamber (Amari et al. 2014; Bortoloti and Baron 2022).

3.2.5.3 Physiological Damage

According to the findings of several studies involving heavy metals, the water content of plants under heavy metal stress decreases. As a result, the toxic impact of metals may be generated indirectly by reducing the amount of water the plant takes in (Ali et al. 2015). On the other hand, Zaier et al. (2010) showed that the water content of *B. juncea* plants saw a significant drop when exposed to lead poisoning. This was the case even in this species, which is thought to be resistant. Several authors have shown that soluble sugars, proline and other amino acids may have a protective effect, controlling osmotic potential and with a more direct detoxifying role of ROS. Both functions are important in detoxification (Keunen et al. 2013). Ali et al. (2015) validated this protective function in *B. napus* plants subjected to Cd stress, as foliar application of 5-aminolevulinic acid enhanced plant resilience and elevated the concentrations of these compounds. It is important to highlight that this investigation was conducted at elevated Cd concentrations (100 and 500 μM), reducing both proline and soluble sugars in Cd-contaminated plants, likely owing to overall metabolic impairment.

Wang et al. (2004) observed a significant reduction in root biomass of *B. juncea* plants at copper doses as low as four μM in their toxicity investigation. In research conducted using lead (Pb), Zaier et al. (2010) observed a reduction in plant biomass when the lead concentration was 200 μM or more significant. This information was based on the same species. Lead concentrations of 50 and 100 μM have been found to substantially impact the growth of *B. napus* plants grown in hydroponic solution (Shakoor et al. 2014). Despite the high toxicity of Cd to plants (Cuypers et al. 2010), both Armas et al. (2015) and Seth et al. (2008) observed an initial enhancement in growth parameters of *B. juncea* plants cultivated in Cd-contaminated soil and hydroponics, respectively.

The majority of the research that has been conducted on the effects of oxidative stress on *Brassica* species has been carried out on mature plants. However, the plants examined were of varying ages. During their research on cadmium toxicity in *B. juncea* seeds, Szollosi et al. (2009) verified that the seeds absorbed significant quantities of these heavy metals and that oxidative stress occurred at an extremely early stage in the plant's growth. Toxic heavy metal ions may induce substantial structural and ultrastructural alterations in the photosynthetic apparatus (Ayyaz et al. 2020).

These heavy metals can, for instance, substitute the magnesium (Mg^{2+}) in chlorophyll molecules, rendering them ineffective for photosynthesis due to the unstable excitation state that disrupts resonance and energy transfer from the pigment antennas to the reaction centres; this will result in disruptions in the electron transport chain and hinder the photosynthetic metabolism (Souri et al. 2019).

Some studies have reported that the decline in the development of *B. juncea* may be linked to insufficient micronutrient uptake. This is because lead (Pb) competes with micronutrients for root absorption sites, resulting in stunted growth and reduced biomass (Sikka et al. 2010). Helal et al. (2016) observed that compared to *B. juncea*, *B. napus* exhibits lower tolerance to heavy metals, with lead consistently decreasing the germination rate, plant biomass and both shoot and root elongation of *B. napus*. Ferreyroa et al. (2017) also found that *B. napus* accumulates and translocates Pb to its aerial parts, adversely affecting plant morphology and photosynthetic functions. The primary subcellular effects of lead include cell wall disintegration, chloroplast disarray and the formation of osmophilic bodies.

References

Achard P, Cheng H, De Grauwe L et al (2006) Integration of plant responses to environmentally activated phytohormonal signals. Science 311(5757):91–94

Acosta-Motos JR, Ortuño MF, Bernal-Vicente A et al (2017) Plant responses to salt stress: adaptive mechanisms. Agronomy 7(1):18

Aggarwal G, Edhigalla P, Walia P et al (2024) A method for screening salt stress tolerance in Indian mustard (*Brassica juncea*)(L.) Czern & Coss at seedling stage. Sci Rep 14(1):12705

Ahmad P, Ozturk M, Gucel S (2012) Oxidative damage and antioxidants induced by heavy metal stress in two cultivars of mustard (*Brassica juncea* L.) plants. Fresenius Environ Bull 21(10):2953–2961

Ahmad P, Hashem A, Abd-Allah EF et al (2015) Role of *Trichoderma harzianum* in mitigating NaCl stress in Indian mustard (*Brassica juncea* L) through antioxidative defense system. Front Plant Sci 6:868

Ahmadi SH, Ardekani JN (2006) The effect of water salinity on growth and physiological stages of eight Canola (*Brassica napus*) cultivars. Irrig Sci 25:11–20

Al Mahmud J, Hasanuzzaman M, Nahar K et al (2017) Relative tolerance of different species of *Brassica* to cadmium toxicity: coordinated role of antioxidant defense and glyoxalase systems. Plant Omics 10(2):107–117

Alamri S, Hu Y, Mukherjee S et al (2020) Silicon-induced postponement of leaf senescence is accompanied by modulation of antioxidative defense and ion homeostasis in mustard (*Brassica juncea*) seedlings exposed to salinity and drought stress. Plant Physiol Biochem 157:47–59

Ali B, Gill RA, Yang S et al (2015) Regulation of cadmium-induced proteomic and metabolic changes by 5-aminolevulinic acid in leaves of *Brassica napus* L. PLoS One 10(4):e0123328

Ali S, Gill RA, Mwamba TM et al (2018) Differential cobalt-induced effects on plant growth, ultrastructural modifications, and antioxidative response among four *Brassica napus* (L.) cultivars. Int J Environ Sci Technol 15:2685–2700

Ali L, Shaheen MR, Ihsan MZ et al (2022) Growth, photosynthesis and antioxidant enzyme modulations in broccoli (*Brassica oleracea* L. var. *italica*) under salinity stress. S Afr J Bot 148:104–111

Amari T, Ghnaya T, Debez A et al (2014) Comparative Ni tolerance and accumulation potentials between *Mesembryanthemum crystallinum* (halophyte) and *Brassica juncea*: metal accumulation, nutrient status and photosynthetic activity. J Plant Physiol 171(17):1634–1644

Andrade Foronda D (2022) Reclamation of a saline-sodic soil with organic amendments and leaching. Environ Sci Proc 16(1):56

Angadi SV, Cutforth HW, Miller PR et al (2000) Response of three *Brassica* species to high temperature stress during reproductive growth. Can J Plant Sci 80(4):693–701

Antunović Dunić J, Mlinarić S, Pavlović I et al (2023) Comparative analysis of primary photosynthetic reactions assessed by OJIP kinetics in three *Brassica* crops after drought and recovery. Appl Sci 13(5):3078

Arif MR, Islam MT, Robin AHK (2019) Salinity stress alters root morphology and root hair traits in *Brassica napus*. Plants 8(7):192

Armas T, Pinto AP, de Varennes A et al (2015) Comparison of cadmium-induced oxidative stress in *Brassica juncea* in soil and hydroponic cultures. Plant Soil 388:297–305

Ayyaz A, Amir M, Umer S et al (2020) Melatonin induced changes in photosynthetic efficiency as probed by OJIP associated with improved chromium stress tolerance in canola (*Brassica napus* L.). Heliyon 6(7):1

Babik I (2005) Nitrogen requirements and fertilization of Brussels sprouts. Veg Crops Res Bull 62:113–126

Babik I (2006) Effect of soil type and irrigation on yield and mineral composition of broccoli. Acta Agrophysica 7(4):793–808

Bac-Molenaar JA, Fradin EF, Becker FF et al (2015) Genome-wide association mapping of fertility reduction upon heat stress reveals developmental stage-specific QTLs in *Arabidopsis thaliana*. Plant Cell 27(7):1857–1874

Balasubramaniam T, Shen G, Esmaeili N, Zhang H (2023) Plants' response mechanisms to salinity stress. Plants 12(12):2253

Bano A, Noreen Z, Tabassum F et al (2022) Induction of salt tolerance in *Brassica rapa* by nitric oxide treatment. Front Plant Sci 13:995837

Barcelo J, Poschenrieder C (1990) Plant water relations as affected by heavy metal stress: a review. J Plant Nutr 13(1):1–37

Barcelo J, Poschenrieder C, Andreu I, Gunse B (1986) Cadmium-induced decrease of water stress resistance in bush bean plants (*Phaseolus vulgaris* L. cv. Contender) I. Effects of Cd on water potential, relative water content, and cell wall elasticity. J Plant Physiol 125(1–2):17–25

Baron KN, Schroeder DF, Stasolla C (2012) Transcriptional response of abscisic acid (ABA) metabolism and transport to cold and heat stress applied at the reproductive stage of development in *Arabidopsis thaliana*. Plant Sci 188:48–59

Batool M, El-Badri AM, Hassan MU et al (2022) Drought stress in *Brassica napus*: effects, tolerance mechanisms, and management strategies. J Plant Growth Regul 2022:1–25

Ben Ammar H, Arena D, Treccarichi S et al (2023) The effect of water stress on the glucosinolate content and profile: a comparative study on roots and leaves of *Brassica oleracea* L. Crops Agron 13(2):579

Benáková M, Ahmadi H, Dučaiová Z et al (2017) Effects of Cd and Zn on physiological and anatomical properties of hydroponically grown *Brassica napus* plants. Environ Sci Pollut Res 24:20705–20716

Bheemanahalli R, Sunoj VJ, Saripalli G et al (2019) Quantifying the impact of heat stress on pollen germination, seed set, and grain filling in spring wheat. Crop Sci 59(2):684–696

Björkman M, Klingen I, Birch ANE et al (2011) Phytochemicals of Brassicaceae in plant protection and human health - influences of climate, environment and agronomic practice. Phytochemistry 72(7):538–556

Blande JD, Holopainen JK, Niinemets Ü (2014) Plant volatiles in polluted atmospheres: stress responses and signal degradation. Plant Cell Environ 37(8):1892–1904

Bortoloti GA, Baron D (2022) Phytoremediation of toxic heavy metals by *Brassica* plants: a biochemical and physiological approach. Environ Adv 8:100204

Bradley FM, Ellis BW, Martin DL (eds) (2009) The organic gardener's handbook of natural pest and disease control. Rodale Inc., New York, USA

Bybordi A (2010) Effects of salinity and N on the growth, photosynthesis and N status of canola (*Brassica napus* L.). Notulae Scientia Biologicae 2(2):92–97

Bytnerowicz A, Olszyk DM, Kats G et al (1987) Effects of SO_2 on physiology, elemental content and injury development of winter wheat. Agric Ecosyst Environ 20(1):37–47

Calatayud A, Barreno E (2004) Response to ozone in two lettuce varieties on chlorophyll a fluorescence, photosynthetic pigments and lipid peroxidation. Plant Physiol Biochem 42(6):549–555

Channaoui S, El Kahkahi R, Charafi J et al (2017) Germination and seedling growth of a set of rapeseed (*Brassica napus*) varieties under drought stress conditions. Int J Environ Agric Biotechnol 2(1):238696

Chao LM, Liu YQ, Chen DY et al (2017) *Arabidopsis* transcription factors SPL1 and SPL12 confer plant thermotolerance at reproductive stage. Mol Plant 10(5):735–748

Chen W, Miao Y, Ayyaz A et al (2022) Purple stem *Brassica napus* exhibits higher photosynthetic efficiency, antioxidant potential and anthocyanin biosynthesis related genes expression against drought stress. Front Plant Sci 13:936696

Clark A (2007) Managing cover crops profitably. In: National SAREOutreach handbook series book 9, 3rd edn. Nat Agric Lab, Beltsville

Cuypers A, Plusquin M, Remans T et al (2010) Cadmium stress: an oxidative challenge. Biometals 23:927–940

de Almeida LMM, Coquemont-Guyot M, Elie N et al (2023) Repeated heat stress events during the reproductive phase impact the dynamic development of seeds in *Brassica napus* L. Plant Sci 327:111559

del Carmen M-BM, Moreno DA, Carvajal M (2013) The physiological importance of glucosinolates on plant response to abiotic stress in *Brassica*. Int J Mol Sci 14(6):11607–11625

Dhaliwal MS (2017) Cole crops. In: Handbook of vegetable crops, 3rd edn. Kalyani Publishers, New Delhi, pp 148–176

Di Gioia F, Rosskopf EN, Leonardi C, Giuffrida F (2018) Effects of application timing of saline irrigation water on broccoli production and quality. Agric Water Manag 203:97–104

Dikšaitytė A, Viršilė A, Žaltauskaitė J et al (2019) Growth and photosynthetic responses in *Brassica napus* differ during stress and recovery periods when exposed to combined heat, drought and elevated CO_2. Plant Physiol Biochem 142:59–72

Dogra S, Chand G, Sinha BK (2018) Physiological and biochemical traits as tools to screen sensitive and resistant genotypes of *Brassica juncea* exposed to drought stress. J Pharmacogn Phytochem 7(5):877–884

Duke JA (1983) Brassica rapa L. In: Handbook of energy crops. Center for New Crops & Plant Products, Purdue Univ., USA

Ebbs SD, Kochian LV (1997) Toxicity of zinc and copper to Brassica species: implications for phytoremediation. J Environ Qual 26(3):776–781

Echer FR, Oosterhuis DM, Loka DA, Rosolem CA (2014) High night temperatures during the floral bud stage increase the abscission of reproductive structures in cotton. J Agron Crop Sci 200(3):191–198

El Sabagh A, Islam MS, Hossain A et al (2022) Phytohormones as growth regulators during abiotic stress tolerance in plants. Front Agron 4:765068

Elferjani R, Soolanayakanahally R (2018) Canola responses to drought, heat, and combined stress: shared and specific effects on carbon assimilation, seed yield, and oil composition. Front Plant Sci 9:1224

El-Hendawy S, Elshafei A, Al-Suhaibani N et al (2019) Assessment of the salt tolerance of wheat genotypes during the germination stage based on germination ability parameters and associated SSR markers. J Plant Interact 14(1):151–163

Eom SH, Baek SA, Kim JK, Hyun TK (2018) Transcriptome analysis in Chinese cabbage (*Brassica rapa* ssp. *pekinensis*) provides the role of glucosinolate metabolism in response to drought stress. Molecules 23(5):1186

Eşiyok D (2012) Kışlık ve Yazlık Sebze Yetiştiriliciği. Meta Basım Matbaacılık, Izmir, Türkiye. ISBN 978-605-87189-0-6

Fahad S, Bajwa AA, Nazir U et al (2017) Crop production under drought and heat stress: plant responses and management options. Front Plant Sci 8:1147

Fahey JW (2016) Brassica: characteristics and properties. In: Encyclopedia of food and health. Elsevier Inc., USA, pp 469–477

FAO (2019) The state of the world's biodiversity for food and agriculture. In: FAO commission on genetic resources for food and agriculture assessments, Rome, p 572

Farooq M, Hussain M, Wakeel A, Siddique KH (2015) Salt stress in maize: effects, resistance mechanisms, and management. A review. Agron Sustain Dev 35:461–481

Fatima A, Umbreen S, Sadia S et al (2024) Mitigation of salinity-induced adverse effects through exogenous application of gibberellic acid in turnip (*Brassica rapa* L.). Cogent Food Agric 10(1):2392042

Feigl G, Kumar D, Lehotai N et al (2013) Physiological and morphological responses of the root system of Indian mustard (*Brassica juncea* L. Czern.) and rapeseed (*Brassica napus* L.) to copper stress. Ecotoxicol Environ Saf 94:179–189

Feng C, Gao H, Zhou Y et al (2023) Unfolding molecular switches for salt stress resilience in soybean: recent advances and prospects for salt-tolerant smart plant production. Front Plant Sci 14:1162014

Ferreyroa GV, Lagorio MG, Trinelli MA et al (2017) Lead effects on *Brassica napus* photosynthetic organs. Ecotoxicol Environ Saf 140:123–130

Frenck G, van der Linden L, Mikkelsen TN et al (2011) Increased [CO_2] does not compensate for negative effects on yield caused by higher temperature and [O_3] in *Brassica napus* L. Eur J Agron 35(3):127–134

Giorno F, Wolters-Arts M, Mariani C, Rieu I (2013) Ensuring reproduction at high temperatures: the heat stress response during anther and pollen development. Plants 2(3):489–506

Goel P, Singh AK (2015) Abiotic stresses downregulate key genes involved in nitrogen uptake and assimilation in *Brassica juncea* L. PLoS One 10(11):e0143645

Golian I, Robak J, Rogowska M et al (2016) Integrated methodology of protecting Italian cabbage. Publishing House Institute of Horticulture, Skierniewice

Gopal R, Rizvi AH (2008) Excess lead alters growth, metabolism and translocation of certain nutrients in radish. Chemosphere 70(9):1539–1544

Hadi F, Ayaz M, Ali S et al (2014) Comparative effect of polyethylene glycol and mannitol induced drought on growth (in vitro) of canola (*Brassica napus*), cauliflower (*Brassica oleracea*) and tomato (*Lycopersicon esculentum*) seedlings. Int J Biosci 4(9):34–41

Han C, Zhen S, Zhu G et al (2017) Comparative metabolome analysis of wheat embryo and endosperm reveals the dynamic changes of metabolites during seed germination. Plant Physiol Biochem 115:320–327

Hannaway DB, Larson C (2004) Forage fact sheet: field mustard (Brassica rapa L. var. rapa). Oregon State Univ, Corvallis

Hasanuzzaman M, Nahar K, Alam MM, Fujita M (2014) Modulation of antioxidant machinery and the methylglyoxal detoxification system in selenium-supplemented *Brassica napus* seedlings confers tolerance to high temperature stress. Biol Trace Elem Res 161:297–307

Hasanuzzaman M, Bhuyan MB, Raza A et al (2020) Selenium in plants: Boon or bane? Environ Exp Bot 178:104170

Hasanuzzaman M, Raihan M, Masud AAC et al (2021) Regulation of reactive oxygen species and antioxidant defense in plants under salinity. Int J Mol Sci 22(17):9326

Hasnain Z, Zafar S, Usman S et al (2023) Elucidating role of melatonin foliar spray in ameliorating adverse effects of drought stress on growth and physio-biochemical attributes of *Brassica rapa* plants. Sci Hortic 321:112336

Heath RL (2008) Modification of the biochemical pathways of plants induced by ozone: what are the varied routes to change? Environ Pollut 155(3):453–463

Helal NM, Shaaban H, Dessoky EDS (2016) Effect of some heavy metals stress on micropropagated plantlets of canola plant (*Brassica napus* L.). Egypt J Exp Biol (Bot) 12(1):67–77

Huang X, Cheng B, Wang Y et al (2023) Effects of fresh-cut and storage on glucosinolates profile using broccoli as a case study. Hortic Plant J 9(2):285–292

Ilyas M, Maqsood MF, Shahbaz M et al (2024) Alleviating salinity stress in canola (*Brassica napus* L.) through exogenous application of salicylic acid. BMC Plant Biol 24(1):611

Iqbal M, Bakht J, Shafi M, Ullah R (2012) Effect of heavy metal and EDTA application on heavy metal uptake and gene expression in different *Brassica* species. Afr J Biotechnol 11(30):7649–7658

Issarakraisila M, Ma Q, Turner DW (2007) Photosynthetic and growth responses of juvenile Chinese kale (*Brassica oleracea* var. *alboglabra*) and Caisin (*Brassica rapa* subsp. *parachinensis*) to waterlogging and water deficit. Sci Hortic 111(2):107–113

Jalil SU, Ansari MI (2020) Physiological role of Gamma-aminobutyric acid in salt stress tolerance. In: Salt and drought stress tolerance in plants: Signaling networks and adaptive mechanisms. Springer International Publishing, Cham, pp 337–350

Jibran R, Hunter D, Dijkwel P (2013) Hormonal regulation of leaf senescence through integration of developmental and stress signals. Plant Mol Biol 82:547–561

Joshi N, Chauhan A, Joshi PC (2009) Impact of industrial air pollutants on some biochemical parameters and yield in wheat and mustard plants. Environmentalist 29:398–404

Kage H, Kochler M, Stützel H (2004) Root growth and dry matter partitioning of cauliflower under drought stress conditions: measurement and simulation. Eur J Agron 20(4):379–394

Kalra A, Goel S, Elias AA (2024) Understanding role of roots in plant response to drought: way forward to climate-resilient crops. Plant Genome 17(1):e20395

Kamran M, Malik Z, Parveen A et al (2020) Ameliorative effects of biochar on rapeseed (*Brassica napus* L.) growth and heavy metal immobilization in soil irrigated with untreated wastewater. J Plant Growth Regul 39:266–281

Kataria S, Baghel L, Guruprasad KN (2017) Pre-treatment of seeds with static magnetic field improves germination and early growth characteristics under salt stress in maize and soybean. Biocatal Agric Biotechnol 10:83–90

Kaur H, Sirhindi G, Bhardwaj R et al (2018) 28-homobrassinolide regulates antioxidant enzyme activities and gene expression in response to salt-and temperature-induced oxidative stress in *Brassica juncea*. Sci Rep 8(1):8735

Kazemi OB, Bandehagh A, Farajzadeh D et al (2024) Effects of Pseudomonas Fluorescens FY32 on canola (*Brassica napus* L.) cultivars under drought stress induced by polyethylene glycol. J Crop Health 76(1):251–260

Keunen ELS, Peshev D, Vangronsveld J et al (2013) Plant sugars are crucial players in the oxidative challenge during abiotic stress: extending the traditional concept. Plant Cell Environ 36(7):1242–1255

Khajeh-Hosseini M, Powell A, Bingham IJ (2003) The interaction between salinity stress and seed vigour during germination of soybean seeds. Seed Sci Technol 31:715–725

Khator K, Shekhawat GS (2020) Nitric oxide mitigates salt-induced oxidative stress in *Brassica juncea* seedlings by regulating ROS metabolism and antioxidant defense system. 3 Biotech 10(11):499

Khodabin G, Tahmasebi-Sarvestani Z, Rad AS et al (2020) The effect of withholding irrigation and foliar application of Zn and Mn on yield and eco-physiological characteristics of rapeseed (*Brassica napus* L.). Iranian J Field Crop Res 18(1):fa85–fa100

Kogan F, Guo W, Yang W (2019) Drought and food security prediction from NOAA new generation of operational satellites. Geomat Nat Haz Risk 2019:2019

Koscielny CB, Hazebroek J, Duncan RW (2018) Phenotypic and metabolic variation among spring *Brassica napus* genotypes during heat stress. Crop Pasture Sci 69(3):284–295

Kourani M, Mohareb F, Rezwan FI, Anastasiadi M, Hammond JP (2022) Genetic and physiological responses to heat stress in *Brassica napus*. Front Plant Sci 13:832147

Kumar V, Sharma PK, Kant S et al (2018) Organic amendments influence mustard (*Brassica juncea*) growth in chromium contaminated soils. J Pharmacogn Phytochem 7(4):2026–2038

La VH, Lee BR, Zhang Q et al (2019) Salicylic acid improves drought-stress tolerance by regulating the redox status and proline metabolism in *Brassica rapa*. Hortic Environ Biotechnol 60:31–40

Latif M, Akram NA, Ashraf M (2016) Regulation of some biochemical attributes in drought-stressed cauliflower (*Brassica oleracea* L.) by seed pre-treatment with ascorbic acid. J Hortic Sci Biotechnol 91(2):129–137

Lee BR, Zaman R, Avice JC et al (2016) Sulfur use efficiency is a significant determinant of drought stress tolerance in relation to photosynthetic activity in *Brassica napus* cultivars. Front Plant Sci 7:459

Lema M, Ali MY, Retuerto R (2019) Domestication influences morphological and physiological responses to salinity in *Brassica oleracea* seedlings. AoB Plants 11(5):plz046

Li N, Cao B, Chen Z, Xu K (2022) Root morphology ion absorption and antioxidative defense system of two Chinese cabbage cultivars (*Brassica rapa* L.) reveal the different adaptation mechanisms to salt and alkali stress. Protoplasma 259(2):385–398

Liu D, Hu LY, Ali B et al (2016) Influence of 5-aminolevulinic acid on photosynthetically related parameters and gene expression in *Brassica napus* L. under drought stress. Soil Sci Plant Nutr 62(3):254–262

Lohani N, Singh MB, Bhalla PL (2020) High temperature susceptibility of sexual reproduction in crop plants. J Exp Bot 71(2):555–568

López-Berenguer C, Martínez-Ballesta MC, García-Viguera C, Carvajal M (2008) Leaf water balance mediated by aquaporins under salt stress and associated glucosinolate synthesis in broccoli. Plant Sci 174(3):321–328

Lu C, Li L, Liu X et al (2023) Salt stress inhibits photosynthesis and destroys chloroplast structure by downregulating chloroplast development-related genes in *Robinia pseudoacacia* seedlings. Plants 12(6):1283

Ma N, Hu C, Wan L et al (2017) Strigolactones improve plant growth, photosynthesis, and alleviate oxidative stress under salinity in rapeseed (*Brassica napus* L.) by regulating gene expression. Frontiers. Plant Sci 8:1671

Mahmoodzadeh H, Bemani M (2008) Influence of salinity at early stage of flowering on the development of male gametophyte in canola (*Brassica napus* L.) cv. symbol. Res J Environ Sci 2(6):415–423

Majid A, Rather BA, Masood A et al (2021) Abscisic acid in coordination with nitrogen alleviates salinity-inhibited photosynthetic potential in mustard by improving proline accumulation and antioxidant activity. Stress 1(3):162–180

Mancilla-Leyton JM, Navarro-Ramos MJ, Muñoz-Vallés S et al (2016) Evaluation of the potential of *Atriplex halimus* stem cuttings for phytoremediation of metal-polluted soils. Ecol Eng 97:553–557

Mittal S, Kumari N, Sharma V (2012) Differential response of salt stress on *Brassica juncea*: photosynthetic performance, pigment, proline, D1 and antioxidant enzymes. Plant Physiol Biochem 54:17–26

Mittler R, Finka A, Goloubinoff P (2012) How do plants feel the heat? Trends Biochem Sci 37(3):118–125

Mondal S, Karmakar S, Panda D et al (2023) Crucial plant processes under heat stress and tolerance through heat shock proteins. Plant Stress 2023:100227

Moradpour M, Abdullah SNA, Namasivayam P (2021) The impact of heat stress on morpho-physiological response and expression of specific genes in the heat stress-responsive transcriptional regulatory network in *Brassica oleracea*. Plants 10(6):1064

Morrison MJ, Stewart DW (2002) Heat stress during flowering in summer *Brassica*. Crop Sci 42(3):797–803

Naaz G, Chauhan KL (2019) Lead tolerance and accumulation potential of *Brassica juncea* L. varieties in imitatively contaminated soil. Res J Life Sci Bionform Pharm Chem Sci 5(2):436–447

Nawaz M, Sun J, Shabbir S et al (2023) A review of plants strategies to resist biotic and abiotic environmental stressors. Sci Total Environ 2023:165832

Nazar R, Umar S, Khan NA (2015) Exogenous salicylic acid improves photosynthesis and growth through increase in ascorbate-glutathione metabolism and S assimilation in mustard under salt stress. Plant Signal Behav 10(3):e1003751

Nouairi I, Ammar WB, Youssef NB et al (2006) Comparative study of cadmium effects on membrane lipid composition of *Brassica juncea* and *Brassica napus* leaves. Plant Sci 170(3):511–519

OGTR (2008) The biology of Brassica napus L. Version 2. Department of Health and Ageing. Australian Government Office of the Gene Technology Regulator

Ondrasek G, Rathod S, Manohara KK et al (2022) Salt stress in plants and mitigation approaches. Plants 11(6):717

Ozturk M, Gemici M, Yılmazer C, Ozdemir F (1992) Alleviation of salinity stress by GA3, KIN and IAA on seed germination of Brassica campestris L. Turk J Bot 17:47–52

Ozturk M, Dogan Y, Baslar S, Mert HH (1997) Preliminary studies on some endangered Brassica species in Turkey. Eucarpia (INRA Rennes) 19:9–10

Ozturk M, Baslar S, Dogan Y, Sakcali S (2006) Alleviation of salinity stress in the seeds of some Brassica species. In: Biosaline agriculture and salinity tolerance. Kluwer- Springer, Netherlands, pp 145–156

Ozturk M, Memon AR, Gucel S, Sakcali MS (2012) Brassicas in Turkey and their possible role in the phytoremediation of degraded habitats. In: The plant family Brassicaceae, contribution towards phytoremediation. Environmental pollution book series. Springer Verlag, New York, pp 265–288

Ozturk M, Akram NA, Turkyilmaz BU, Ashraf M (2022) Introduction and application of organic fertilizers as protectors of our environment. Camb Scholars Publ, Cambridge, UK, pp 1–514

Ozturk M, Bhat RA, Ashraf M et al (2023a) Phytohormones and stress responsive secondary metabolites. Academic Press-Elsevier, Oxford, UK, pp 1–287

Ozturk M, Altay V, Nazish M et al (2023b) Halophyte plant diversity and public health. Springer, New York

Ozturk M, Altay V, Keskin M (2025a) Bioengineering of halophytes in Turkiye: a case study on the Leguminous Halophytes and their economic potential. In: Sustainable utilisation and bioengineering of halophytes. Springer Nature Singapore, Singapore, pp 169–182

Ozturk M, Gul A, Rafique K et al (2025b) Nature conservation, air pollution and plant diversity across various ecosystems and the impact of climate change. In: Forests and carbon sequestration. Cambridge Scholars Publishing, UK, pp 330–422

Pakdel H, Hassani SB, Ghotbi-Ravandi AA, Bernard F (2020) Contrasting the expression pattern change of polyamine oxidase genes and photosynthetic efficiency of maize (*Zea mays* L.) genotypes under drought stress. J Biosci 45:1–12

Pavlović I, Mlinarić S, Tarkowská D et al (2019) Early *Brassica* crops responses to salinity stress: a comparative analysis between Chinese cabbage, white cabbage, and kale. Front Plant Sci 10:450

Pillai AJ, Walia P (2024) Heat stress in indian mustard (*Brassica juncea* L.): a critical review of impacts and adaptation strategies. Plant Cell Biotechnol Mol Biol 25(5–6):1–11

Pirasteh-Anosheh H, Rahimpour B, Mohammadi H et al (2023) Induced salinity tolerance by salicylic acid through physiological manipulations. In: Phytohormones and stress responsive secondary metabolites. Elsevier, New York, pp 99–110

Pirasteh-Anosheh H, Ozturk M, Gul A et al (2025) Salinity stress tolerance in plants: antioxidant defense mechanisms and latest developments. In: Exogenous priming and engineering of plant metabolic and regulatory genes. Academic Press, Amsteram, Netherlands, pp 279–304

Pokharel M, Chiluwal A, Stamm M et al (2020) High night-time temperature during flowering and pod filling affects flower opening, yield and seed fatty acid composition in canola. J Agron Crop Sci 206(5):579–596

Raihan MRH, Nahar K, Nowroz F et al (2023) Oilseed brassica responses and tolerance to salt stress. In: Oilseed crops-uses, biology and production. IntechOpen, London, United Kingdom

Ramzan T, Shahbaz M, Maqsood MF et al (2023) Phenylalanine supply alleviates the drought stress in mustard (*Brassica campestris*) by modulating plant growth, photosynthesis, and antioxidant defense system. Plant Physiol Biochem 201:107828

Ranjit SL, Manish P, Penna S (2016) Early osmotic, antioxidant, ionic, and redox responses to salinity in leaves and roots of Indian mustard (*Brassica juncea* L.). Protoplasma 253:101–110

Raza A (2021) Eco-physiological and biochemical responses of rapeseed (*Brassica napus* L.) to abiotic stresses: consequences and mitigation strategies. J Plant Growth Regul 40(4):1368–1388

Raza MAS, Shahid AM, Saleem MF et al (2017) Effects and management strategies to mitigate drought stress in oilseed rape (*Brassica napus* L.): a review. Zemdirbyste-Agriculture 104(1):85–94

Raza MM, Ullah S, Abbas T et al (2019a) Alleviation of salinity stress in maize using silicon nutrition. Notulae Botanicae Horti Agrobotanici Cluj-Napoca 47(4):1340–1347

Raza A, Razzaq A, Mehmood SS et al (2019b) Impact of climate change on crops adaptation and strategies to tackle its outcome: a review. Plants 8(2):34

Ren Y, Wang W, He J et al (2020) Nitric oxide alleviates salt stress in seed germination and early seedling growth of pakchoi (*Brassica chinensis* L.) by enhancing physiological and biochemical parameters. Ecotoxicol Environ Saf 187:109785

Rocco M, Arena S, Renzone G et al (2013) Proteomic analysis of temperature stress-responsive proteins in *Arabidopsis thaliana* rosette leaves. Mol BioSyst 9(6):1257–1267

Rosendahl J, Erlewein A, Hecheltjen A (2022) UNCCD COP 14: mirroring soil and land's growing relevance at the interface of climate and biodiversity; discussion focus on drought and land tenure. In: International yearbook of soil law and policy 2020/2021. Springer International Publishing, Cham, pp 211–224

Rucker KS, Kvien CK, Holbrook CC, Hook JE (1995) Identification of peanut genotypes with improved drought avoidance traits. Peanut Sci 22(1):14–18

Saibi W (2021) Oxidative stress and antioxidant defense in brassicaceae plants under abiotic stresses. J Plant Sci 2021:244

Salehi B, Quispe C, Butnariu M et al (2021) Phytotherapy and food applications from Brassica genus. Phytother Res 35(7):3590–3609

Salehi-Lisar SY, Bakhshayeshan-Agdam H (2016) Drought stress in plants: causes, consequences, and tolerance. In: Drought stress tolerance in plants, vol 1: physiology and biochemistry. Springer International Publishing, Cham, pp 1–16

Sánchez-Rodríguez E, Rubio-Wilhelmi M, Cervilla LM et al (2010) Genotypic differences in some physiological parameters symptomatic for oxidative stress under moderate drought in tomato plants. Plant Sci 178(1):30–40

Santo A, Mattana E, Frigau L et al (2017) Effects of NaCl stress on seed germination and seedling development of *Brassica insularis* Moris (Brassicaceae). Plant Biol 19(3):368–376

Sardar H, Ramzan MA, Naz S et al (2023) Exogenous application of melatonin improves the growth and productivity of two broccoli (*Brassica oleracea* L.) cultivars under salt stress. J Plant Growth Regul 42(8):5152–5166

Schiessl S (2020) Regulation and subfunctionalization of flowering time genes in the allotetraploid oil crop *Brassica napus*. Front Plant Sci 11:605155

Schreiner M, Beyene B, Krumbein A, Stutzel H (2009) Ontogenetic changes of 2-propenyl and 3-indolylmethyl glucosinolates in *Brassica carinata* leaves as affected by water supply. J Agric Food Chem 57(16):7259–7263

Seth CS, Chaturvedi PK, Misra V (2008) The role of phytochelatins and antioxidants in tolerance to Cd accumulation in *Brassica juncea* L. Ecotoxicol Environ Saf 71(1):76–85

Shah SH, Islam S, Mohammad F et al (2023) Gibberellic acid: a versatile regulator of plant growth, development and stress responses. J Plant Growth Regul 42:7352–7373

Shahrasbi S, Pirasteh-Anosheh H, Emam Y et al (2021) Elucidating some physiological mechanisms of salt tolerance in Brassica napus L. seedlings ınduced by seed priming with plant growth regulators. Pak J Bot 53(2):367–377

Shakoor MB, Ali S, Hameed A et al (2014) Citric acid improves lead (Pb) phytoextraction in *Brassica napus* L. by mitigating Pb-induced morphological and biochemical damages. Ecotoxicol Environ Saf 109:38–47

Shawon RA, Kang BS, Lee SG et al (2020) Influence of drought stress on bioactive compounds, antioxidant enzymes and glucosinolate contents of Chinese cabbage (*Brassica rapa*). Food Chem 308:125657

Shehzad J, Mustafa G, Arshad H et al (2023) Morpho-physiological and biochemical responses of *Brassica* species toward lead (Pb) stress. Acta Physiol Plant 45(1):8

Shen J, Guo MJ, Wang YG et al (2020) Humic acid improves the physiological and photosynthetic characteristics of millet seedlings under drought stress. Plant Signal Behav 15(8):1774212

Sheoran AR, Lakra N, Saharan BS et al (2024) Enduring drought: effects and strategies for *Brassica* crop resilience. Acad Biol 2(3):1

Sikka R, Nayyar V, Sidhu SS (2010) Bioaccumulation of lead by Indian mustard in a loamy sand soil artificially contaminated with lead: impact on plant growth and uptake of metal. Commun Soil Sci Plant Anal 41(10):1257–1270

Singh BK, Singh SP, Shekhawat K et al (2019) Comparative analysis for understanding salinity tolerance mechanism in Indian mustard (*Brassica juncea* L.). Acta Physiol Plant 41:1–14

Singh J, Sharma PC, Singh V (2021) Breeding mustard (*Brassica juncea*) for salt tolerance: problems and prospects. Brassica Breed Biotechnol 3:69–85

Sitch S, Cox PM, Collins WJ, Huntingford C (2007) Indirect radiative forcing of climate change through ozone effects on the land-carbon sink. Nature 448(7155):791–794

Soni S, Kumar A, Sehrawat N et al (2021) Effect of saline irrigation on plant water traits, photosynthesis and ionic balance in durum wheat genotypes. Saudi J Biol Sci 28(4):2510–2517

Souri Z, Cardoso AA, da-Silva CJ et al (2019) Heavy metals and photosynthesis: recent developments. In: Photosynthesis, productivity and environmental stress. Wiley Online Library, NJ, USA, pp 107–134

Szollosi R, Varga IS, Erdei L, Mihalik E (2009) Cadmium-induced oxidative stress and antioxidative mechanisms in germinating Indian mustard (*Brassica juncea* L.) seeds. Ecotoxicol Environ Saf 72(5):1337–1342

Tomer YS, Parkash G (1989) Effect of long term SO_2 exposures on *Brassica campestris* L. Indian J Ecol 16(2):111–114

Toosi AF, Baki BB, Azizi M (2014) Effect of drought stress by using PEG 6000 on germination and early seedling growth of *Brassica juncea* var. Ensabi. Agronomy LVII:360

Torabi B, Ardestani FG (2013) Effect of salt and drought stresses on germination components in canola (*Brassica napus* L.). Int J Agric Crop Sci 5(15):1642

Uçarlı C (2020) Effects of salinity on seed germination and early seedling stage. Abiotic Stress Plants 211(10):5772

USDA-ARS (2017) Germplasm resources information network (GRIN) taxonomy for plants. Taxon: Brassica napus L. United States Department of Agriculture. Agricultural Research Service, Beltsville Area

USDA-ERS (2014) Soybeans & oil crops-Canola. United States Department of Agriculture. Economic Research Service

Wang SH, Yang ZM, Yang H et al (2004) Copper-induced stress and antioxidative responses in roots of *Brassica juncea* L. Botan Bull Acad Sinica 2022:45

Wang L, Zuo Q, Zheng J et al (2022) Salt stress decreases seed yield and postpones growth process of canola (*Brassica napus* L.) by changing nitrogen and carbon characters. Sci Rep 12(1):17884

Waraich EA, Hussain A, Ahmad Z et al (2022) Foliar application of sulfur improved growth, yield and physiological attributes of canola (*Brassica napus* L.) under heat stress conditions. J Plant Nutr 45(3):369–379

Wei J, Li H, Huang X et al (2024) Elucidating the regulatory role of long non-coding RNAs in drought stress response during seed germination in leaf mustard. PeerJ 12:e17661

Wu QS, Zou YN, Liu W et al (2010) Alleviation of salt stress in citrus seedlings inoculated with mycorrhiza: changes in leaf antioxidant defense systems. Plant Soil Environ 56(10):470–475

Yang X, Lu M, Wang Y et al (2021) Response mechanism of plants to drought stress. Horticulturae 7(3):50

Yousuf PY, Ahmad A, Aref IM et al (2016) Salt-stress-responsive chloroplast proteins in *Brassica juncea* genotypes with contrasting salt tolerance and their quantitative PCR analysis. Protoplasma 253:1565–1575

Yu X, Wang H, Lu Y et al (2012) Identification of conserved and novel microRNAs that are responsive to heat stress in *Brassica rapa*. J Exp Bot 63(2):1025–1038

Zaier H, Mudarra A, Kutscher D et al (2010) Induced lead binding phytochelatins in *Brassica juncea* and *Sesuvium portulacastrum* investigated by orthogonal chromatography inductively coupled plasma-mass spectrometry and matrix assisted laser desorption ionization-time of flight-mass spectrometry. Anal Chim Acta 671(1–2):48–54

Zandalinas SI, Mittler R, Balfagón D et al (2018) Plant adaptations to the combination of drought and high temperatures. Physiol Plant 162(1):2–12

Zhang SS, Yang H, Ding L et al (2017) Tissue-specific transcriptomics reveals an important role of the unfolded protein response in maintaining fertility upon heat stress in *Arabidopsis*. Plant Cell 29(5):1007–1023

Zhang Q, Gong M, Xu X et al (2022) Roles of auxin in the growth, development, and stress tolerance of horticultural plants. Cells 11(17):2761

Zhou R, Kan X, Chen J et al (2019) Drought-induced changes in photosynthetic electron transport in maize probed by prompt fluorescence, delayed fluorescence, P700 and cyclic electron flow signals. Environ Exp Bot 158:51–62

Zhu J, Cai D, Wang J et al (2021) Physiological and anatomical changes in two rapeseed (*Brassica napus* L.) genotypes under drought stress conditions. Oil. Crop Sci 6(2):97–104

Zou M, Yuan L, Zhu S et al (2017) Effects of heat stress on photosynthetic characteristics and chloroplast ultrastructure of a heat-sensitive and heat-tolerant cultivar of wucai (*Brassica campestris* L.). Acta Physiol Plant 39:1–10

Chapter 4
Brassica's: Mycorrhizal Interactions and Development Related to Other Organisms

4.1 Mycorrhizal Interactions

In the current era of global climate change, agriculture is under extreme pressure due to variety of abiotic stresses (extreme temperature conditions, prolonged drought situations, unexpected flooding, soil salinization, luxurious use of agrochemicals, depleting micro flora, metal and metalloid toxicity and war or warlike situations) and biotic stress factors (pathogens, pests and weeds), all seriously challenging the global food security (Dervash et al. 2024; Lahlali et al. 2025; Jiang et al. 2025; Zhang et al. 2025; Manghwar and Zaman 2024; Lin et al. 2023). Abiotic and biotic stresses significantly disrupt the vital physiological processes like net photosynthetic rate, uptake of essential and beneficial nutrients, water uptake and overall health of the plants and lead to serious yield penalties (Ashraf et al. 2009, 2012; Hakeem et al. 2013; Hasanuzzaman et al. 2019; Ozturk and Gul 2020; Akhtar et al. 2024; Wang et al. 2025a, b).

Current scenarios of frequently changing global climatic conditions leads to prolonged exposure to various stresses which not only weakens plant resilience but also limits the efficiency of resource use such as carbon assimilation, water uptake and limited access to essential nutrients, thereby posing a serious threat to agricultural sustainability and global food security (Janni et al. 2024; Oishy et al. 2025; Ozturk et al. 2025a). Prolonged exposure to stress conditions compromises plant resilience by inducing oxidative damage, impairing cellular repair mechanisms and reducing the growth efficiency (Gill and Tuteja 2010; Singh 2022).

Keeping in view the future global food demand, there will be nearly 10 billion more people on earth by 2050 (about 3 billion more mouths to feed than there were in 2010); therefore, the cumulative impact of climate-induced stress on crop performance will pose a critical threat to agricultural sustainability. Without adaptive strategies and the use of sustainable measures, these challenges will intensify

M. Ozturk et al., *Climate-Resilient* Brassicas,
https://doi.org/10.1007/978-3-032-12709-9_4

existing vulnerabilities in food systems, jeopardising food security, particularly in climate-sensitive regions (Abdullahi 2023).

Mycorrhizal interactions are fundamental to sustainable agriculture, as they significantly strengthen plant tolerance to abiotic and biotic stresses (Dervash et al. 2024; Wang et al. 2024; Ahmed et al. 2025; Mythili et al. 2025). Mycorrhizal fungi are symbiotic organisms that form mutualistic associations with the roots of most terrestrial plants, playing a vital role naturally in improving nutrient uptake, maintaining soil health and enhancing stress resilience. Through these functions, they significantly influence plant health, promote robust growth and increase adaptability to a wide range of environmental stresses (Rigobelo 2024; Ahmed et al. 2025; Ozturk et al. 2025b).

Mycorrhizal fungi form vast hyphal networks that spread into the soil, significantly increasing the root architecture's reach for water and nutrient absorption. These fungal networks improve the uptake of essential nutrients such as nitrogen (N), phosphorus (P), potassium (K) and zinc (Zn), thus reducing reliance on synthetic fertilisers and supporting robust plant growth, even in nutrient-poor soils in a sustainable manner (Etesami et al. 2021; Hawkins et al. 2023; Sheikh-Assadi et al. 2023; Bate-Weldon et al. 2024; Ahmed et al. 2025). Among these, arbuscular mycorrhizal fungi (AMF) are the most widespread soil microorganisms with the potential to make symbiotic relationship with the root system of majority of plants (except some) and have been shown to enhance nutrient uptake (N, P, K, Zn, etc.), soil moisture retention, improvement of soil structure and resilience to various environmental stresses in wide range of plants (Jiménez et al. 2023; Salomon and Watts-Williams 2024; Wu et al. 2024; Valenzuela-Aragon et al. 2025; Li et al. 2025; Ghorui et al. 2025).

The interaction of *Brassica* spp. with AMF is particularly intriguing, as the *Brassicas* (mustard, cabbage, broccoli and canola) have long been regarded as non-mycorrhizal (non-symbiotic) or only weakly so (Sharma et al. 2023; Trautwig et al. 2023). However, emerging research reveals an association of the members of Brassicaceae with AMF for growth enhancement and stress resilience (Table 4.1; Fig. 4.1). Sharma et al. (2023) have explored the Brassicaceae family, known for its non-mycorrhizal status and glucosinolate-based defence system, exhibiting a unique trade-off between symbiosis and defence. Glucosinolates, particularly their toxic breakdown products (isothiocyanates), inhibit AM. However, AM colonisation paradoxically enhances glucosinolate production in rare AM-hosting Brassicales species like *Moringa oleifera* and *Tropaeolum* spp. PHR1, a transcription factor that regulates phosphate starvation responses, also activates glucosinolate biosynthesis in non-host *Arabidopsis thaliana.* Recent microbiome studies have revealed that Brassicaceae may rely on alternative microbial mutualists to compensate for the loss of AM symbiosis, with glucosinolates influencing microbiome composition, suggesting glucosinolates play a key role in Brassicaceae mycorrhizal avoidance and defence strategy.

Shah et al. (2024) have investigated the combined effect of fullerenol nanoparticles (FNPs) and AMF in alleviating lead stress in *Brassica napus*. The plants exposed to 200 μM $PbCl_2$ show stunted growth (15%–25% reduction), decreased

Table 4.1 Mycorrhizal association for sustainable agriculture and stress resilience in Brassicaceae

Plant species	Mycorrhizal association	Stress	Response	Reference
Brassica napus	AMF	Pb	Foliar FNPs (3 mM) + AMF inoculation significantly mitigated Pb toxicity by boosting antioxidant enzymes (CAT, SOD, APX and GR by 37%–200%) and improving nutrient uptake	Shah et al. (2024)
Alliaria petiolata	AMF	–	It reveals a unique, non-symbiotic pathway where AMF inadvertently aid invasion by improving nutrient access in garlic mustard	Trombley et al. (2025)
B. napus	AMF	Cd	Biochar + AMF combination not only alleviated Cd stress but also optimised nutrient uptake and stress response mechanisms	Yin et al. (2024)
B. oleracea	AMF	Zn deficiency	AMF colonisation reduced oxidative stress markers and boosted antioxidant defence and cysteine/glutathione as well as upregulated Zn transporter genes (*BoZIP1* and *BoNRAMP1*), thus enhancing Zn acquisition	Arifuzzaman et al. (2024)
B. oleracea var. *alboglabra*	AMF	–	Bamboo biochar + AMF enhances kale growth, chlorophyll and photosynthetic pigments	Jatuwong et al. (2024)
B. rapa	AMF	Cr	Si + AMF mitigated Cr-induced oxidative damage, restored growth and improved physiological and biochemical functions	Shah et al. (2024)

photosynthesis (38%) and elevated oxidative stress (62%–103% rise in ROS and MDA). However, foliar FNPs (3 mM) + AMF inoculation significantly mitigated lead toxicity by boosting antioxidant enzymes (CAT, SOD, APX, GR by 37%–200%) and improving nutrient uptake. This combination also restored photosynthetic efficiency and gas exchange. Notably, FNPs and AMF synergistically reduced lead accumulation while enhancing plant resilience to lead stress. The research findings suggest that carbon-based nanoparticles combined with AMF can effectively counteract heavy metal stress in crops, offering a sustainable solution for contaminated soils (Shah et al. 2024; Tonelli et al. 2024).

Trombley et al. (2025) noted that AMF typically suppress Brassicaceae growth, but the invasive garlic mustard (*Alliaria petiolata*) defies expectations by thriving with AMF inoculation by nearly doubling in size in lab and field conditions. Unlike classic symbiosis, AMF did not form nutrient-exchanging arbuscules in roots but instead boosted nitrogen availability, enhancing plant growth. This effect is correlated with reduced investment in defensive glucosinolates, suggesting a trade-off between defence and AMF-mediated growth. Contrary to assumptions that garlic

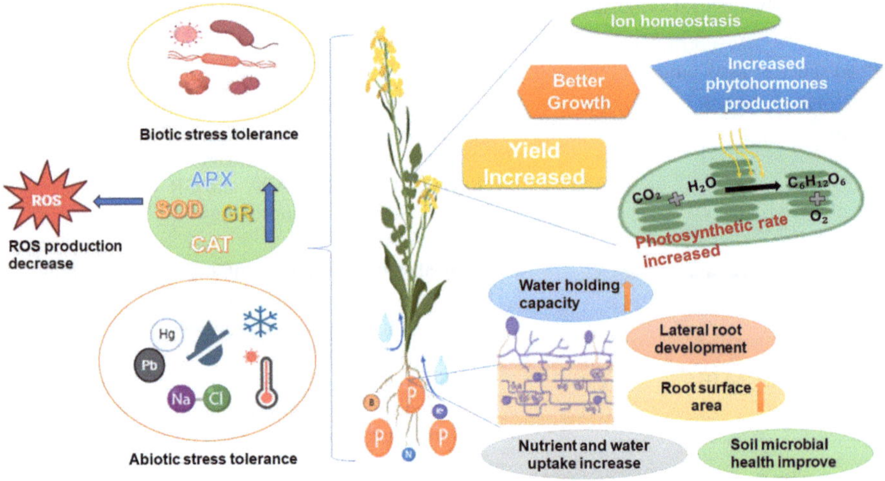

Fig. 4.1 Schematic representation of AMF interaction with Brassicaceae

mustard inhibits AMF, the present research reveals a unique, non-symbiotic pathway where AMF inadvertently aid invasion by improving nutrient access. The findings challenge traditional views on AMF-Brassicaceae interactions and highlight overlooked mechanisms in invasion ecology. For agriculture, this raises questions about AMF's role in weed management, as invasive mustards may exploit AMF-rich soils (Trombley et al. 2025). The effects of biochar and AMF for enhancing *B. napus* tolerance to cadmium, stress (200–300 mg/kg), have been investigated, and it has been revealed that the combined application of 5% BC + 10 g AMF (BC1A) proved most effective, significantly improving plant growth while reducing cadmium uptake (Yin et al. 2024). At 200 mg/kg cadmium, BC1A increased fresh weight by 35.5%, reduced cadmium content by 21.2%, boosted antioxidant enzymes (SOD by 30.6%, POD by 73.4%) and lowered oxidative stress markers (MDA by 40.8%, H_2O_2 by 70%); the soluble sugars, crucial for stress resilience, rose by 38%. Similar trends have been observed at 300 mg/kg Cd, with fresh weight increasing by 32.8% and Cd content decreasing by 16%. Transcriptomic analysis has revealed that biochar upregulated genes linked to peroxidase activity, photosynthesis and MAPK signalling, enhancing cadmium resistance. The AMF activated pathways related to peroxisomes, phytohormone signalling and carotenoid biosynthesis, mitigating the cadmium toxicity in *B. napus*. Furthermore, the biochar + AMF combination not only alleviated cadmium stress but also optimised nutrient uptake and stress response mechanisms. These findings highlight biochar + AMF as a promising strategy for phytoremediation in heavily cadmium-contaminated soils, offering insights for ecological restoration and sustainable agriculture in polluted environments (Hakeem et al. 2015; Ozturk et al. 2015a, b).

Further field trials could validate scalability for real-world applications (Yin et al. 2024). Arifuzzaman et al. (2024) have demonstrated that AMF can mitigate zinc deficiency in broccoli, a typically non-mycorrhizal crop. Under zinc-deficient conditions, AMF-inoculated plants showed improved growth, photosynthesis and zinc uptake compared to non-inoculated plants. AMF colonisation reduced oxidative stress markers (cell death, electrolyte leakage, H_2O_2) and boosted antioxidant defences (POD, APX, SOD) and sulphur metabolites (cysteine, glutathione) in roots. Notably, AMF upregulated zinc transporter genes (*BoZIP1* and *BoNRAMP1*), enhancing zinc acquisition. For the first time, it is revealed that AMF can alleviate zinc deficiency in broccoli by improving nutrient uptake, reducing oxidative damage and activating stress-responsive pathways. The results suggest potential for microbiome-assisted strategies to enhance crop resilience in zinc-deficient soils, extending to other vegetable crops (Arifuzzaman et al. 2024). Jatuwong et al. (2024) have evaluated the effect of three biochar types-bamboo (BB-char), corn cob (CC-char) and coffee grounds (CG-char) for their effects on soil quality and growth of Chinese kale (*B. oleracea* var. *alboglabra*). Biochar improved soil porosity and texture, with 5% BB-char + AMF enhancing kale growth, chlorophyll and photosynthetic pigments more effectively as compared to the 10% applications. Soil analyses after 42 days showed biochar amendments elevated pH, organic matter and carbon levels. Additionally, biochar-AMF combinations increased phenolic compounds in leaves, improving plant health and nutrient cycling (Jatuwong et al. 2024). Shah et al. (2024) have investigated the role of silicon (Si; 300 mg kg^{-1} potash silica) and AMF (*Rhizophagus irregularis*) in mitigating chromium (0 and 300 mg kg^{-1}) toxicity in *Brassica rapa*. Co-application of Si and AMF has markedly enhanced plant defence responses by boosting antioxidant enzyme activities, potassium citrate and GSH levels, glyoxalase system function and associated gene expression. The synergistic treatment also restricted Cr uptake and accumulation in roots and shoots more effectively than either treatment alone. Overall, the combined use of Si and AMF mitigated chromium-induced oxidative damage, restored growth and improved physiological and biochemical functions in *B. rapa* (Shah et al. 2024).

Guo et al. (2023) have examined the effects of AMF and *B. napus* (rape) straw return on maize growth and heavy metal dynamics in a lead-zinc mining area in Southwest China. AMF significantly enhanced rape straw decomposition, nutrient release and soil nutrient availability. AMF + SR improved maize height and biomass by 32%–35% and reduced cadmium and lead accumulation in both soil and maize by 20%–30% and 18%–25%, respectively. Strong positive correlations were observed between soil and maize cadmium and lead levels, while negative correlations were found between heavy metals and maize growth. Overall, AMF played a key role in promoting nutrient cycling, improving plant nutrition and reducing heavy metal uptake in maize, highlighting its potential for sustainable crop production in contaminated soils.

Floc'h et al. (2022) detected 49 AMF variants in both rhizosphere and bulk soils after 10 years of monoculture, challenging assumptions about AMF survival without host plants. The study compared three cropping systems (canola monoculture, 2-year and 3-year rotations), finding that AMF communities persisted regardless of rotation frequency. Network analysis uncovered novel microbial interactions, particularly between AMF (*Funneliformis mosseae* and *Rhizophagus iranicus*) and bacterial cohorts in bulk soil, suggesting non-specific AMF-microbe relationships may support AMF survival without host plants. These findings reshape understanding of AMF ecology, demonstrating their ability to maintain diverse communities in non-host crop systems through alternative microbial partnerships. The results open new research avenues into AMF subsistence mechanisms and their ecological roles beyond traditional plant symbioses, with potential implications for sustainable cropping systems.

4.2 Potential Biotic Stresses in the *Brassicas*

Biotic stresses, including fungi, bacteria, viruses and insects, utilise different molecular mechanisms as virulence strategies to harm plants, ultimately leading to damage. Approximately 25% of crops are lost due to diseases and insect attacks. Plants can defend themselves against biotic stresses; however, when they lack adequate information about the pathogen, they may develop disease symptoms. This disrupts their fundamental processes, impairing growth and development and ultimately reducing crop yields, causing economic losses (Kayum et al. 2016; Dervash et al. 2024).

The infection of *Brassica oleracea* leads to a series of physiological and molecular changes that disrupt cellular functions, ultimately resulting in host cell death. At the infection sites, distinct changes appear on the leaf surface, including bright regions where the wax layer has been removed, indicating the beginning of the pathogen attack. This is accompanied by an increase in ROS production and lipid peroxidation, which are key indicators of the plant's defence response (Macioszek et al. 2024). As the infection progresses, the accumulation of peroxidase and phenolic compounds in the cell wall leads to necrotic areas, where the centre becomes brown and is surrounded by a chlorotic zone, eventually causing the host cell to die (Macioszek et al. 2020).

Chloroplasts, which are essential for energy production and carbon assimilation, play a crucial role in plant defence during pathogen attacks. Upon infection, chloroplasts participate in the defence process by generating ROS, synthesising phytohormones like jasmonic acid (JA) and salicylic acid (SA) and producing secondary metabolites such as sinapic acid, which contribute to the plant's immune response (Kretschmer et al. 2019). However, this defence mechanism can lead to an oxidative burst, damaging plant cells, including electrolyte leakage, lipid peroxidation, pigment oxidation and ultimately tissue necrosis (Abdollahi et al. 2024). The pathogen's attack profoundly disrupts key photosynthetic processes. Infected leaves show downregulation of genes involved in light reactions, such as LHCb and LHCa, as well as genes crucial for the Calvin cycle, including Rubisco small subunit and glyceraldehyde-3-phosphate dehydrogenase. These disruptions impair the photosynthetic machinery and hinder the plant's ability to synthesise the energy needed for survival. Additionally, structural changes in the nucleus, including local swelling of the nuclear envelope and loss of electron density in the nucleoplasm, underscore the extensive cellular damage (Macioszek et al. 2020). In addition, at the molecular level, downregulation of LOX2, a chloroplast lipoxygenase critical for JA accumulation, contributes to the plant's weakened defence in response to salicylic acid-dependent gene activities, which promotes host cell death (Vidhyasekaran 2015).

4.3 Interactions with Pathogen and Pest Organisms

Numerous biotic stressors also affect the rapeseed-mustard crop group. More than 50 insect pests have been found to be connected to it (Root 1973; Lamb 1989; Li et al. 2024). Among them, cabbage butterfly (*Pieris brassicae*), pollen beetle (*Brassicogethes aeneus*), aphid (*Lipaphis erysimi*), painted bug (*Bagrada cruciferarum*), termite (*Odontotermus obesus*), mustard sawfly (*Athalia proxima*), diamondback moth (*Plutella xylostella*), Bihar hairy caterpillar (*Spilosoma obliqua*) and pea leaf miner (*Chromatomyia horticola*) are the important pests (Hansen 2004; Ahuja et al. 2011; CFIA 2014).

The most serious insect pests of oilseed brassicas, particularly in Europe and North America, are those belonging to the order Coleoptera; in contrast, aphids cause the most serious damage in India and New Zealand. Other insect orders are also known to be important pests in India; notable among these are *Meligethes aeneus*, *Phyllotreta cruciferae*, *P. striolata*, *P. undulata*, *Psylliodes chrysocephala*, *Entomoscelis americana*, *Ceutorhynchus assimilis* and *C. quadri*. All these belong to the order Coleoptera; other important insect pests include *Lygus elisus*, *L. lineolaris* and *Bagrada cruciferarum* from the order Hemiptera; *Brevicoryne brassicae* and *Lipaphis erysimi* from the order Homoptera; *Diacrisia obliqua*, *Mammestra configurata*, *Pieris brassicae* and *Plutella xylostella* from the order Lepidoptera; *Dasineura brassicae*, *Delia radicum* and *Phytomyza horticola* from the order Diptera and *Athalia proxima* from the order Hymenoptera (Root 1973; Lamb 1989; Ahuja et al. 2011). Some insect pests affecting *Brassica* plants may also play a significant role in transmitting other diseases, including viral infections. Several such viral infections affecting specific *Brassica* species have been reported, including cauliflower mosaic virus, turnip yellow mosaic virus and turnip mosaic virus (Raybould et al. 1999; Spence et al. 2007; Quezada-Martinez et al. 2021).

The poor and inconsistent harvest of rapeseed-mustard generally results in 90% yield loss, largely caused by diseases. In India, the oilseed crop known as rapeseed mustard is currently shown to be afflicted by over 22 diseases. The important diseases affecting the productivity are damping-off and seedling blight (*Rhizoctonia solani*, *Sclerotium rolfsii*, *Macrophomina phaseolina* and *Fusarium* spp.), sclerotinia rot (*Sclerotinia sclerotiorum*), white rust (*Albugo candida*), powdery mildew (*Erysiphe cruciferarum*), Alternaria blight (*Alternaria brassicae*), downy mildew (*Hyaloperonospora parasitica*), club root (*Plasmodiophora brassicae*), Fusarium wilt (*F. oxysporum* conglutinans), stem blight (*Nigrospora oryzae*), root rot (*Sclerotium rolfsii*), bacterial rot (*Xanthomonas campestris*), bacterial stalk rot (*Erwinia carotovora*), mosaics and phyllody (*Mycoplasma* like organism) (Tewari and Mithen 1999; Oilseeds 2006; Jabeen 2020; Priyamedha et al. 2020).

The clubroot disease is widespread among many crop plants worldwide, including *Brassica* cultivars, and has a significant negative impact on production (Dixon 2009). Large-scale screening studies, particularly on *B. rapa* and *B. oleracea*, report disease-resistant cultivars (Piao et al. 2009; Peng et al. 2014; Zhang et al. 2015). Some researchers have investigated the resistance of *B. oleracea* to clubroot disease and powdery mildew, revealing its frequent resistance (Carlsson et al. 2004; Monot and Silué 2009). Sclerotinia stem rot, a fungal disease, has been reported to cause yield losses of up to 70% in winter rapeseed (Koch et al. 2007; Quezada-Martinez et al. 2021). However, resistance to this disease has been identified in certain *Brassica* cultivars, particularly *B. oleracea* and *B. napus* (Mei et al. 2011, 2013; Taylor et al. 2015). Blackleg disease, caused by *Leptosphaeria maculans*, is one of the most prevalent diseases affecting rapeseed plants, primarily cultivated in Canada, Europe and Australia (West et al. 2001; Quezada-Martinez et al. 2021). Resistance to blackleg disease has been identified in certain *B. napus* and *B. rapa* lines (Delourme et al. 2006; Rimmer 2006; Yu et al. 2005, 2008; Light et al. 2011). *Xanthomonas campestris* pv. *campestris*, the pathogen that causes black rot, is the most significant pathogen of *B. oleracea* (Vicente et al. 2001). It can affect up to 50% of the crop and cause serious economic damage (Singh et al. 2011). Several studies have reported the presence of black rot-resistant lines in *B. oleracea* and *B. rapa* (Lema et al. 2012, 2015; Saha et al. 2016; Ribeiro da Silva et al. 2020). A resistance to the *Albugo candida* pathogen, which causes white rust, has been identified in *B. napus* and *B. rapa* cultivars (Panjabi-Massand et al. 2010; Awasthi et al. 2012). Furthermore, *B. napus*, *B. oleracea* and *B. fruticulosa* genotypes have been shown to exhibit high resistance to white leaf spot, caused by *Pseudocercosporella capsellae* (Gunasinghe et al. 2014, 2017).

Since weeds battle with the crops for soil moisture and nutrients, they constitute one of the main factors preventing rapeseed-mustard output from being sustained. As a result, effective weed control is essential to reduce weed-related damage, along with boosting the yield of the crop. The major weeds of rapeseed-mustard are (*Asphodelus tenuifolius*, *Amaranthus viridis*, *Argemone maxicana*, *Convolvulus arvensis*, *Chenopodium murale*, *C. album*, *Carthamus oxycantha*, *Medicago hispida*, *Pluchea lanceolata*, *Trianthema portulacastrum/monogyna*), sedges (*Cyperus rotundus*), grasses (*Phalaris minor*, *Cynodon dactylon*, *Avena ludoviciana*, *Echinochloa colona*) and parasitic (*Orobanche aegyptiaca*) (Sharma et al. 2015).

There are several interconnected molecular pathways involved in the intricate interactions between plants and pathogens. The key to interpreting these processes is identifying the sets of genes transcribed in the pathogen and the host during infection, along with the way in which certain genes impact the expression of other genes in the pathways. The pathogen must avoid being detected by the plant's defences. Plants use external obstacles like cuticles, wax and trichomes to provide a passive primary line of defence to keep pathogens and insects away. They can also produce substances to keep away from diseases. At the initial step of pathogen detection, defence-related genes are activated by pathogen recognition receptors (PRRs), which identify pathogen-associated molecular patterns (PAMPs), leading to the initiation of the plant defence mechanism referred to as PAMP-triggered immunity

(PTI). Moreover, the pathogen release effectors/elicitors pass through PTI defence responses in order to enter the plant cell. These effectors are subsequently recognised by plant receptors, primarily but not exclusively via nucleotide-binding leucine-rich repeat, or NRL-receptors, which trigger the effector-triggered immunity (ETI) defence responses. Brassicaceae, like various other plant families, differ regarding the quantity and structural characteristics of PRRs and NLR receptors. Of the 59 PRRs with known ligands, 28 have been successfully identified in *A. thaliana*; in *B. napus*, only one is known; this single known receptor of *B. napus* perceives MAMP ligands (like oligogalacturonides and chitin) derived from fungal cell wall and proteinaceous effectors/elicitors (Phyto cytokine-like peptides). The hypersensitivity reactions (HRs) are triggered by ETI, which also cause programmed cell death (PCD) in affected and non-affected nearby cells. The R-gene-mediated reaction of a plant to an invading pathogen can sometimes initiate an additional local level immunity called systemic acquired resistance. SAR further interplays with different pathways, providing protection to other plant parts. Although both kinds of plant defences typically result in a similar response, ETI is believed to be more resistant to invasion by pathogens (Das et al. 2024).

References

Abdollahi H, Nasiri J, Mohajer S (2024) A comprehensive review on oxidative stress and ROS-related resistance strategies in compatible interaction between *Erwinia amylovora* and host plants. J Plant Growth Regul 2024:1–24

Abdullahi M (2023) Land degradation, causes, implications and sustainable management in arid and semi arid regions: a case study of Egypt. Egypt J Soil Sci 63:659–676

Ahmed N, Li J, Li Y et al (2025) Symbiotic synergy: how arbuscular mycorrhizal fungi enhance nutrient uptake, stress tolerance, and soil health through molecular mechanisms and hormonal regulation. IMA Fungus 16:e144989

Ahuja I, Rohloff J, Bones AM (2011) Defence mechanisms of brassicaceae: implications for plant-insect interactions and potential for integrated pest management. In: Sustainable agriculture, vol 2. Springer, pp 623–670

Akhtar K, Ul Ain N, Vara Prasad PV et al (2024) Physiological, molecular, and environmental insights into plant nitrogen uptake, and metabolism under abiotic stresses. Plant Genome 17(2):e20461

Arifuzzaman M, Bagchi R, Hasan MR et al (2024) Arbuscular mycorrhizal fungi induce Zn uptake and antioxidant efficiency in broccoli (*Brassica oleracea* L.) exposed to Zn deficiency. Plant Stress 11:100415

Ashraf M, Ozturk M, Athar HR (eds) (2009) Salinity and water stress: improving crop efficiency. Tasks for vegetation science 44. Springer, New York

Ashraf M, Ozturk M, Ahmad M, Aksoy A (eds) (2012) Crop production for agricultural improvement. Springer, New York

Awasthi RS, Nashaat NI, Kolte SB et al (2012) Screening of putative resistant sources against Indian and exotic isolates of *Albugo candida* inciting white rust in rapeseed-mustard. J Oilseed Brassica 3:27–37

Bate-Weldon MPA, Edmondson JL, Field KJ (2024) Impact of zinc on arbuscular mycorrhizal-mediated nutrient acquisition in urban horticulture. Iscience 27:8110580

Carlsson M, von Bothmer R, Merker A (2004) Screening and evaluation of resistance to downy mildew (*Peronospora parasitica*) and clubroot (*Plasmodiophora brassicae*) in genetic resources of Brassica oleracea. Hereditas 141:293–300

CFIA (2014) Canadian food inspection agency. https://inspection.gc.ca/plants/plants-with-noveltraits/applicants/directive-94-08/biology-documents/1330965093062/1330987674945

Das LS, Kundu A, Podder S (2024) Impact of biotic stresses on the Brassicaceae family and opportunities for crop improvement by exploiting genotyping traits. Planta 259:97

Delourme R, Chèvre AM, Brun H et al (2006) Major gene and polygenic resistance to Leptosphaeria maculans in oilseed rape (Brassica napus). Eur J Plant Pathol 114:41–52

Dervash M, Yousuf A, Bhat MA, Ozturk M (2024) Soil organisms: deciphering the life beneath our feet, Springer briefs in microbiology. Springer. ISBN 978-3-031-66293-5

Dixon GR (2009) The occurrence and economic impact of Plasmodiophora brassicae and clubroot disease. J Plant Growth Regul 28:194–202

Etesami H, Jeong BR, Glick BR (2021) Contribution of arbuscular mycorrhizal fungi, phosphate-solubilizing bacteria, and silicon to P uptake by plant. Front Plant Sci 12:699618

Floc'h J-B, Hamel C, Laterrière M et al (2022) Long-term persistence of arbuscular mycorrhizal fungi in the rhizosphere and bulk soils of non-host *Brassica napus* and their networks of co-occurring microbes. Front Plant Sci 13:828145

Ghorui M, Chowdhury S, Burla S (2025) Recent advances in the commercial formulation of arbuscular mycorrhizal inoculants. Front Ind Microbiol 3:1553472

Gill SS, Tuteja N (2010) Reactive oxygen species and antioxidant machinery in abiotic stress tolerance in crop plants. Plant Physiol Biochem 48:909–930

Gunasinghe N, You MP, Banga SS, Barbetti MJ (2014) High level resistance to Pseudocercosporella capsellae offers new opportunities to deploy host resistance to effectively manage white leaf spot disease across major cruciferous crops. Eur J Plant Pathol 138:873–890

Gunasinghe N, You MP, Banga SS et al (2017) Outstanding host resistance will resolve the threat from white leaf spot disease (*Pseudocercosporella capsellae*) to oilseed and vegetable Brassica spp.crops. Australas Plant Pathol 46:137–146

Guo J, Chen J, Li C et al (2023) Arbuscular mycorrhizal fungi promote the degradation of the fore-rotating crop (*Brassica napus* L.) straw, improve the growth of and reduce the cadmium and lead content in the subsequent maize. Agronomy 13(3):767

Hakeem KR, Parvaiz A, Ozturk M (eds) (2013) Crop improvement-new approaches and modern techniques, vol XXVII. Springer, New York

Hakeem KR, Sabir M, Ozturk M, Mermut A (2015) Soil remediation and plants: prospects and challenges. Academic Press/Elsevier, London/New York

Hansen LM (2004) Economic damage threshold model for pollen beetles (*Meligethes aeneus* F.) in spring oilseed rape (*Brassica napus* L.) crops. Crop Prot 23:43–46

Hasanuzzaman M, Nahar K, Ozturk M (2019) Ecophysiology, abiotic stress responses and utilization of halophytes. Springer Nature Singapore, Singapore

Hawkins H-J, Cargill RIM, Nuland MEV et al (2023) Mycorrhizal mycelium as a global carbon pool. Curr Biol 33(11):R560–R573

Jabeen N (2020) Agricultural, economic and societal importance of Brassicaceae plants. In: The plant family Brassicaceae: biology and physiological responses to environmental stresses. Springer, pp 45–128

Janni M, Maestri E, Gullì M et al (2024) Plant responses to climate change, how global warming may impact on food security: a critical review. Front Plant Sci 14:1297569

Jatuwong K, Aiduang W, Kiatsiriroat T et al (2024) Effects of biochar and arbuscular mycorrhizal fungi on soil health in chinese kale (*Brassica oleracea* var. alboglabra L.) cultivation. Microbiol Res 15(1):404–421

Jiang Z, van Zanten M, Sasidharan R (2025) Mechanisms of plant acclimation to multiple abiotic stresses. Commun Biol 8:655

Jiménez EM, Ferrol N, Corradi N et al (2023) The potential of arbuscular mycorrhizal fungi to enhance metallic micronutrient uptake and mitigate food contamination in agriculture: prospects and challenges. New Phytol 242(4):1441–1447

Kayum MA, Kim HT, Nath UK et al (2016) Research on biotic and abiotic stress related genes exploration and prediction in *Brassica rapa* and *B. oleracea*: a review. Plant Breed Biotechnol 4(2):135–144

Koch S, Dunker S, Kleinhenz B et al (2007) A crop loss-related forecasting model for sclerotinia stem rot in winter oilseed rape. Phytopathology 97:1186–1194

Kretschmer M, Damoo D, Djamei A, Kronstad J (2019) Chloroplasts and plant immunity: where are the fungal effectors? Pathogens 9(1):19

Lahlali R, Laasli S-E, Ait Barka E (2025) Plant responses to biotic and abiotic stresses: from cellular to morphological changes—series II. Agronomy 15(1):229

Lamb RJ (1989) Entomology of oilseed *Brassica* crops. Annu Rev Entomol 34:211–229

Lema M, Velasco P, Soengas P et al (2012) Screening for resistance to black rot in Brassica oleracea crops. Plant Breed 131:607–613

Lema M, Cartea ME, Francisco M et al (2015) Screening for resistance to black rot in a Spanish collection of *Brassica rapa*. Plant Breed 134:551–556

Li Z, Costamagna AC, Beran F, You M (2024) Biology, ecology, and management of flea beetles in *brassica* crops. Annu Rev Entomol 69(1):199–217

Li X, Wu Y, Huang C et al (2025) Inoculation with arbuscular mycorrhizal fungi in the field promotes plant colonization rate and yield. Eur J Agron 164:127503

Light KA, Gororo NN, Salisbury PA (2011) Usefulness of winter canola (*Brassica napus*) race-specific resistance genes against blackleg (causal agent *Leptosphaeria maculans*) in southern Australian growing conditions. Crop Pasture Sci 62:162–168

Lin F, Li X, Jia N et al (2023) The impact of Russia-Ukraine conflict on global food security. Glob Food Sec 36:100661

Macioszek VK, Gapińska M, Zmienko A et al (2020) Complexity of *Brassica oleracea-Alternaria brassicicola* susceptible interaction reveals downregulation of photosynthesis at ultrastructural, transcriptional, and physiological levels. Cells 9(10):2329

Macıoszek VK, Piotrowska-Niczyporuk A, Kononowicz AK (2024) *Alternaria brassicicola*-induced ROS accumulation during black spot disease differentially affects antioxidant efficiency, phenolic content and susceptibility of *Brassica species*. Plant Pathol 2024:2024

Manghwar H, Zaman W (2024) Plant biotic and abiotic stresses. Life 14(3):372

Mei J, Qian L, Disi JO et al (2011) Identification of resistant sources against *Sclerotinia sclerotiorum* in Brassica species with emphasis on *B. oleracea*. Euphytica 177:393–399

Mei J, Ding Y, Lu K et al (2013) Identification of genomic regions involved in resistance against *Sclerotinia sclerotiorum* from wild *Brassica oleracea*. Theor Appl Genet 126:549–556

Monot C, Silué D (2009) Comparison of the resistance and susceptibility to downy mildew [*Hyaloperonospora parasitica*, constant. (Pers. Ex Fr)] of nine *Brassica olearacea* accessions in laboratory, seedbed, and field screens. Euphytica 169:413–419

Mythili M, Ramalakshmi A, Gopal NO (2025) Arbuscular mycorrhizal fungi for sustainable agriculture ecosystem. In: Encyclopedia of green materials. Springer, Singapore

Oilseeds WA (2006) Growing western Canola: an overview of canola production in western Australia. Oil Seeds Industry Association of Western Australia, Perth

Oishy MN, Shemonty NA, Fatema SI et al (2025) Unravelling the effects of climate change on the soil-plant-atmosphere interactions: a critical review. Soil Environ Health 3(1):100130

Ozturk M, Gul A (2020) Climate change and food security with emphasis on wheat. Academic Press/Elsevier Inc, London/New York

Ozturk M, Ashraf M, Aksoy A, Ahmad MSA (eds) (2015a) Phytoremediation for green energy. Springer, New York

Ozturk M, Ashraf M, Aksoy A, Ahmad MSA (eds) (2015b) Plants, pollutants & remediation. Springer, New York

Ozturk M, Gul A, Rafique K et al (2025a) Nature conservation, air pollution and plant diversity across various ecosystems and the ımpact of climate change. In: Forests and carbon sequestration. Cambridge Scholars Publishing, pp 330–422

Ozturk M, Hanif A, Gul A, Altay V (2025b) Environmental health: an urgent matter. In: Allelopathy. Apple Academic Press, pp 49–66

Panjabi-Massand P, Yadava SK, Sharma P et al (2010) Molecular mapping reveals two independent loci conferring resistance to *Albugo candida* in the east European germplasm of oilseed mustard *Brassica juncea*. Theor Appl Genet 121:137–145

Peng G, Falk KC, Gugel RK et al (2014) Sources of resistance to *Plasmodiophora brassicae* (clubroot) pathotypes virulent on canola. Can J Plant Pathol 36:89–99

Piao Z, Ramchiary N, Lim YP (2009) Genetics of clubroot resistance in Brassica species. J Plant Growth Regul 28:252–264

Priyamedha RB, Kumar A et al (2020) Genetics and genomic approaches for disease resistance in *Brassicas*. In: *Brassica* improvement: molecular, genetics and genomic perspectives. Springer International Publishing, Cham, pp 147–157

Quezada-Martinez D, Addo Nyarko CP, Schiessl SV, Mason AS (2021) Using wild relatives and related species to build climate resilience in Brassica crops. Theor Appl Genet 134(6):1711–1728

Raybould AF, Maskell LC, Edwards M-L et al (1999) The prevalence and spatial distribution of viruses in natural populations of *Brassica oleracea*. New Phytol 141:265–275

Ribeiro da Silva AL, Secchi Candian J, Ramalho do Rego E et al (2020) Screening cabbage cultivars for resistance to black rot under field conditions. Am Soc Hortic Sci 30:448–455

Rigobelo EC (2024) Mycorrhiza. In: Microbial services for cereal crops. Springer, Cham

Rimmer SR (2006) Resistance genes to *Leptosphaeria maculans* in *Brassica napus*. Can J Plant Pathol 28:S288–S297

Root RB (1973) Organization of a plant-arthropod association in simple and diverse habitats: the fauna of collards (*Brassica oleracea*). Ecol Monogr 43(1):95–124

Saha P, Kalia P, Sharma M, Singh D (2016) New source of black rot disease resistance in *Brassica oleracea* and genetic analysis of resistance. Euphytica 207:35–48

Salomon MJ, Watts-Williams SJ (2024) Realising the promise of arbuscular mycorrhizal fungal biofertilisers through more applied research. New Phytol 246(3):811–813

Shah T, Khan H, Ali A et al (2024) Silicon and arbuscular mycorrhizal fungi alleviate chromium toxicity in *Brassica rapa* by regulating Cr uptake, antioxidant defense expression, the glyoxalase system, and secondary metabolites. Plant Physiol Biochem 206:108286

Sharma P, Meena PD, Singh D (2015) Present and emerging biotic stress challenges of oilseed Brassica. In: ICAR sponsored winter school on new paradigms in diseases management: conventional and molecular approaches for rapeseed-mustard production. ICAR-Directorate of Rapeseed-Mustard Research, Bharatpur, p 122

Sharma A, Sinharoy S, Bisht NC (2023) The mysterious non-arbuscular mycorrhizal status of Brassicaceae species. Environ Microbiol 25(5):917–930

Sheikh-Assadi M, Khandan-Mirkohi A, Taheri MR et al (2023) Arbuscular mycorrhizae contribute to growth, nutrient uptake, and ornamental characteristics of statice (*Limonium sinuatum* [L.] Mill.) subject to appropriate inoculum and optimal phosphorus. Horticulturae 9(5):564

Singh D (2022) Juggling with reactive oxygen species and antioxidant defense system—a coping mechanism under salt stress. Plant Stress 5:100093

Singh D, Dhar S, Yadava DK (2011) Genetic and pathogenic variability of Indian strains of *Xanthomonas campestris* pv. campestris causing black rot disease in crucifers. Curr Microbiol 63:551–560

Spence NJ, Phiri NA, Hughes SL et al (2007) Economic impact of turnip mosaic virus, cauliflower mosaic virus and beet mosaic virus in three Kenyan vegetables. Plant Pathol 56:317–323

Taylor A, Coventry E, Jones JE, Clarkson JP (2015) Resistance to a highly aggressive isolate of sclerotinia sclerotiorum in a *Brassica napus* diversity set. Plant Pathol 64:932–940

Tewari JP, Mithen RF (1999) Diseases. In: Biology of Brassica coenospecies. Elseiver Science B.V., pp 375–393

Tonelli FMP, Roy A, Ozturk M, Murthy HCA (2024) Nanotechnology-based sensors for detection of environmental pollution. Elsevier

Trautwig AN, Jackson MR, Kivlin SN, Stinson KA (2023) Reviewing ecological implications of mycorrhizal fungal interactions in the Brassicaceae. Front Plant Sci 14:1269815

Trombley J, Celenza JL, Frey SD, Anthony MA (2025) Arbuscular mycorrhizal fungi boost development of an invasive Brassicaceae. Plant Cell Environ 48(7):4928–4937. https://doi.org/10.1111/pce.15508

Valenzuela-Aragon B, Cardinale M, Rolli E et al (2025) The role of arbuscular mycorrhizal fungi in abiotic stress management in viticulture under climatic shifts. Plant Stress 16:100863

Vicente JG, Conway J, Roberts SJ, Taylor JD (2001) Identification and origin of *Xanthomonas campestris* pv. Campestris races and related pathovars. Phytopathology 91:492–499

Vidhyasekaran P (2015) Plant hormone signaling systems in plant innate immunity, vol 2. Springer, Dordrecht, pp 117–122

Wang Q, Liu M, Wang Z et al (2024) The role of arbuscular mycorrhizal symbiosis in plant abiotic stress. Front Microbiol 14:1323881

Wang D, Jewaria PK, Xiao J (2025a) Photosynthetic adaptation in poplar under abiotic and biotic stress: integrating molecular, physiological, and biotechnological perspectives. Int J Plant Biol 16(2):42

Wang L, Ju C, Han C et al (2025b) The interaction of nutrient uptake with biotic and abiotic stresses in plants. J Integr Plant Biol 67(3):455–487

West JS, Kharbanda PD, Barbetti MJ, Fitt BDL (2001) Epidemiology and management of *Leptosphaeria maculans* (phoma stem canker) on oilseed rape in Australia, Canada and Europe. Plant Pathol 50:10–27

Wu Y, Chen C, Wang G (2024) Inoculation with arbuscular mycorrhizal fungi improves plant biomass and nitrogen and phosphorus nutrients: a meta-analysis. BMC Plant Biol 24:960

Yin C, Lei W, Wang S et al (2024) Biochar and arbuscular mycorrhizal fungi promote rapid-cycling *Brassica napus* growth under cadmium stress. Sci Total Environ 953:176034

Yu F, Lydiate DJ, Rimmer SR (2005) Identification of two novel genes for blackleg resistance in Brassica napus. Theor Appl Genet 110:969–979

Yu F, Lydiate DJ, Rimmer SR (2008) Identification and mapping of a third blackleg resistance locus in Brassica napus derived from B. Rapa subsp. sylvestris. Genome 51:64–72

Zhang H, Feng J, Zhang S et al (2015) Resistance to *Plasmodiophora brassicae* in *Brassica rapa* and *Brassica juncea* genotypes from China. Plant Dis 99:776–779

Zhang H, Lang Z, Zhu JK, Wang P (2025) Tackling abiotic stress in plants: recent insights and trends. Stress Biol 5:8

Chapter 5
Molecular Aspects of *Brassica*

5.1 Introduction

The Brassicaceae family includes economically important crops such as *Brassica napus* (rapeseed), *Brassica juncea* (mustard) and *Brassica oleracea* (cabbage), which play a crucial role in the global agriculture sector. The largest producers of *Brassica* vegetables are India and China, with 44 and 18 million tonnes, respectively. Currently, *Brassica napus*, which evolved 7600 years ago, is becoming the second largest oil crop. Numerous extreme morphotypes of *Brassica* crops have been cultivated, including *B. juncea* var. *tumida* (tuber mustard), *B. oleracea* var. *italica* (broccoli), *B. rapa* ssp. *rapa* (enlarged roots in turnip), *B. rapa* ssp. *pekinen sis* (leafy head in Chinese cabbage), *B. oleracea* var. *botrytis* (thickened inflorescence in cauliflower), *B. juncea* spp. *napiformis* (root mustard), *B. rapa* var. *rapifera* (rutabaga), *B. oleracea* var. *capitata* (cabbage) and *B. oleracea* var. *gongyloides* (enlarged stems in kohlrabi). *B. juncea*, also known as "Indian mustard", significantly contributes to the production of edible oils. *B. juncea* is an amphitetraploid ($2n = 36$) crop developed by crossing between *B. rapa* ($2n = 20$) and *B. nigra* ($2n = 16$) (Xue et al. 2020). In India, 26.2 million hectares of the land area are used for cultivating oilseed crops. Approximately 34 million tonnes of edible seed oil are needed to meet the demand, and between 8 and 10 million tonnes of oil are produced from rapeseed and mustard in India. These crops exhibit remarkable genetic diversity, which makes them valuable crops for understanding the molecular mechanisms for plant growth and stress responses (Tan et al. 2024; Shaw et al. 2021). Understanding the molecular pathways in *Brassica*, including key metabolic pathways, hormone signalling and genetic regulation of different genes involved in different functions like growth and development, flowering, seed germination and stress responses, can enhance the ability to manipulate these pathways for improved varieties of high yield and stress resilience. Key regulatory components, such as ROS signalling, hormone signalling, accumulation of osmo-protectants and transcription factors, contribute to stress adaptation in plants.

M. Ozturk et al., *Climate-Resilient* Brassicas,
https://doi.org/10.1007/978-3-032-12709-9_5

5.2 Genomic Features of *Brassica*

The low spectrum of angiosperm genome size is determined by the Brassicaceae family, which has around 3350 species and almost 340 genera. Due to genome evolution, Brassicaceae is particularly interesting despite significant chromosomal alteration; comparatively limited range (0.16 pg < 1C < 1.95 pg) of DNA content was preserved. For Brassicaceae, no DNA concentrations greater than 1C = 1.95 pg have been documented (Plant DNA C-values Database (www.rbgkew.org.uk/cval/homepage.html). The currently available allopolyploids, *B. juncea*, *B. napus* and *B. carinata*, which were created via hybridisation between *B. rapa* × *B. oleracea*, *B. nigra* × *B. oleracea* and *B. nigra* × *B. oleracea*, appear to have seen additional reductions in DNA content. *B. carinata* (0.97 pg), *B. juncea* (0.92 pg) and *B. napus* (0.92 pg) DNA content are proportionately lower than the total DNA content of their putative parental species (Johnston et al. 2005). The six most well-known species associated with *Brassica*, an agriculturally important genus, include three diploids, *B. rapa* (A genome, *n* = 10), *B. nigra* (B genome, *n* = 8) and *B. oleracea* (C genome, *n* = 9), and three allotetraploid species, *B. juncea* (AB genome, *n* = 18), *B. napus* (AC genome, *n* = 19) and *B. carinata* (BC genome, *n* = 17). The "Triangle of U" (Fig. 5.1) represents the genetic connection between these six representative individuals. The publication of the complete genome sequence of *B. rapa* not only demands investigation into the roles of its genes, but also serves as an essential source for studying the evolution of the "Triangle of U" and the polyploidization of

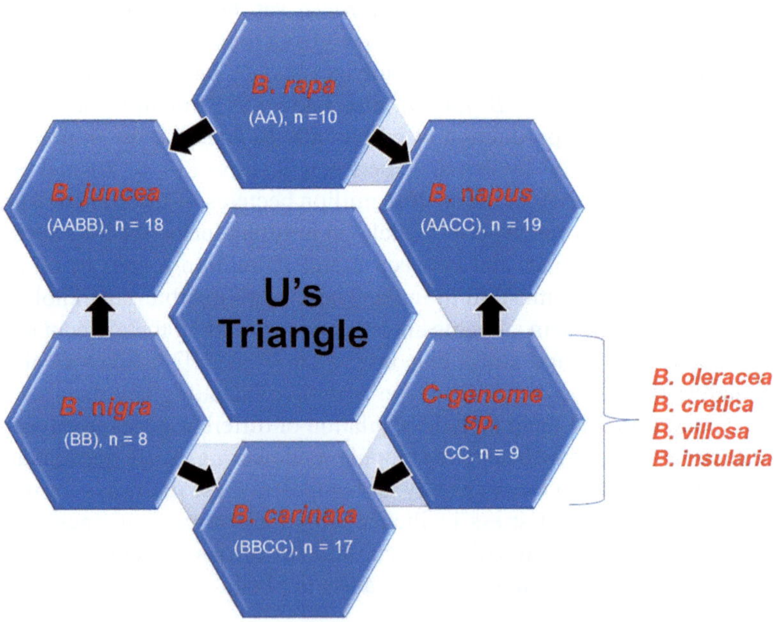

Fig. 5.1 Triangle of U representation of the genetic relation between *Brassica* individuals

Brassicaceae species. Least fractionated (LF), more fractionated (MF1) and most fractionated (MF2) are the three subgenomes that resulted from the additional three diploid members of the tPCK family in "Triangle of U" created by whole-genome triplication (WGT) of their ancestors.

The WGT enhanced the genomic variety that contributes to the many morphotypes in *Brassica* species, whereas LF is the least variable subgenome. The development of the "Triangle of U" species genome comparison tool made it possible to investigate the differences in polyploid formation and the importance of subgenome dominance to agronomic development and domestication (Wu et al. 2022). In 2016, 119 *B. oleracea* and 199 *B. rapa* accessions were sequenced for the purpose of investigating the domestication of *Brassicas* that produce tubers and leafy heads. Additionally, the considerable genomic flexibility of *Brassica* species makes it possible to domesticate crops having a very high degree of phenotypic variety. Parallel domestication of comparable characteristics in many *Brassica* species includes characteristics like expanded root/stem and leafy head. *Brassica* provides a perfect platform to study the domestication of crops and artificial selection because of all these characteristics (Cheng et al. 2016). New research on plant development may benefit from naturally present genetic variants in *Brassica*, especially when examining the consequences of alleles lacking distinct characteristics in *Arabidopsis*. Research on the function and expression of duplicated genes in *Brassica* species can help clarify how genome replication contributes to the genus's phenotypic diversity. Additionally, the research on polyploidy-related fast genome alterations is made possible by *Brassica* species. Chromosomal doubling after hybridising the diploid species allows for the resynthesis of each of the amphidiploid species: *B. napus*, *B. juncea* and *B. carinata*. This produces fully homozygous polyploid lines that shouldn't separate, yet offspring produced by self-pollination of individual resynthesised polyploids have been found to differ in phenotypic and DNA fragment restriction (Paterson et al. 2001). The genomes of *B. rapa* (ArAr) and *B. oleracea* (CoCo) experienced whole-genome triplication, according to relative genome mapping, chromosomal segment evaluation and a whole-genome sequencing study. Pan-genome evaluation can now accurately represent within-species genetic diversity, especially structural variations (SVs), because of the rapid development of DNA sequencing and assemblage platforms, which have allowed investigators to obtain a previously unattainable number of high-quality genomes. *Brassica* is frequently used for agricultural diversification, which makes it a great model system to investigate evolution caused by polyploidization. The development of *Brassica* species genomes has been studied by genome sequencing and comparative genome analysis, which additionally demonstrated conserved features. The relationships of repetitive elements (REs), known to be essential to epigenetic mechanisms and therefore perform an essential part in plant genome adaptation and evolution, have not been well understood, despite the fact that asymmetric sub-genome retention of genes and chromosomal rearrangement in *Brassica* species have been extensively documented. The standard genome sequences have been constructed with estimated genome sizes of *B. rapa* (AA) (58%), *B. oleracea* (CC) (86%), and their derived allotetraploid, *B. napus* (AACC) (75%). The core components of each genome are concealed, but some of the non-assembled genome segments differ among these three genome sequences (Waminal et al. 2016). Duplication has been a major concern in function and evolution investigations of genes and gene expression among Brassicaceae due to current relative genomic research. About 40 million years ago, a paleopolyploid occurrence took place close to the Brassicaceae's genesis. The discovery of wide genome duplication and the preservation of a sizable repertoire of duplicated gene pairs (DGPs) within the *A. thaliana* sequence led to the conclusion that the organism was paleopolyploid. Despite having a similar structure and sequence, DGPs in the Brassicaceae vary between and within lineages in terms of their expression, cis-regulation and presence or absence.

5.2.1 Advances in Sequencing Technologies: Genome Assemblies and Annotations

More targeted studies employing plant lines and genotypes promoting climate adaptation are additionally becoming necessary as a result of climate change. The easy accessibility of data like this is highly valuable in the contemporary environment of constantly changing artificial intelligence, technological advances and chemistry. Large collections of data must be assembled, annotated, represented or visualised in order to constitute an aggregation with biological information. Traditional conservation genetics has given way to restoration genomics due to quick developments in genome sequencing and bioinformatic techniques. The total GC content, repetition number, the percentage of duplicated reads and the reliability of the sequence information ought to be evaluated before assembly. Data analysis tools like FastQC (www.bioinformatics.babraham.ac.uk/projects/fastqc) are a good place where you begin. Several alternative programs and scripts are capable of being used to trim poor-quality data and readings generated by PCR duplications. After an assembly has been finished effectively, its quality and many assemblies can be compared through various techniques. A genome sequence must be annotated using appropriate biological data in order to be fully utilised. This data may include microRNA and epigenetic changes, gene models and functional details such as gene ontology (GO) or "Kyoto encyclopaedia of genes and genomes" (KEGG) pathways. The annotation procedure can be logically separated into two stages: a "computational phase" in which preliminary gene and transcript hypotheses are made by combining multiple lines of evidence gathered from various genomes or from transcriptome information particular to a species. Using a set of guidelines established within the annotation pipeline, all (often conflicting) data is subsequently combined into a gene annotation in phase second that is the "annotation phase." Furthermore, because preliminary sequences of genomes are currently being generated at a rapid pace, genome publication is crucial. Conventional databases that offer accessibility to genomes and meta-information, like the National Centre for Biotechnology Information's (NCBI) or the European Molecular Biology Labs' (EMBL) ENSEMBL, are presently not able to annotate and curate all incoming genomes. As a result, NCBI already offers to submit annotations and draft genome sequences created by users. The completed genome and any pertinent metadata must be posted, like as a BioProject on NCBI, in addition to every bit of unprocessed information, so that other users can enhance the assembly and its annotation (Ekblom and Wolf 2014).

5.3 Molecular Pathways in *Brassica*

5.3.1 Key Metabolic Pathways: Glucosinolate Biosynthesis, Lipid Metabolism

Glucosinolates are nitrogen and sulphur-containing secondary metabolites that have a prime occurrence in the order Capparales and are exclusively present in the family Brassicaceae. Usually under normal conditions, glucosinolates show a stable chemical nature, whereas enzyme myrosinase hydrolyses glucosinolates and results in various products like thiocyanates, isothiocyanates, nitriles, oxazolidines and epithionitriles upon cell or tissue injury. The final products produced upon hydrolysis are determined by various variables like pH, availability of proteins that interact with myrosinase and ferrous ions (del Carmen et al. 2013). Glucosinolates biosynthesis mainly involves three three-step processes: (a) amino acids (Met, Phe) side chain elongation, (b) core structure development and (c) secondary modification of side chain. Branched chain amino acid aminotransferase 4 (BCAT4) initiates chain elongation by deaminating Met and produces 2-oxo acid. Further, 2-ketoacid is formed by aliphatic chain elongation of 2-oxo acid that is carried out by a cluster of enzymes, including methylthioalkylmalate (MAMS), isopropylmalate isomerase (IPMI) and isopropylmalate dehydrogenase (IPMDH). Finally, BCAT3 transaminates and produces chain-elongated Met, which enters the next step of glucosinolate biosynthesis. The second step includes core glucosinolate synthesis, where CYP79F1 and CYP79A2 catalyse methionine derivatives and phenylalanine derivatives, respectively, to form aldoximes. Aldoximes are converted into aci-nitro compounds by CYP83A1 and CYP83B1, which further produce S-alkyl-thiohydroximates, a glutathione conjugate by glutathione-S-transferases (GST). Later on, SUR1, a carbon-sulfur lyase, produce thiohydroximates from S-alkyl-thiohydroximates. S-Glycosylation of thiohydroximates is carried out by UGT74C1 and UGT74B1 and forms desulfoglucosinolates. A group of sulfotransferases (SOTs) add sulphate to desulfoglucosinolates and produce glucosinoate molecules ready for side chain modifications. Side chain modification of aliphatic GSL is done by flavin-monooxygenase glucosinolate S-oxygenase (FMO), 4-methylsulfinyl butyl GSL, which is further modified by 2-oxoglutarate-dependent dioxygenase (GS-OH) and 2-hydroxy-3-butenyl GSL. Furthermore, modification in Benzenic GSL carried out by GS-OH results in 2-hydroxy-2-phenylethyl GSL (Kitainda and Jez 2021) (Fig. 5.2).

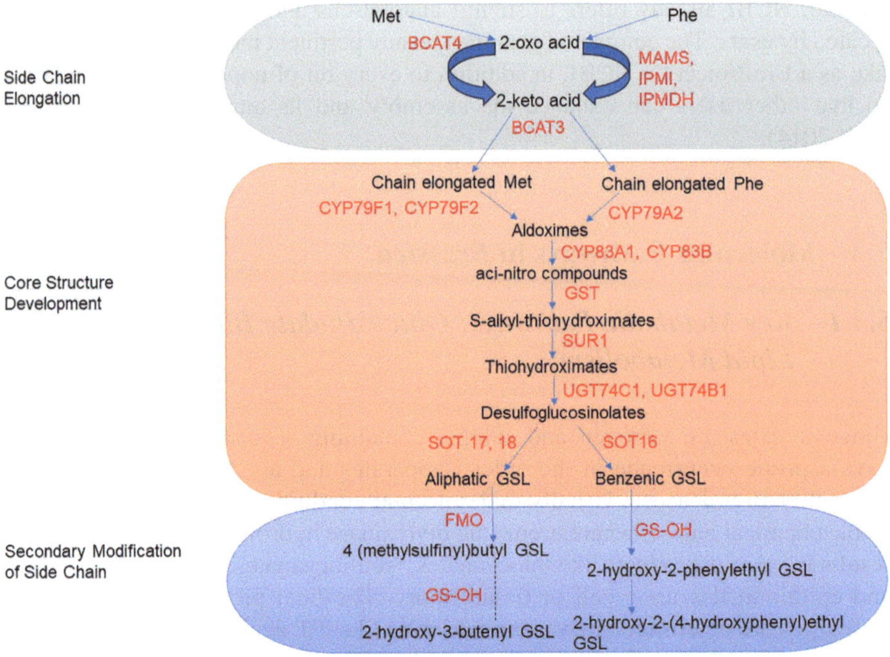

Fig. 5.2 Glucosinolates biosynthesis pathways in Brassicaceae

Multiple organelles and a multi-step reaction are involved in lipid metabolism, including biosynthesis and catabolism of fatty acids and glycerolipids. FA is a primary pathway for carbon to enter lipids because FAs are the building blocks of all lipids, and the carbon component of FAs comes from phosphoenolpyruvate (PEP) and pyruvate of the glycolytic pathway. Pyruvate is transported into the plastid and decarboxylated by the pyruvate dehydrogenase complex to produce acetyl-coenzyme A (acetyl-CoA). The FA synthase complex delivers the malonyl component to acyl-carrier proteins (ACPs) for the production of long-chain FAs, primarily C16:0, C18:0 and C18:1, after acetyl-CoA carboxylase (ACCase) carboxylates acetyl-CoA into malonyl-CoA. The primary enzyme that controls the length of the FA chain is fatty acyl-ACP thioesterase. Following production, FAs get esterified in the plastid or ER using an enzyme called FA desaturase (Manan et al. 2016). The acyl-CoAs are used for membrane lipid synthesis or TAGs once they have been transferred from the plastid to the ER. The acyl-CoA-dependent Kennedy route is the method for assembling FAs into TAGs, which are important storage lipids in oilseeds. Phosphatidic acid (PA) is created in this process by G3P acyltransferase (GPAT) and lysophosphatidic acid acyltransferase (LPAAT), respectively, incorporating acyl-CoAs into the sn-1 and sn-2 locations of glycerol-3-phosphate (G3P). Phosphatidic acid phosphatase (PAP) then removes a phosphate group from PA, and acyl-CoA: diacylglycerol acyltransferase (DGAT) acylates the resultant

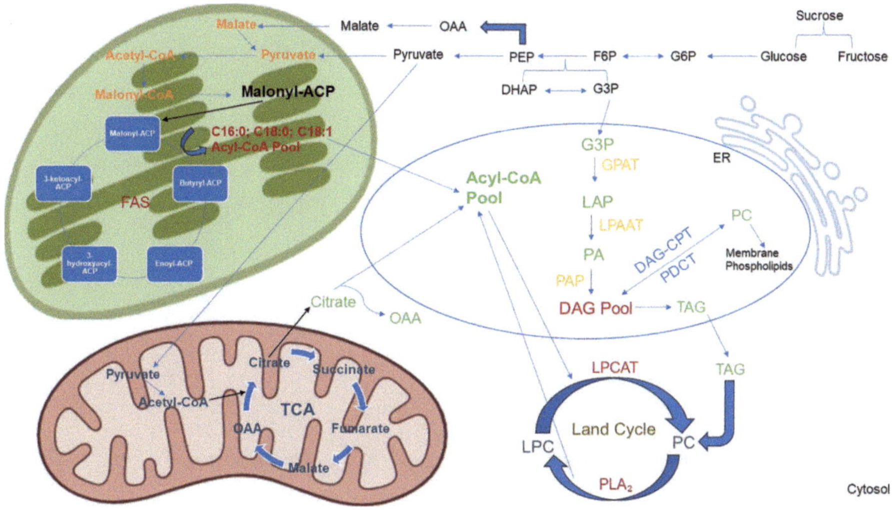

Fig. 5.3 Schematic representation of fatty acids and TAG synthesis in oilseed plants

diacylglycerol (DAG) on the sn-3 position. This completes the de novo production of TAGs; DGATs are recognised as the primary enzymes that cause the accumulation of oil. *Brassica napus* studies showed that the isoforms of DGAT2s exhibit extremely varied activity against 18:3. In addition to this, it has small to substantial activity towards 22:1, with varying substrate specificities (Miklaszewska et al. 2021) (Fig. 5.3).

5.3.2 Genetic Regulation of Flowering Time, Seed Development and Stress Responses

Plants include different developmental stages in the whole life cycle, and among those, the most important is flowering. For the best development of harvestable leaves and seeds, crops must flower at the right time. Delay in flowering also affects the seed yield, while early bolting stunts vegetative growth and yields low-quality leaves and seeds. Plant leaves perceive temperature, day duration and internal development signal, also florigen, a long-distance signal molecule involved in floral transition, move through phloem to the SAM (shoot apical meristem). Florigen produces FT protein complexed with FD transcription factor at SAM and further initiates a signal to activate LFY and AP1 (meristem identity genes). Expression of FT is regulated by various factors that include the age of the plant, GA levels, thermosensory and photoperiodic pathways. Day length is measured by the photoperiodic pathway, which then transmits the data to the CONSTANS (CO) TF to detect seasonal variations. FT expression is enhanced by elevated CO protein activity during ideal

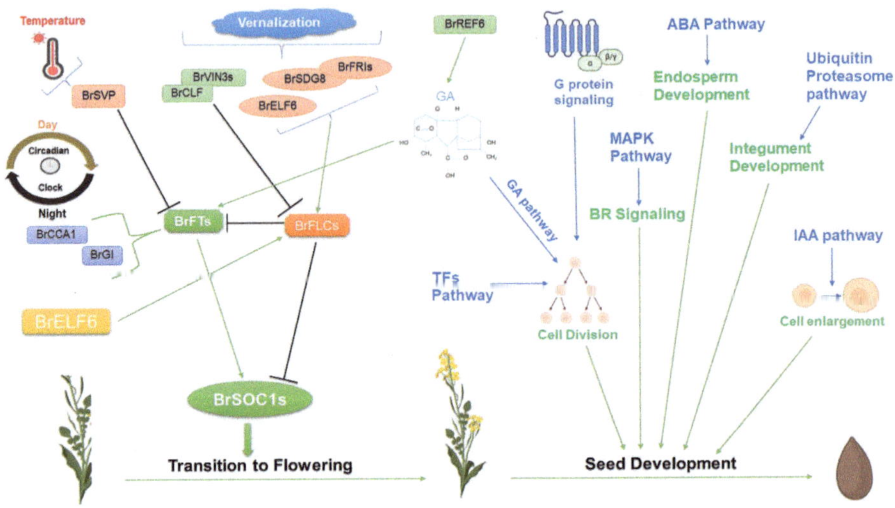

Fig. 5.4 Regulation of flowering transition in *Brassica rapa*. (Adopted from Li et al. 2024)

long-day conditions, which results in increased flowering. The primary method that GA encourages flowering is by triggering FT expression in leaves during long days. By substantially promoting the expression of SOC1 and LFY, GA may also speed up flowering during short days. When the ideal temperature is reached, the thermosensory pathway triggers FT. A number of microRNAs (miRNAs) and SQUAMOSA PROMOTER BINDING-LIKE (SPL) TFs also contribute to the rise in flower competence that occurs with plant age. The premature flowering problem can be prevented by suppressing FT and LFY with the action of these miRNAs and TFs. By modifying the period of flowering transition of various cultivars, we can enhance crop tolerance and versatility through molecular breeding, which may be facilitated by recognising essential genes and QTL that regulate flowering time (Li et al. 2024) (Fig. 5.4). Along with flowering, seed development is also an important event in a plant's life cycle. In order to create the suppression subtractive cDNA libraries of the *B. napus*, genes associated with the development of seeds have been examined at various pattern development and maturity stages. Five tested genes were detected tissue-specifically in root, stem, leaf, flower, bud, pod and embryo tissues using RT-PCR. The major functions of the differentially expressed genes, as determined by the COG functional classification database, are signal transmission, post-transcriptional alterations, lipid and amino acid metabolism and other processes. The five differentially expressed genes detected by RT-PCR results showed that two of them were implicated in lipid and protein metabolisms and seed development, two of them might be related to signal transduction, and one of them did not correspond to the homologous sequences that had been identified at the time and turned out to be probably a new gene (Huang et al. 2011).

5.4 Stress Physiology and Molecular Mechanisms

5.4.1 *Abiotic Stress Responses: Drought, Salinity and Temperature Extremes*

The hazardous scenario imposed by climate change highlights the requirement to create agricultural crops which are resilient to stress while enhancing quality and output. A number of abiotic stresses have affected *Brassica* yield parameters. A variety of stresses drastically lowers the productivity of crops and production of biomass, which eventually impacts crop survival. Plant viability is hampered by salinity imbalances that affect cellular ions, leading to osmotic stress and ion toxicity. Reactive oxygen species (ROS), including singlet oxygen, hydroxyl radical ($OH^{\cdot}$), hydrogen peroxide (H_2O_2) and superoxide anion ($O_2^{\bullet}$), are produced in response to salinity stress, which results in oxidative stress. ROS production causes alteration to various metabolic pathways and damages biomolecules (nucleic acid, protein and lipids), also leading to an enzymatic imbalance. Photosynthetic capacity, area of leaf, relative water content, stomatal conductance, cell membrane stability, ROS generation and chlorophyll damage are some of the activities that are adversely affected. Researchers further discovered that rapeseed during salt stress exhibits a reduction in shoot and root length, leaf area and biomass. It has been discovered that the rapeseed crop can withstand salt exposure to a moderate level. Under varying saline concentrations, a notable influence was noted on the percent seed germination, shoot length, growth parameters, protein and oil content of *Brassica* (Raza 2021).

Rapeseed is particularly vulnerable to drought stress, and it is predicted that changes in the global environment that cause severe and protracted drought in some areas of the biosphere will reduce rapeseed production. Plants typically respond to drought stress in a variety of ways, including physiological, biochemical, molecular and cellular alterations meant to validate their presence. Drought physiologically results in changes to respiration, transpiration, leaf turgor, chlorophyll levels and osmotic changes, which together reduce photosynthesis and stomata closure. Production of proteins and the accumulation of stress-related metabolites may biochemically represent immune cell homeostasis. Researchers found that irrigation during primary vegetative growth, flowering, and at the time of seed production resulted in the maximum seed output in rapeseed. The productivity of rapeseed is significantly impacted by drought during pod filling or flowering, as the reproductive stage is most sensitive. Research indicates that the flowering stage and stem elongation are particularly vulnerable to dryness, which may result in decreased yields. The productivity decline from drought during the silique growth period was found to be 21%, but at the flowering stage it was 30%. Similar to this, a decrease in yield of about 30% was expected when drought stress persisted throughout the stages of pod formation and blooming. Similarly, increased temperature also affects various physiological processes, including photosynthesis, carbon assimilation and causes oxidative injury. It has been reported that heat stress during flowering alters gametogenesis, embryo sac variation, fertilisation, and endosperm and embryo development, hence significantly reducing seed output. Water stress and high temperatures frequently happen at the same time. Furthermore, *Brassica* is sensitive to low temperatures and freezing in the early morning during winter. Intracellular ice crystals are formed due to very low temperature, which affects new seedlings, flowering and seed formation. The positive association between biomass and grain harvest revealed that dry mass is the constantly changing component that drives grain development. In order to achieve increased output during temperature stress, increased translocation of dry mass to the grain could prove an appropriate approach (Raza 2021).

5.4.2 *Role of Transcription Factors, MicroRNAs and Epigenetic Modifications in Stress Adaptation*

Rapid change in climate resulting in various abiotic stresses that pose a serious threat to the *Brassica* crops production worldwide. Under various stressors, plants activate a complicated signalling cascade to produce distinct defence mechanisms. Plants that are subjected to stress are encouraged to activate protein kinase cascades, maintain the activation of particular ion channels, or produce phytohormones, TFs, miRNAs, ROS, and other chemicals. Plant cell sensors that detect stressors are present in the cell membrane or cell wall. When the cell membrane receives signals, the cells attempt to respond by generating ROS, calcium ions, inositol phosphate, and other substances within the cell. Furthermore, several TFs are activated or suppressed by protein kinase-mediated phosphorylation, interact selectively with cis-acting regions in the promoters of stress-responsive genes and regulate their expression for defence against stress. Various TFs and microRNAs play an important role in stress adaptations. In *B. napus*, 2167 TFs from five distinct families were found out of which 7.6% are novel and exclusive to *B. napus*. Seventy percent of the differentially expressed genes found to respond to various abiotic stressors, including 62 BnAP2/EREBPs, 32 BnbZIPs, 61 BnMYBs, 40 BnNACs and 27 BnWRKYs, were responsive to heat stress. The responses of WRKY, MYB and Alfin-like TFs to various abiotic stressors in *B. rapa* have been extensively described. BrWRKY22, BrWRKY44 showed increased expression under cold stress. In comparison to the control, BrWRKY17, 57 and 58 exhibited increased expression during salt stress conditions, whereas BrWRKY51, 65, 98 and 104 exhibited the greatest expression following drought stress treatment. *B. rapa*'s viability assurance gene and other BZR TFs demonstrated resilience to ABA, cold and drought stresses. It has been documented that various Alfine-like TFs (BoAL1, 4, 5, 6, 7, 8, 9, 10 and 12) in *B. oleracea* react to abiotic stresses. Auxin treatment considerably decreased the expression of six MYB genes, while ABA treatment substantially boosted the expression of these genes (BrMYB172, BrMYB208, BrMYB210, BrMYB140, BrMYB137 and BrMYB229). Despite having no reaction to cold stress, BrMYB261 (an ortholog of AtMYB28) displayed elevated expression in response to osmotic stress. BrMYB210, BrMYB137, BrMYB88, BrMYB154 and BrMYB222 all showed major responses to both cold and osmotic stress. Numerous genes and TFs have been shown to have variable expression in response to various biotic stressors. The levels of expression of TF-BrAL2, 3, 4, 7, 9, 10, 13, 14 and 15 were elevated in *Brassica* spp. upon infection with *Fusarium oxysporum* f. sp. *conglutinans*, and BrAL2, 3, 7, 9, 13, 14 and 15 have been detected when exposed to both biotic and abiotic stresses. Furthermore, the chitinase genes (BrCLP1, BrCLP2 and BrCLP3) respond to infections caused by *Pectobacterium carotovorum* subsp. *carotovorum*. In contrast, two genes (BoAL8 and BoAL12) that resembled Alfin were activated in *B. oleracea* following infection with *P. carotovorum* subsp. *carotovorum*. Nevertheless, six BrWRKY genes (BrWRKY4, 65, 72, 97, 133 and 141) exhibited noticeably increased expression upon infection with *F. oxysporum* f. sp. *conglutinans*. Nearly 180 times more BrWRKY141 was expressed following contamination with *P. carotovorum* subsp. *Carotovorum* (Kayum et al. 2016). In *B. juncea*, *E. cruciferarum* triggers the LAR and SAR pathways, resulting in assistance in the selection of putative candidate genes such as PR1, PR2 and PR3 for developing resistance against disease (Ali et al. 2017).

Furthermore, via controlling gene expression, non-coding RNAs contribute to resilience against stress in addition to their other functions in various biological processes. miRNAs, the most well-characterised small RNAs in defence mechanisms, are among other small RNAs. The small RNA population of *B.* spp. is not well understood, despite the fact that numerous studies have found both conventional and unique miRNAs across a variety of model plants, including rice and *Arabidopsis*, when exposed to various abiotic stressors. In *B. rapa*, it was discovered that five conserved miRNA families and four of the new ones responded to temperature stress and the associated putative genes seemed either negatively or positively regulated. Two types of *B. napus* (YN171 and 93275) were found to have 22 methylase genes and six demethylase genes. Evaluation of gene expression at various stages of development in 22 tissues indicated that such genes have distinct regulatory functions in development, along with the variations in methylation patterns seen at different contexts (CHG, CHH and CG) and sequence positions (gene areas and up-/downstream regions) (Kourani et al. 2022). Apart from TFs and miRNAs, epigenetic controls are crucial in adjusting how plants express their genes in response to specific external stimuli. Epigenetics has emerged as a promising subject for comprehending the intricate processes underlying stress-related adaptability. Histone modification, DNA methylation, and microRNA-mediated DNA methylation are examples of epigenetic phenomena. In response to cell conditions, epigenetic modifications regulate the nuclear gene expression processes during plant developmental phases. At CpCpGpG sites, *B. napus* salt stress caused a rise in the cytosine methylation process, accompanied by de novo methylation and demethylation processes. Nevertheless, during heat treatment, it was observed that the sensitive genotypes of *B. napas* had higher levels of DNA methylation than the intolerant ones. When *Brassica* sp. was subjected to heat stress, the tolerant cultivar displayed hypomethylation while the sensitive cultivar displayed hypermethylation (Rehman and Tanti 2020) (Table 5.1).

Table 5.1 Some examples of epigenetic applications and results used to determine stress-related adaptations in certain *Brassica* species

Plant spp.	Transcription factors	Stress	Result	Reference
Brassica rapa ssp. *pekinensis*	BrbZIPs	Drought	Improve plant tolerance	Eom et al. (2018)
B. napus	BnERF039 and BnERF245	Salt stress	Improve salt tolerance	Owji et al. (2017)
B. napus	BnERF227, BnERF228, BnERF234, BnERF134, BnERF132, BnERF176, and BnERF235	*Leptosphaeria maculans* and *L. biglobosa*	Enhance resistance against fungal infection	Owji et al. (2017)
B. napus	*BnMRD107*	Osmotic stress	Improved resistance to osmotic stress	Li et al. (2021a)
B. rapa L.	*DREB1* and *DREB2*	Drought and heat	Improve resistance	Gong et al. (2024)
microRNAs				
B. napus	miR398	Cd stress	Involved in a pathway for detoxification of superoxide radicals	Fu et al. (2019)
B. rapa	*mir168a*	Salinity stress	Salinity tolerance is achieved by regulating melatonin levels, preventing osmolytes from building up, and preserving ion homeostasis	Shamloo-Dashtpagerdi et al. (2022)
B. napus	miR169n	Drought	Improved ability to withstand drought stress in seedlings and osmotic stress in the post-germination stage	Li et al. (2021b)
B. napus	miR159, miR5139, and miR39	*Sclerotinia sclerotiorum* infection	Increase expression of stress response genes	Regmi et al. (2021)

5.5 Genetic Resources and Biotechnological Interventions

5.5.1 *Germplasm Diversity and Conservation Strategies*

Conserving and propagating *Brassica* genetic material is crucial to accelerating global studies, and plant genetic resource centres or gene banks are responsible for ensuring a sufficient supply of germplasm. Additionally, considering the number of genes available for *Brassica* crops is becoming smaller because of ongoing breeding for fewer specific traits and the general decline in biological diversity brought on by climate change, gene banks are a great option for conserving plants' genetic resources, such as cultivars, crop wild relatives (CWR) and landraces. Gene banks are primarily responsible for maintaining the genetic variety of particular plant varieties and distributing them to farmers and plant investigators under the current agricultural environment (Subramanian et al. 2023). The national agricultural enhancement initiatives require an extensive genetic foundation with germplasm of various genetic origins, primarily from the centres of crop variety. Therefore, plant investigation, collection and introduction efforts are continuously used to increase the ancestral material of *Brassica* in order to further develop its genetic resources. Germplasm should ideally be assessed for disease and pest responses at the hot spot sites, as well as for key morpho-agronomic variables across various agroclimatic circumstances. Any effort aimed at improving crops has made use of germplasm that has been maintained and conserved. This means that it can be stored for 40–60 years at −18 to −20 °C. To conserve *Brassica* genetic material, various storage methods such as in-vitro collection, DNA collection and cryopreservation are also utilised; however, they are rarely utilised. Additional study on protected genetic material is necessary in order to comprehend the features and attributes of seeds that are kept in storage, as well as to enhance the storage procedures that are already in use. Additional research on *Brassica* species that have been conserved in gene banks focuses on the changes in biochemistry which take place in seeds over an extended period of preservation and the metabolite screening of collections. Research on the effects of membrane permeability and seed lipid peroxidation in Brassicaceae members after prolonged storage on germination and vigour revealed that higher membrane permeability was elevated in *Brassica* seeds, resulting in lower germination rates and vigour loss. According to data, preserved *Brassica* is dispersed randomly among nations and is heavily focused on a limited number of species. A great way to maintain *Brassica* biodiversity for scientific, breeding and commercial uses is through plant gene banks (Singh and Sharma 2007).

5.5.2 CRISPR/Cas9 and Genome Editing Application in Brassica

Molecular breeding of crops requires genome editing. Reverse genetics is used to understand gene function by using genome sequencing, gene editing and mutagenesis-like techniques. Nowadays, it combines with modern gene editing techniques like CRISPR-Cas9, RNAi and omics technologies to gain insights into the gene functions and is applied to improve the crop traits for better agricultural productivity (Ben-Amar et al. 2016; Gurumurthy et al. 2016). The genome editing method mediated by CRISPR (clustered regularly interspaced short palindromic repeats) and Cas9 has been used extensively as a potent tool for altering desirable endogenous genes. Research has demonstrated that, via altering the biochemical route in *Brassica*, the CRISPR/Cas9 system can be applied to create beneficial mutant crops with an agronomically appropriate phenotype. In order to create stable, hereditary knockdown variants in important crops, CRISPR/Cas9 technologies have the ability to concurrently alter many genes in addition to a particular gene. The possible application of CRISPR/Cas9 to simultaneously alter genes in a polyploid species. The specific genes are rapeseed continue to vary as CRISPR/Cas9 technological applications advance; however, the primary emphasis is on the genetic enhancement of economically significant agricultural properties (such as yield, nutritional value and stress tolerance). The primary objective of rapeseed modification is to enhance production and nutritional value, which is easily achieved via CRISPR/Cas9. The extremely poor transformation performance of *Brassica rapa* var. *rapa* has prevented the editing of genes from being detected. Using CRISPR/Cas9 gene-editing technology, 20 BrrTCP4b-edited plants were produced, increasing the quality of leaf trichomes and enhancing regeneration capacity (Liu et al. 2022).

5.5.3 *Transgenic Approaches for Improved Traits: Biofortification, Pest Resistance and Herbicide Tolerance*

The term "biofortification" or "biological fortification" describes agricultural crops which have been produced and cultivated utilising traditional plant breeding, current biotechnology approaches and agricultural approaches to produce nutrition-enriched food crops that have higher accessibility to the people of the world. There are three primary methods for biofortifying vital micronutrients into agricultural plants: transgenic, conventional and agronomic. These methods involve the application of fertilisation techniques, crop breeding and biotechnology, respectively. Agronomical, conventional and transgenic methods mostly target primary crops such as rice, sorghum, common beans, wheat, potatoes, sweet potatoes, maize and tomatoes. While transgene and agronomic methods have been used to biofortify barley, soybean, lettuce, carrot, canola and mustard, transgenic and breeding methods have been used to biofortify cassava, cauliflower, and banana. Transgenic methods are aimed at more crops, but breeding methods have made biofortification more widely used in practice (Garg et al. 2018). The expression of the FAD3 enzyme caused the transgenic mustard to produce gamma linoleic acid (Hong et al. 2002).

Cultivar tolerance and chemical treatment were thought to be more successful approaches to disease management of insect pests. Chemical treatments are expensive and pose health or environmental risks in the cultivation of *Brassica*. Adopting cultivars that resist has been seen as the most cost-effective and ecologically responsible approach to pest control. As genetic engineering has advanced, heterologous genes are currently used to improve the majority of commercially valuable *Brassica* species. Toxic proteins to fungal infections include the pathogenesis-related (PR) protein. Glucan, a key component of pathogenic fungi's cell walls, can be hydrolysed by glucanase, a plant defence mechanism. By introducing a tomato class I basic glucanase, transgenic plants of *B. juncea* were found to be tolerant to *Alternaria brassicae*. Protein taken from separate transformants stopped *A. brassicae*'s hyphal development by 15–54%. Comparing transgenic strains to wild type, the total number, size and dissemination of lesions induced by *A. brassicae* were limited during polyhouse bioassay (Mondal et al. 2021). Broad herbicides are typically employed to suppress weeds in mustard fields, killing the mustard plants at the same time. Therefore, some optimism for weed management, especially for noxious weeds like *Orobanche aegyptiaca*, is provided by the creation of transgenic Indian mustard with an herbicide-tolerance trait. While the transgenic mustard plants will not be impacted by the herbicide-tolerance characteristic expressed by the transgene, the weedy plants will be killed by a spraying of the target herbicide.

5.6 System Biology and Functional Genomics

Bioinformatics is a multidisciplinary field of science and technology that involves the use of computers to analyse the huge biological data and information. Ben Hesper and Paulien Hogeweg created bioinformatics in 1970 as "the study of informatic processes in biotic systems". Recent developments in genomic and proteomic studies have generated huge biological data that require sophisticated tools to analyse and conclude some information. These tools can help in the identification of candidate genes, understanding the functional and metabolic importance of the genes, genome assembly and annotation, three-dimensional structure variations, evolutionary history and gene expression. These tools help in the prediction of traits that are related to high yield, involved in stress tolerance and disease resistance, and ultimately helping in improving quality, productivity and stress resilience of *Brassica* crops. A study of the complete genome, transcriptome, metabolome and proteome clears the path for all future research of any organism and is important to interpret the molecular information. Proteins are very important molecules that regulate the functions of cells/organs, and information about protein function and structure is deposited in specialised databases like PDB (Protein Data Bank). There are different tools to analyse the proteins. To study the evolutionary history of proteins/genes/organisms, PHYLIP (Phylogeny Inference Package) and MEGA (Molecular Evolutionary Genetics Analysis) are used. To simulate the protein's 3-dimensional structure, SWISS-MODEL, Phyre, MODELLER (MDLR), AlphaFold Protein Structure Database and I-TASSER (Iterative Threading ASSEmbly Refinement) are commonly employed (de Almeida et al. 2022). ShinyGO provides a graphical gene-set enrichment tool for animals and plants, useful for gene ontology. Multi-omics datasets studies resulted in analysis of genetic material like transcriptome-wide association studies (TWAS), genome-wide association studies (GWAS), expression quantitative trait locus mapping (eQTL) and identification of candidate genes that are involved in desirable traits. Due to the complex and special evolutionary history of *Brassica*, it makes a polyploid genome and more complex sequence composition than diploid crops. Different databases have been constructed to compile the molecular information of *Brassica*, like BnPIR: *Brassica napus* pan-genome information resource (www.cbi.hzau.edu.cn/bnapus/), BnVIR: *Brassica napus* variation information resource (www.yanglab.hzau.edu.cn/BnVIR), BnTIR: *Brassica napus* transcriptome information resource (www.yanglab.hzau.edu.cn/BnTIR) and *Brassica* database (BRAD) (www.brassicadb.org/brad/). These databases and tools are used for predicting protein functions and structure, predicting conserved domain gene structure and annotating proteins. The bioinformatics analysis of the *Brassica* genome and proteome identifies the genes involved with high yield, seed germination and growth that can be further used to enhance the quality and quantity of the crop (Mu et al. 2022). The advancement of digital technologies and artificial intelligence has led to the development in the use of automated robots in agricultural fields, which has changed the time-consuming and labour-intensive practices. In order to irrigate plants at specific times, the irrigation system uses a NodeMCU ESP32 microcontroller to gather environmental information like humidity, temperature and moisture in the soil via detectors. It additionally features an infrared passive sensor to identify trespassers near the farmland and alert the cultivator to any dangerous circumstances, like extremely high or low temperatures. Unpredictable events and a constantly changing climate of the different diseases severely affect the crop yield and grain quality, so early disease detection is crucial for crop protection. Machines are trained by using different datasets to recognise the different disease conditions and crop monitoring, with the help of ML (machine learning) and DL (deep learning) techniques (Singla et al. 2024).

References

Ali S, Mir ZA, Tyagi A et al (2017) Identification and comparative analysis of *Brassica juncea* pathogenesis-related genes in response to hormonal, biotic and abiotic stresses. Acta Physiol Plant 39:268

Ben-Amar A, Daldoul S, Reustle G et al (2016) Reverse genetics and high throughput sequencing methodologies for plant functional genomics. Curr Genomics 17(6):460–475

Cheng F, Sun R, Hou X et al (2016) Subgenome parallel selection is associated with morphotype diversification and convergent crop domestication in *Brassica rapa* and *Brassica oleracea*. Nat Genet 48:1218–1224

de Almeida PV, de Souza GI, Monteiro CR et al (2022) Protein structural bioinformatics: an overview. Comput Biol Med 147:105695

del Carmen M, Moreno DA, Carvajal M (2013) The physiological importance of glucosinolates on plant response to abiotic stress in *Brassica*. Int J Mol Sci 14(6):11607–11625

Ekblom R, Wolf JB (2014) A field guide to whole-genome sequencing, assembly and annotation. Evol Appl 7(9):1026–1042

Eom SH, Baek S-A, Kim JK, Hyun TK (2018) Transcriptome analysis in chinese cabbage (*Brassica rapa* ssp. *pekinensis*) provides the role of glucosinolate metabolism in response to drought stress. Molecules 23(5):1186

Fu Y, Mason AS, Zhang Y et al (2019) MicroRNA-mRNA expression profiles and their potential role in cadmium stress response in *Brassica napus*. BMC Plant Biol 19:570

Garg M, Sharma N, Sharma S et al (2018) Biofortified crops generated by breeding, agronomy, and transgenic approaches are improving lives of millions of people around the world. Front Nutr 5:12

Gong Y, Qiu Z, Hou W et al (2024) Characteristics of rapeseed (*Brassica rapa* L.) genome DREB family demonstrate their roles in stress. Plant Stress 13:100536

Gurumurthy CB, Grati MH, Ohtsuka M et al (2016) CRISPR: a versatile tool for both forward and reverse genetics research. Hum Genet 135:971–976

Hong H, Datla N, Reed DW et al (2002) High-level production of γ-linolenic acid in *Brassica juncea* using a Δ6 desaturase from pythium irregular. Plant Physiol 129(1):354–362

Huang JY, Jie ZJ, Wang LJ et al (2011) Analysis of the differential expression of the genes related to *Brassica napus* seed development. Mol Biol Rep 38:1055–1061

Johnston JS, Pepper AE, Hall AE et al (2005) Evolution of genome size in Brassicaceae. Ann Bot 95(1):229–235

Kayum MA, Kim HT, Nath UK et al (2016) Research on biotic and abiotic stress related genes exploration and prediction in *Brassica rapa* and *B. oleracea*: a review. Plant Breed Biotechnol 4(2):135–144

Kitainda V, Jez JM (2021) Structural studies of aliphatic glucosinolate chain-elongation enzymes. Antioxidants 10(9):1500

Kourani M, Mohareb F, Rezwan FI et al (2022) Genetic and physiological responses to heat stress in *Brassica napus*. Front Plant Sci 13:832147

Li J, Lin K, Zhang S et al (2021a) Genome-wide analysis of myeloblastosis-related genes in *Brassica napus* L. and positive modulation of osmotic tolerance by BnMRD107. Front Plant Sci 12:678202

Li J, Duan Y, Sun N et al (2021b) The miR169n-NF-YA8 regulation module involved in drought resistance in *Brassica napus* L. Plant Sci 313:111062

Li N, Yang R, Shen S, Zhao J (2024) Molecular mechanism of flowering time regulation in *Brassica rapa*: similarities and differences with *Arabidopsis*. Hortic Plant J 2024:2024

Liu Y, Zhang L, Li C et al (2022) Establishment of *Agrobacterium*-mediated genetic transformation and application of CRISPR/Cas9 genome-editing system to *Brassica rapa* var. *rapa*. Plant Methods 18:98

Manan S, Chen B, She G et al (2016) Transport and transcriptional regulation of oil production in plants. Crit Rev Biotechnol 37(5):641–655

Miklaszewska M, Zienkiewicz K, Inchana P, Zienkiewicz A (2021) Lipid metabolism and accumulation in oilseed crops. OCL 28:50

Mondal HA, Paul B, Gurung A, Mallick M (2021) Factors involved in enhancing host susceptibility towards aphid clonal propagation on leaf foliage of *Arabidopsis*. Curr Sci 121(8):1080–1089

Mu H, Wang B, Yuan F (2022) Bioinformatics in plant breeding and research on disease resistance. Plants 11(22):3118

Owji H, Hajiebrahimi A, Seradj H, Hemmati S (2017) Identification and functional prediction of stress responsive AP2/ERF transcription factors in *Brassica napus* by genome-wide analysis. Comput Biol Chem 71:32–56

Paterson AH, Lan T, Amasino R et al (2001) *Brassica* genomics: a complement to, and early beneficiary of, the *Arabidopsis*sequence. Genome Biol 2(reviews1011):1

Raza A (2021) Eco-physiological and biochemical responses of rapeseed (*Brassica napus* L.) to abiotic stresses: consequences and mitigation strategies. J Plant Growth Regul 40:1368–1388

Regmi R, Newman TE, Kamphuis LG et al (2021) Identification of *Brassica napus* small RNAs responsive to infection by a necrotrophic pathogen. BMC Plant Biol 21:366

Rehman M, Tanti B (2020) Understanding epigenetic modifications in response to abiotic stresses in plants. Biocatal Agric Biotechnol 27:101673

Shamloo-Dashtpagerdi R, Lindlöf A, Tahmasebi S (2022) Evidence that miR168a contributes to salinity tolerance of *Brassica rapa* L. via mediating melatonin biosynthesis. Physiol Plant 174(5):e13790

Shaw RK, Shen Y, Wang J et al (2021) Advances in multi-omics approaches for molecular breeding of black rot resistance in *Brassica oleracea* L. Front Plant Sci 12:742553

Singh R, Sharma SK (2007) Evaluation, maintenance, and conservation of germplasm. Adv Bot Res 45:465–481

Singla A, Nehra A, Joshi K et al (2024) Exploration of machine learning approaches for automated crop disease detection. Curr Plant Biol 2024:100382

Subramanian P, Kim SH, Hahn BS (2023) *Brassica* biodiversity conservation: prevailing constraints and future avenues for sustainable distribution of plant genetic resources. Front Plant Sci 14:1220134

Tan Z, Han X, Dai C et al (2024) Functional genomics of *Brassica napus*: progress, challenges, and perspectives. J Integr Plant Biol 66(3):484–509

Waminal NE, Sampath Perumal SP, Lee JongHoon LJ et al (2016) Repeat evolution in *Brassica rapa* (AA), *B. oleracea* (CC), and *B. napus* (AACC) genomes. Plant Breed Biotechnol 4(2):107–122

Wu J, Liang J, Lin R et al (2022) Investigation of *Brassica* and its relative genomes in the post-genomics era. Hortic Res 25(9):uhac182

www.bioinformatics.babraham.ac.uk/projects/fastqc. Accesssed 15 Feb 2025

www.brassicadb.org/brad/. Accesssed 15 Feb 2025

www.cbi.hzau.edu.cn/bnapus/. Accesssed 15 Feb 2025

www.rbgkew.org.uk/cval/homepage.html. Accesssed 15 Feb 2025

www.yanglab.hzau.edu.cn/BnTIR. Accesssed 15 Feb 2025

Xue JY, Wang Y, Chen M et al (2020) Maternal inheritance of U's triangle and evolutionary process of *Brassica* mitochondrial genomes. Front Plant Sci 11:805

Chapter 6
Propagation and Production of *Brassica*

6.1 Introduction

Brassicas, encompassing vital crops like *Brassica juncea*, *B. napus* and *B. rapa*, are integral to global agriculture, serving as a primary source of edible oils, vegetables and fodder. This chapter deals with the propagation and production of Brassicas, highlighting their significance in sustainable farming systems and addressing food security challenges. Propagation techniques, including seed propagation and vegetative methods like tissue culture, are explored,` with a focus on improving efficiency and scalability. Furthermore, strategies for managing biotic and abiotic stresses—major constraints in *Brassica* cultivation—are presented, incorporating conventional breeding to tissue culture and CRISPR techniques. Emerging trends, such as genetic improvements, advanced propagation technologies and their role in mitigating climate change impacts, are discussed, highlighting the scope for innovation. This chapter offers a comprehensive guide to *Brassica* cultivation, addressing challenges and proposing solutions for enhanced productivity and sustainability. This chapter aims to provide a resource for researchers, agriculturists and policymakers to optimise *Brassica* production systems while contributing to global food security.

One or two of the three genome types A, B and C are found in *Brassica* species. *B. rapa* (genotype AA), B. *juncea* (AABB), *B. nigra* (BB), *B. oleracea* (CC), *B. napus* (AACC) and *B. carinata* (BBCC) are among them. According to the "triangle of U" theory, which illustrates the relationships between the six *Brassica* species, the triangle's three vertices, which stand for the three fundamental genomes, AA, BB and CC, respectively, are occupied by the diploid species *B. rapa*, *B. nigra* and *B. oleracea*. Conversely, the neo tetraploid genomes of *B. juncea*, *B. napus* and *B. carinata*, which are all allotetraploids, are located at the three corners of the triangle as AABB, AACC and BBCC, respectively (Bhattacharya 2019). The Brassicaceae family comprises a large number of economically significant plants

M. Ozturk et al., *Climate-Resilient* Brassicas,
https://doi.org/10.1007/978-3-032-12709-9_6

that humans have significantly cultivated and modified, particularly those belonging to the genus *Brassica*, which includes turnip, rutabaga, kohlrabi, napa cabbage, kale, cabbage, broccoli and Brussels sprouts. White mustard, horseradish and radish are some of the other significant crops of Brassicaceae. Certain members of the family are regarded as invasive in areas outside of their natural habitat, while several species like basket-of-gold, candytuft and honesty are planted as ornamentals. According to Zheng and Liu (2022), the output of main *Brassica* crops, such as broccoli, mustard seeds, cauliflower, cabbage and seeds of rape, amounted to over 186 million tonnes in 2022, spanning around 44 million hectares, and had a combined worth of over $82 billion. Oilseeds, veggies and condiments are the three main categories into which Brassicas are divided according to the parts they are used. Economically important members of Brassicaceae are Brussels sprouts (*B. oleracea* var. *gemmifera*), collard (*B. oleracea* var. *acephala*), turnip (*B. rapa* var. *rapa*), cabbage (*B. oleracea* var. *capitata*), brown mustard (*B. juncea*), bok choy (*B. rapa* var. *chinensis*), napa cabbage (*B. rapa* var. *pekinensis*), kohlrabi (*B. oleracea* var. *gongylodes*), broccoli (*B. oleracea* var. *italica*), cauliflower (*B. oleracea* var. *botrytis*), rape (*B. napus* var. *napus*), kale (*B. oleracea* var. *acephala*) and rutabaga (*B. napus* var. *napobrassica*). In addition to being cultivated as a leaf food, additionally, Abyssinian mustard or cabbage is also used to produce biofuel as well as an oilseed crop. Perennial herbs like elongated mustard have the capacity to serve as a source of genes to enhance the nutritional value of rape crops. Twiggy turnip, sometimes known as Mediterranean cabbage, is an herb which may be used as a gene source to help kale become resistant to pests. In addition to being cultivated for their leaves, stems and seeds, Indian mustard, also known as green mustard, is utilised in phytoremediation. Rapeseed, commonly known as canola, is used to make biofuel, feed for animals and vegetable oil for consumption by humans. Spinach mustard, also known as spoon mustard, or rosette bok choy, is a leaf used as human food. Black mustard is an annual weedy plant. Ground seeds are commonly used as a spice. Wild cabbage is an important human and animal food crop plant. Tender green or mustard spinach is also used as human food. Chinese cabbage, also known as field mustard or turnip mustard, has leaves/roots and seeds (as a vegetable oil) that are utilised for livestock feed and human consumption. In addition to being utilised as food for humans, brown mustard may also be a source of genes that could improve the quality of kale. Sahara mustard, Asian mustard and African mustard are annual herbs that are frequently seen as invasive weeds (Petruzzello 2023). Among the highest-grown and consumed crop types worldwide is *Brassica*. The past few decades have seen an increase in its output. *Brassica* crops have a major positive impact on human health through consumption. *Brassica* vegetables are great plant-based providers of macronutrients that are necessary for daily dietary requirements. Taste, fragrance and pungency are among the sensory qualities that are strongly associated with some of their metabolites. *Brassica* crops offer a wide variety of macronutrients, such as proteins, soluble sugars, amino acids, organic acids and dietary fibres, but with the exception of those utilised to produce oil, they generally are not considered important sources of energy due to their small carbohydrate and fat concentration. Growing research indicates that because

Brassica vegetables have special phytonutrients that promote health, eating them may lower the risk of developing chronic illnesses, particularly some forms of cancer (Xiaomeng et al. 2025). In contrast to other vegetables, brassica veggies have the largest concentrations of healthy substances called glucosinolates (GLSs). A side chain made of amino acids, a sulfonated oxime moiety and a β-d-thioglucose group make up GLSs. The most prevalent aliphatic GLS in broccoli is glucoraphanin (GLR), which accounts for 20 to 80% of the total GLS content, 1-methoxy-glucobrassicin (5–30%) and glucobrassicin (10–40%). Moreover, additional indolyl GLSs include 4-methoxy-glucobrassicin (1–5%) (Sun et al. 2021).

6.2 Propagation Methods: Tissue Culture and Micropropagation

Seed is the traditional method of propagation for production, where seedlings are cultivated in beds or modules before being moved to field locations. Nevertheless, stem and shoot cuttings can be used to vegetatively grow certain *B. olerace*a species, such as tronchuda. This technique involves taking the stem and side branch cuttings from plants that are 5 weeks old, rooting them and then transplanting them as regular cuttings. The following are some benefits of employing micropropagation techniques: The method can be used to keep stocks free of disease after first eradicating viruses. Managing certain high-value lines, such as highly self-incompatible lines for hybrid production, is made easy using this method (Basak et al. 2012).

6.2.1 Tissue Culture and Micropropagation

Most *Brassica* species have well-reported requirements for in vitro regeneration. Sterilised seeds were put in culture tubes on MS media (Murashige and Skoog 1962) for seed germination. Media preparation includes MS (dehydrated plant tissue culture media kit) media from Himedia was used for tissue culture experiments. Composition of media (Murashige and Skoog 1962) is given in Table 6.1.

Along with the above ingredients, sucrose (30 g/L) and agar (8 g/L) were added prior to autoclaving the media, with the pH of media set at 5.8. Eight to 12 seeds were cultured inside each tube. Sterile surgical blades were utilised for cutting seedling leaves about 2–5 mm from 6–7-day-old plants into 12–15 pieces. Four leaf segments were then carefully inserted into each tube, which held sterile MS media containing different quantities and combinations of growth regulators such as $AgNO_3$, NAA (1-naphthaleneacetic acid) and 2, 4-D (2-dichlorophenoxyacetic acid). Explants were inoculated for 20–50 days before the calli reached the appropriate size for transferring to regenerating medium. They were then in an aseptic manner transferred with forceps from their original tubes to fresh tubes that contained sterile induction media. The laminar airflow cabinet was used as the location

Table 6.1 Media composition

MS basal media	
Ingredients	Concentration (mg/l)
Potassium nitrate	1900
Ammonium nitrate	1650
Calcium chloride.$2H_2O$	440
Magnesium sulphate	180.69
Potassium phosphate monobasic	170
Manganese sulphate.$2H_2O$	16.9
Boric acid	6.2
Potassium iodide	0.83
Molybdic acid.$2H_2O$	0.25
Zinc sulphate.$7H_2O$	8.6
Cooper sulphate.$5H_2O$	0.025
Cobalt chloride.$6H_2O$	0.025
$FeSO_4.7H_2O$	27.8
$Na_2EDTA.2H_2O$	37.3
Myo-inositol	100
Thiamine hydrochloride	0.1
Pyridoxine hydrochloride	0.5
Nicotinic acid (free acid)	0.5
Glycine (free base)	4
Total g/L	4.54

to perform the transfer. The subcultured media employed were MS (Murashige and Skoog 1962) medium that contained various concentrations and mixtures of $AgNO_3$, NAA and BAP (6-benzylaminopurine). In order to maintain calli and organogenesis, multiple subcultures were additionally carried out at intervals of 15 to 20 days and maintained at the same temperature as previously stated. The calli that had been subcultured kept growing and developing into shoots. After growing to a length of two to three centimetres, these shoots were separated from one another and cultivated once again separately on bottles filled with sterile root formation medium in order to stimulate rooting. Later, these plantlets were transferred into pots for acclimatisation and hardening and shifted to a polyhouse or greenhouse for further development (Basak et al. 2012). Regeneration of plant methods using direct organogenesis has been reported in numerous *Brassica* species. The emergence of a new organ, firstly a protrusion, at a location where only a parent structure is present, is known as organogenesis. The initial material used in regeneration was portions of roots of *B. juncea* (Sharma et al. 1990), stems (Barfield and Pua 1991), petioles (Ghnaya et al. 2007), hypocotyls of *B. napus* (Phogat et al. 2000), shoot tips and cotyledons of *B. oleracea* (Ravanfar et al. 2009, 2013, 2014). On the other hand, adventitious shoot and/or root production from callus is known as indirect organogenesis (inductive) (Krikorian 1995). Additionally, because somatic embryogenesis takes a long time, it might not be particularly popular. Genotype plays a crucial role in the embryogenic frequency of the majority of *Brassica* species employed in

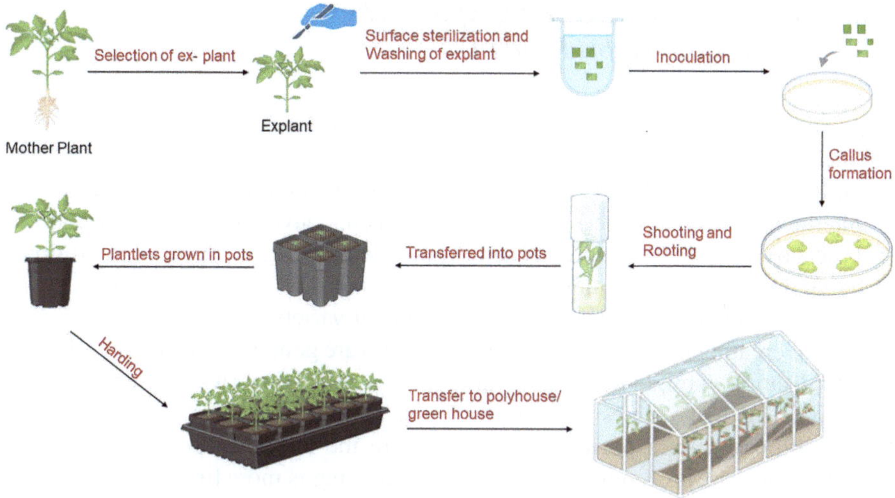

Fig. 6.1 Steps involved in plant tissue culture

organogenesis. Hypocotyls of *B. napus* have been used to promote somatic embryogenesis (Kohlenbach et al. 1982), immature cotyledons (Turgut et al. 1998) and protoplast-derived colonies of cabbage (Choi et al. 1996). Growth media, environmental conditions, explant sources, explant age, plant growth regulators and genetics are some of the major aspects that affect the success and results of tissue culture procedures (Fig. 6.1).

6.2.2 Breeding and Genetic Improvement

One effective method for increasing output and creating cultivars that are resistant to biotic (disease and insect) and abiotic stresses is the creation of genetic diversity-focused crop breeding. In breeding, separate progeny were selected for specific desirable features, resulting in hybridisation. Agricultural techniques have undergone a number of changes, such as the invention and introduction of selective herbicides, fungicides and pesticides, as well as the enhancement of fertiliser dosage for improved growth. Crops with high production value have the ability to change the composition of crops. The Brassicaceae family's genetic variety during the last several centuries has produced numerous crops that are now widely consumed, including oilseed rape (*B. napus*), field mustard (*B. campestris*), Ethiopian mustard (*B. carinata*), cabbage (*B. oleracea*), turnip (*B. rapa*) and mustard greens (*B. juncea*). Over the past 2500 years, plant breeding has resulted in the production of more than six agricultural crops, each of which differs greatly from its wild ancestor in terms of shape, colour and nutrients in order to serve various decorative and/or consumption needs (Fahey 2016).

6.2.3 *Conventional Breeding Approaches (Selection, Hybridisation)*

In conventional breeding, novel plant varieties are created despite the assistance of advanced molecular breeding technologies (Ozturk et al. 2018, 2024; Rehman et al. 2020; Ruqia et al. 2020). The traditional breeding method does not violate the law of inheritance. Using a selective breeding approach, conventional breeding chooses traits according to their exceptional achievement on specific attributes. In traditional breeding, breeders use the methods of crossing (hybridisation) to combine desired features from many, but typically produce a novel variety from related plants. The process of creating hybrids by mating parents who are genetically different is known as hybridisation. It might involve the same species, individuals with different genetic makeup or distinct species and can be either natural or man-made. Intraspecific hybridisation and intraspecific hybridisation are the two most prevalent forms of hybridisation. The procedure of conventional breeding is more time-consuming and often takes over 10 years to produce a new cultivar. Traditional breeding relies too much on phenotypic appearances to select better varieties. As genotype by environment interactions have a significant impact on phenotypes, chosen cultivars are therefore not error-free. The most popular method of crop development is still selection, which is the oldest method. Through years of selection amid the natural biological variety that existed, people progressively but methodically selected traits that made plants more useful. Since selection is focused on phenotype, a field-grown plant to a land race may have similar physical characteristics yet have a very different genetic makeup (Acquaah 2015). Conventional breeding for self-pollinated crops includes pure-line selection and backcross breeding. One of the oldest techniques for breeding plants is selection within the same population or pure line selection. It is used frequently in self-pollinated crops. Crossing an F1 hybrid with one of the homozygous parents to get offspring who are exactly like the parent is known as backcrossing. It is a method for producing hybrids with traits which are more desirable than those of the original parents. On the other hand, recurrent selection—the method that involves reselecting favourable traits generation after generation to raise their frequency solely through crossings among high-performing individuals—is a component of traditional breeding for cross-pollinated crops. In maize, recurrent selection was first and extensively used; later, it was also utilised for cereals, millet, and soybeans. Heterozygous is the source population of recurrent selection. Open-pollinated, single- or double-crossed F1 hybrids, interbred offspring of specific inbred lines, synthetic or composite are some possible varieties. Recurrent selection is of three types; simple recurrent selection, recurrent selection for combining ability, and reciprocal recurrent selection. Furthermore, crops which show self-pollination and cross-pollination are both subject to mass selection. Mass selection is used to increase the frequency of favourable features and boost core population performance (Lamichhane and Thapa 2022).

6.2.4 *Advances in Molecular Breeding and Biotechnology*

The uses of molecular marker systems have altered the future development of plants due to advances in technology (Ozturk et al. 2018, 2024; Rehman et al. 2020; Ruqia et al. 2020). Advances in sequencing technology have been directed towards crop improvement through the development of various techniques and applications of molecular markers (Fig. 6.2). These techniques are currently being used on a variety

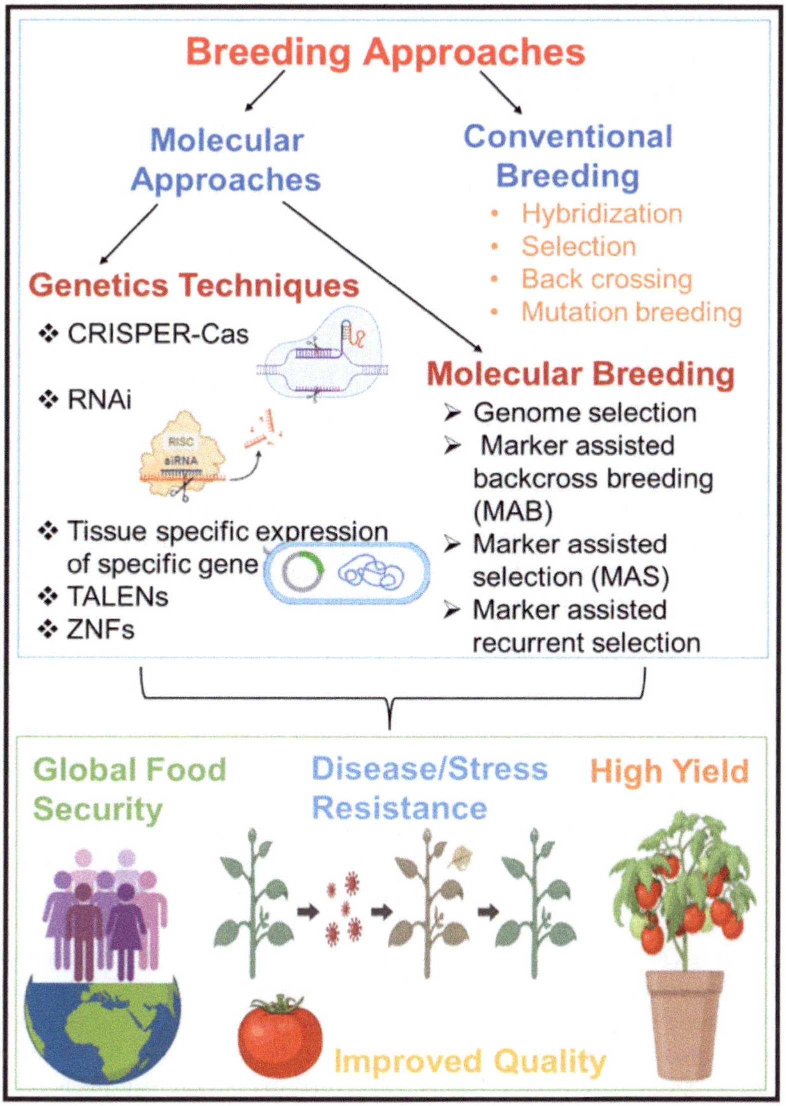

Fig. 6.2 Breeding approaches

of crops and show promise, especially for oilseed crops, in regard to the total output of food and non-food products as well as the nutritional quality of the oils. Increasing total oil yield and its quality, which encompasses a variety of factors, is one of the goals in this context. Molecular breeding (MB), genetic engineering (GE) or recombinant DNA technology, allele mining, linkage and association mapping, genome-wide selection (GWS), gene pyramiding, marker-assisted recurrent selection (MARS) and next-generation sequencing (NGS) are some of the molecular biology techniques used for crop improvement. In contrast, biotechnological methods such as plant tissue culture, direct DNA manipulation for gene transfer, Agrobacterium-mediated gene transfer, mutagenesis, RNA interference, next-generation sequencing, bioinformatics tools and nanotechnology are crucial for improving crops and preventing disease (Rathore et al. 2018).

Ideal DNA markers for effective usage in marker-assisted breeding should satisfy the following requirements, depending on the species and application: an elevated degree of polymorphism, uniform dispersion throughout the entire genome (not grouped in specific areas), expression co-dominance (to differentiate heterozygotes from homozygotes), clearly defined allelic characteristics (to facilitate the easy identification of the many alleles), there is only one copy and no pleiotropic impact, low cost of use (or economical genotyping and marker development), simple automation and assay/detection, high availability (unrestricted use) and suitability for multiplication or duplication (to allow for the accumulation and sharing of data throughout laboratories), unique to each genome (particularly in polyploids), and no negative impact on phenotypic (Adlak et al. 2019). Numerous DNA molecular markers have been created and effectively used in breeding and genetics research for a range of agricultural crops. PCR-based molecular markers and hybridisation-based molecular markers are the two primary types of DNA markers, which comprise many sets of markers. The molecular marker known as restriction fragment length polymorphism (RFLP) is based on hybridisation. Additionally, PCR-based molecular markers include single nucleotide polymorphism (SNPs), micro-satellites or short tandem repeats (STRs), single-strand conformation polymorphism (SSCP), random amplified polymorphic DNAs (RAPD), amplification length polymorphism (ALP), simple sequence repeat (SSRs), amplified fragment length polymorphism (AFLP), sequence characterised amplified regions (SCARs), cleaved amplified polymorphic sequences (CAPS), sequence tagged sites (STS), single polymorphic amplification test (SPLAT), variable number of tandem repeats (VNTRs), DNA amplification fingerprinting (DAF) and diversity arrays technology (DArT). SNP markers have become the most widely used among these. Recombination may result in the loss of random DNA markers (RDMs), which make up the majority of these marker types. On the other hand, if candidate characteristic genes are identified, it is possible to create and apply DNA markers (also known as functional markers, or FMs) that are either inside or closely associated with the gene in crop varieties. It is generally acknowledged that DNA markers have the potential to be useful instruments for crop improvement in plants. Compared to conventional phenotypic and biochemical markers in plants, molecular markers offer a number of benefits. DNA markers are mostly used in agricultural research for quantitative trait loci (QTL) mapping, map-based gene cloning, hybrid testing, genetic diversity analysis, mapping of mutations, cultivar identification and "purity" assessment. Examples include mapping key genes, characterising transformants, marker-assisted pyramiding, marker-assisted selection (MAS), marker-assisted backcross breeding (MAB) and more (Sood and Chauhan 2023). Genetic transformation mediated by *Agrobacterium tumefaciens* represents one of the dependable methods being employed to produce robust transgenic lines. In contrast, alternative methods, including sonication, electroporation and particle bombardment (biolistics), are employed to temporarily express the foreign DNA. In this regard, the idea of the pangenome-which encompasses all of the genetic data present in the species' accessions drawn a lot of attention. Since the reference genome uses all of the genetic variation found in the species, it has the potential to significantly speed up molecular breeding when combined alongside the resequencing data of all accessible accessions of that species. Along with other crops, the development of pangenomes has also been done for *Brassica* (Bayer et al. 2019).

6.2.5 CRISPR and Gene-Editing Applications for Yield Improvement and Stress Tolerance

Plant biologists can now alter nearly every desired trait utilising genome editing technologies, which can also provide a useful approach for the functional genomics of the crop. First-generation genome editing technologies include zinc finger nucleases (ZFNs), site-specific nucleases (SSNs), meganucleases and transcription activator-like effector nucleases (TALENs). Whereas the second-generation methods for genome editing, clustered regularly interspaced short palindromic repeats/Cas9 (CRISPR/Cas9). Designing, cloning and/or gene-free editing techniques are all part of CRISPR. The Cas9 protein, which creates ds cuts, and a tiny gRNA molecule, which points Cas9 towards a particular DNA region for cleavage, are both components of the CRISPR/Cas9 system. After interacting with mature gRNA and inducing modifications in conformation, Cas9 is activated. Cas causes a break in both strands three nucleotides upstream of the PAM sequence (5′ NGG 3′) when it binds to the area of interest DNA. If a DNA template that is homologue to the flanking site of the DSB is present, the damage is going to be fixed using the homologous-directed repair (HDR) route or non-homologous end joining (NHEJ). The cut and repair mechanism can modify the structure of the protein or eliminate its function entirely by removing or adding nucleotides to the original sequence. Depending on various repair processes, genome editing either overexpresses or eliminates certain genes. Many research investigations have demonstrated the effectiveness of DNA editing, primarily CRISPR/Cas9, for enhancing resistance to a range of stresses (Rahman et al. 2022). Cas9 belongs to the class 2 type II CRISPR system and is an RNA-guided endonuclease. Trans-activating crRNA and CRISPR RNA combine in order to generate a two-RNA structure, which instructs Cas9 to cause DSB in the area of interest DNA. Another RNA-guided endonuclease, Cas12, representing the class 2 type V CRISPR system, was additionally investigated. A single crRNA guides the majority of Cas12 endonucleases, although some also require a second tracrRNA. Previously known as CRISPR/Cpf1, CRISPR/Cas12a has been modified for genome editing. In contrast to Cas9, Cas12a generates DSBs having staggered overhangs of 4-5 nt and cuts target DNA distal to T-rich PAM sequences under the guidance of a single crRNA. Like Cas12a, Cas12b favours PAMs that are rich in T (Li et al. 2022). A sgRNA, which is derived from a two-RNA structure (crRNA and tracrRNA), instructs Cas12b to create a DSB with staggered ends, in contrast to Cas12a. The use of additional Cas12 nucleases in genome editing is currently being investigated. Cas12e (formerly known as CasX) and Cas12j (previously known as CasΦ) are capable of altering eukaryotic genomes. Cas12f (formerly known as Cas14), which is approximately 50% smaller than Cas9/Cas12a, is of special significance (Liu et al. 2019; Pausch et al. 2020). In most cases, the cell's natural DNA repair system fixes the nick and makes gene editing easier. Depending on whether a template of DNA that is homologous to the flanking site of the DSB is present, most cuts are going to be fixed either via the homologous-directed repair (HDR) route or via non-homologous end joining (NHEJ). As a technique for improving traits, regulating genes, developing virus resistance and creating mutant libraries, the use of CRISPR/Cas9 for genome modification has revolutionised agriculture. The cutting-edge genome editing technologies controlled by CRISPR-Cas are growing rapidly every day. Functional annotation of abiotic stress-responsive genes and regulatory genomic regions may be significantly facilitated by the use of CRISPR-mediated technologies in forward genetic approaches (Singh et al. 2020). Given that polyploid rapeseed has four or six homoeologous gene copies, making it especially complex, the potential to introduce multiple mutations simultaneously using the CRISPR/Cas9 approach offers substantial benefits. Through the use of CRISPR/Cas9 to eliminate gene copy clusters, advances have been achieved in understanding the complexity of polyploidy in *B. napus* (Lohani et al. 2020). CBE that introduced specific C-to-T transition mutations close to the PAM region by combining CRISPR-Cas9 with an inhibitor of uracil glycosylase and APOBEC1. Three sgRNAs were utilised to target the ALS, RGA, and IAA7 genes. Plants with ALS base-editing showed great herbicide resistance, whereas plants with RGA or IAA7 base-editing showed reduced height of the plant (Cheng et al. 2021). Mutant Bnalpat2 and Bnalpat5 seeds had a wrinkled appearance, a lower oil content (24–39%) and reduced larger oil bodies, fully developed seeds also showed an accumulation of carbohydrate and protein complexes (Zhang et al. 2019). A number of copies of the BnaFAD2 gene were targeted using the CRISPR/cas9 method, which might alter FA compositions in oilseed crops, with the goal to create novel rapeseed varieties with improved FA profiles. Bnafad2.A5/Bnafad2.C5 double mutants contain a lot more oleic acid than the wild type. However, the oleic concentration of the Bnafad2.A5 and Bnafad2.C5 single mutants similarly increased by 63% (Huang et al. 2020). *B. napus* is susceptible to the *Verticillium longisporum* (Vl43) (hemibiotrophic fungus). In order to produce Vl43-resistant plants, the CRISPR/Cas9 method was utilised to delete CRT1a in *B. napus*. The BnaC09.CRT1 double mutant exhibited a more serious reduced-sensitivity phenotype when compared to BnaA09. CRT1a mutants, however, the combined BnaC09.CRT1a/BnaA09.CRT1a mutants demonstrated decreased susceptibility to Vl43 (Pröbsting et al. 2020). In rapeseed, pod fracture during the time of maturity frequently results in reduced yield, accounting for 20% of the overall yield and up to 50% in unfavourable weather circumstances. Majorly enhanced pod-shattering resistance was demonstrated by the BnA03.ind/BnC03.ind mutants, bnjag.a08 mutants, and the mutants' combining mutagenesis at five homeologs of BnSHP1 and BnSHP2. These genes were useful for breeding for resistance to pod-shattering (Zaman et al. 2021). Homozygous mutants were made for BnaTT2 in yellow-seeded mutants and resulted in a better lipid profile and enhanced amount of oil (Xie et al. 2020). Through the use of editing methods like homology-directed modification of genes or prime editing, in addition to expanding beyond CRISPR/Cas9 to include additional CRISPR-associated systems, it will be possible to more effectively and accurately enhance variety within the *Brassica* genome and make highly accurate and stated modifications in this significant oilseed crop. Given their complexity, many traits like yield and disease tolerance should necessitate editing many separate loci. In order to accurately modify several connected or unconnected loci inside an individual cell and hence inside the same regenerated plant, multiplex gene editing entails an ongoing management of reagents. Additionally, the next development in gene-editing technology is expected to be the de novo combination of gene-edited features, similar to introgression breeding but with much shorter timeframes (Table 6.2).

6.3 Conclusion and Future Perspectives

Productivity needs to be increased in order to meet the expanding market requirement for oil and other products of *Brassica*, and micropropagation (a good alternative to seed propagation) may prove a viable may to address this problem. These consist of simple and quick multiplication of certain genotypes produced by the traditional breeding procedure. Propagation of parents of hybrids and asexually produced crops is also done using micropropagation. *Brassica* members are a valuable, multipurpose crop, hence necessitating greater care, safeguarding, and advancement. Therefore, the importance of *Brassica* has to stimulate scientists' curiosity, learn more concerns about its potential, and reinvigorate experts' interest (decreased by crop decline due to numerous diseases and decreased production that restricts *Brassica* revenue). A number of investigations have made it possible to successfully employ the molecular mechanism governing advantageous features (oil, protein content, production, and disease tolerance). Attempts to maximise various desirable qualities are currently ongoing. In order to prevent the high input/low yield in production, which could encourage producers to grow their crops, approaches to management need to be updated. So that production sustainability might be assured, and revenue could be attained.

Table 6.2 Several studies have shown the utility of genome editing assays, especially CRISPR/Cas9, for improving *Brassica* spp.

Plant	Technique	Gene	Response	References
Brassica napus	*CRISPER-CAS*	*BnaFAD2*	Develop novel variations in fatty acid profiles, oleic acid content increased inoleic and linolenic acid contents decreased	Huang et al. (2020)
Brassica oleracea var. capitata	*CRISPR-CAS*	BoSRK, BoMS1	BoMS1 gene results in a male-sterile mutant, and the BoSRK3 gene helps in self-incompatibility suppression	Ma et al. (2019a)
Brassica oleracea var. capitata	*CRISPR-CAS*	*BoPDS*	Albino-phenotype transgenic shoot in T0 Basta-resistant lines was observed	Ma et al. (2019b)
Brassica napus	*CRISPR-CAS*	*BnaA9. WRKY47*	Boron content in tissues was found to be enhanced in *BnaA9.WRKY47* overexpressing plants	Feng et al. (2020)
Brassica napus	*CRISPR-CAS*	BnFAD2 and BnFAE1	Oleic acid content (70–80%) content significantly increased, whereas polyunsaturated fatty acids content decreased	Shi et al. (2022)
Brassica rapa	*CRISPR-CAS*	*eIF(iso)4E*	Conferred resistance to turnip mosaic virus (TuMV)	Lee et al. (2023)
Brassica campestris	*CRISPR-CAS*	Bra003491, Bra007665, and Bra014410	Mutations in Bra003491 were stable and inherited by T1 and T2 generations	Xiong et al. (2019)
Broccoli	*CRISPR-CAS*	BroMYB28	Increased glucoraphanin content	Kim et al. (2019)
Brassica napus	*CRISPR-CAS*	*BnALS*	Herbicide-resistant	Wu et al. (2020)
Broccoli	*CRISPR-CAS*	BoC3H4	Enhanced salt stress tolerance	Jiang et al. (2019)
Broccoli	*CRISPR-CAS*	BoGL5	Mutants lacked cuticular waxes	Liu et al. (2023)
Brassica napus	*CRISPR-CAS*	*ALS*, *RGA* and *IAA7*	Improve plant architecture and enhance the resistance to herbicides in order to control weeds	Cheng et al. (2021)
Broccoli	*CRISPR-CAS*	BoMAM1	Led to an increase in SF content	Cao et al. (2021)
Rapeseed	*CRISPR-CAS*	BnaMAX1	Overall yield has been improved because BnaMAX1 expressing plants have increased branching with more siliques and semi-dwarf height	Zheng et al. (2020)
Brassica oleracea	*CRISPR-CAS*	*myb28*	A decrease in methionine-derived compounds like glucosinolate and glucoraphanin has been noticed due to the downregulation of A-GSL in *myb28* mutants	Neequaye et al. (2021)

References

Acquaah G (2015) Conventional plant breeding principles and techniques. In: Advances in plant breeding strategies: breeding, biotechnology and molecular tools. Springer, Cham. https://doi.org/10.1007/978-3-319-22521-0_5

Adlak T, Tiwari S, Tripathi MK et al (2019) Biotechnology: an advanced tool for crop improvement. Curr J Appl Sci Technol 33(1):1–11

Barfield DG, Pua EC (1991) Gene transfer in plants of *Brassica juncea* using *Agrobacterium tumefaciens* mediated transformation. Plant Cell Rep 10:308–314

Basak H, Biswas BK, Azad MAK et al (2012) Micropropagation of mustard (*Brassica* spp.) from leaf explants. Thai J Agric Sci 45(2):75–81

Bayer PE, Golicz AA, Tirnaz S et al (2019) Variation in abundance of predicted resistance genes in the *Brassica oleracea* pangenome. Plant Biotechnol J 17(4):789–800

Bhattacharya S (2019) Brassica-aphid interaction: challenges and prospects of genetic engineering for integrated aphid management. Physiol Mol Plant Pathol 108:101442

Cao H, Liu R, Zhang J et al (2021) Improving sulforaphane content in transgenic broccoli plants by overexpressing MAM1, FMOGS-OX2, and Myrosinase. Plant Cell Tissue Organ Cult (PCTOC) 146:461–471

Cheng H, Hao M, Ding B et al (2021) Base editing with high efficiency in allotetraploid oilseed rape by A3A-PBE system. Plant Biotechnol J 19:87–97

Choi PS, Soh WY, Liu JR (1996) Somatic embryogenesis and plant regeneration in cotyledonary explant cultures of Chinese cabbage. Plant Cell Tissue Organ Cult 44:253–256

Fahey JW (2016) Brassica: characteristics and properties. In: Encyclopedia of food and health. Academic, Oxford, pp 469–477

Feng Y, Cui R, Wang S et al (2020) Transcription factor BnaA9. WRKY47 contributes to the adaptation of *Brassica napus* to low boron stress by up-regulating the boric acid channel gene BnaA3. NIP5; 1. Plant Biotechnol J 18(5):1241–1254

Ghnaya AB, Charles G, Branchard M (2007) Rapid shoot regeneration from thin cell layer explants excised from petioles and hypocotyls in four cultivars of *Brassica napus* L. Plant Cell Tissue Organ Cult 92:25–30

Huang HB, Cui TT, Zhang LL et al (2020) Modifications of fatty acid profile through targeted mutation at BnaFAD2 gene with CRISPR/Cas9-mediated gene editing in Brassica napus. Theor Appl Genet 133:2401–2411

Jiang M, Miao L, Zhang H, Zhu X (2019) Over-expression of a transcription factor gene BoC3H4 enhances salt stress tolerance but reduces Sclerotinia stem rot disease resistance in broccoli. J Plant Growth Regul 39:1162–1176

Kim Y-C, Cha A, Hussain M et al (2019) Impact of *Agrobacterium*-infiltration and transient overexpression of BroMYB28 on glucoraphanin biosynthesis in broccoli leaves. Plant Biotechnol Rep 14:373–380

Kohlenbach HW, Wenzel G, Hoffmann F (1982) Regeneration of *Brassica napus* plantlets in cultures from isolated protoplasts of haploid stem embryos as compared with leaf protoplasts. Z Pflanzenphysiol 105:131–142

Krikorian AD (1995) Hormones in tissue culture and micropropagation. In: Plant hormones: physiology, biochemistry and molecular biology. Springer, Dordrecht, pp 774–796

Lamichhane S, Thapa S (2022) Advances from conventional to modern plant breeding methodologies. Plant Breed Biotechnol 10(1):1–4

Lee YR, Siddique MI, Kim DS et al (2023) CRISPR/Cas9-mediated gene editing to confer turnip mosaic virus (TuMV) resistance in Chinese cabbage (*Brassica rapa*). Hortic Res 10(6):uhad078

Li J, Yu X, Zhang C et al (2022) The application of CRISPR/Cas technologies to *Brassica* crops: current progress and future perspectives. aBIOTECH 3:146–161

Liu J, Orlova N, Oakes B et al (2019) CasX enzymes comprise a distinct family of RNA-guided genome editors. Nature 566(7743):218–223

Liu Y, Wei M, Liu Y et al (2023) Functional characterization of BoGL5 by an efficient CRISPR/Cas9 genome editing system in broccoli. Sci Hortic 319:112136
Lohani N, Jain D, Singh MB, Bhalla PL (2020) Engineering multiple abiotic stress tolerance in canola, *Brassica napus*. Front Plant Sci 11:3
Ma C, Zhu C, Zheng M et al (2019a) CRISPR/Cas9-mediated multiple gene editing in *Brassica oleracea* var. *capitata* using the endogenous tRNA-processing system. Hortic Res 2019:6
Ma C, Liu M, Li Q et al (2019b) Efficient BoPDS gene editing in cabbage by the CRISPR/Cas9 system. Hortic Plant J 5(4):164–169
Murashige T, Skoog F (1962) A revised medium for rapid growth and bio assays with tobacco tissue cultures. Physiol Plant 15(3):1
Neequaye M, Stavnstrup S, Harwood W et al (2021) CRISPR-Cas9-mediated gene editing of MYB28 genes impair glucoraphanin accumulation of *Brassica oleracea* in the field. CRISPR J 4(3):416–426
Ozturk M, Altay V, Hakeem KR, Akçiçek E (2018) Molecular aspects. In: Liquorice: from botany to phytochemistry. Springer, pp 41–43
Ozturk M, Kamili AN, Altay V, Rohela GK (2024) Molecular aspects of mulberry. In: Mulberry: from botany to phytochemistry. Springer, Cham, pp 69–99
Pausch P, Al-Shayeb B, Bisom-Rapp E et al (2020) CRISPR-CasΦ from huge phages is a hypercompact genome editor. Science 369(6501):333–337
Petruzzello M (2023) List of economically important members of the family Brassicaceae. Encyclopedia Britannica
Phogat SK, Burma PK, Pental D (2000) High frequency regeneration of *Brassica napus* varieties and genetic transformation of stocks containing fertility restorer genes for two cytoplasmic male sterility systems. J Plant Biochem Biotechnol 9:73–79
Pröbsting M, Schenke D, Hossain R et al (2020) Loss of function of CRT1a (calreticulin) reduces plant susceptibility to Verticillium longisporum in both Arabidopsis thaliana and oilseed rape (Brassica napus). Plant Biotechnol J 18:2328–2344
Rahman M, Zulfiqar S, Raza MA et al (2022) Engineering abiotic stress tolerance in crop plants through CRISPR genome editing. Cells 11(22):3590
Rathore JP, Rashid M, Sharma A et al (2018) Biotechnology and breeding approaches to increase disease resistances in cabbage. J Pharmacogn Phytochem 7(4):2667–2671
Ravanfar SA, Aziz MA, Kadir MA et al (2009) Plant regeneration of *Brassica oleracea* subsp. *italica* (broccoli) cv. Green Marvel as affected by plant growth regulators. Afr J Biotechnol 8:2523–2528
Ravanfar SA, Aziz MA, Shabanimofrad M, Samarfard S (2013) Greenhouse evaluation on the performance of heat tolerant transgenic broccoli and genetic diversity analysis using inter simple sequence repeat (ISSR) markers. Electron J Biotechnol 16(5):1–10
Ravanfar SA, Aziz MA, Rashid AA, Shahida S (2014) In vitro adventitious shoot regeneration from cotyledon explant of *Brassica oleracea* subsp. *italica* and *Brassica oleracea* subsp. *capitata* using TDZ and NAA. Pak J Bot 46:329–335
Rehman R, Tarıq A, Haq N et al (2020) Mutations leading to a better tomorrow in the production of improved crop varieties. Cambridge Scholar
Ruqia S, Haq SN, Ilyas M et al (2020) Plant metabolomics: role in crop improvement. Cambridge Scholar
Sharma KK, Bhojwani SS, Thorpe TA (1990) Factors affecting high frequency differentiation of shoots and roots from cotyledon explants of *Brassica juncea* (L.) Czern. Plant Sci 66:247–253
Shi J, Ni X, Huang J et al (2022) CRISPR/Cas9-mediated gene editing of BnFAD2 and BnFAE1 modifies fatty acid profiles in *Brassica napus*. Genes 13(10):1681
Singh A et al (2020) Improvement of crop's stress tolerance by gene editing CRISPR/CAS9 system. In: Sustainable agriculture in the era of climate change. Springer, Cham
Sood R, Chauhan A (2023) Applications of genetic engineering and DNA markers in plant breeding. Res Rev Genet Plant Breed 1:45

Sun J, Wang Y, Pang X et al (2021) The effect of processing and cooking on glucoraphanin and sulforaphane in brassica vegetables. Food Chem 360:130007

Turgut K, Barghchi M, Scott R (1998) Efficient shoot regeneration and somatic embryogenesis from immature cotyledons of *Brassica napus* L. Plant Breed 117:503–504

Wu J, Chen C, Xian G et al (2020) Engineering herbicide-resistant oilseed rape by CRISPR/Cas9-mediated cytosine base-editing. Plant Biotechnol J 18(9):1857

Xiaomeng Z, Qiong J, Xin J et al (2025) *Brassica* vegetables-an undervalued nutritional goldmine. Hortic Res 12(2):uhae302

Xie T, Chen X, Guo T et al (2020) Targeted knockout of BnTT2 homologues for yellow-seeded *Brassica napus* with reduced flavonoids and improved fatty acid composition. J Agric Food Chem 68(20):5676–5690

Xiong X, Liu W, Jiang J et al (2019) Efficient genome editing of *Brassica campestris* based on the CRISPR/Cas9 system. Mol Gen Genomics 294:1251–1261

Zaman QU, Wen C, Yuqin S et al (2021) Characterization of SHATTERPROOF homoeologs and CRISPR-Cas9-mediated genome editing enhances pod-shattering resistance in Brassica napus L. CRISPR J 4:360–370

Zhang K, Nie L, Cheng Q et al (2019) Effective editing for lysophosphatidic acid acyltransferase 2/5 in allotetraploid rapeseed (*Brassica napus* L.) using CRISPR-Cas9 system. Biotechnol Biofuels 12:1–8

Zheng Q, Liu K (2022) Worldwide rapeseed (*Brassica napus* L.) research: a bibliometric analysis during 2011–2021. Oil Crop Sci 7(4):157–165

Zheng M, Zhang L, Tang M et al (2020) Knockout of two Bna MAX 1 homologs by CRISPR/Cas9-targeted mutagenesis improves plant architecture and increases yield in rapeseed (*Brassica napus* L.). Plant Biotechnol J 18(3):644–654

Chapter 7
Phytochemical Constituents, Pharmacological Activities and Human Health

7.1 Phytochemical Constituents

Brassica vegetables are characterised by their low fat content and high vitamin (C, E, B, A and K) and carotenoid content. They are also renowned for their antioxidant properties, which include catalase, superoxide dismutase (SOD) and peroxidase, as well as sulphur-containing glucosinolates, flavonoids, anthocyanins, terpenes and coumarins (Cartea et al. 2011; Kapusta-Duch et al. 2012; Fahey 2016; Sanlier and Guler 2018). These vegetables are also important for their mineral content. Notably, their low levels of oxalic and phytic acids, which bind calcium (Ca^{2+}), increase the bioavailability of Ca^{2+} (Cartea et al. 2011; Sanlier and Guler 2018). Some studies have also reported that these vegetables accumulate essential minerals, including phosphorus (P), sulphur (S), chlorine (Cl), calcium (Ca), iron (Fe), manganese (Mn), copper (Cu), strontium (Sr), potassium (K) and zinc (Zn) (Jahangir et al. 2009; Cartea et al. 2011; Sanlier and Guler 2018). They also contain alkaloids, cardiac glycosides, coumarins, phlobatannins, quinines, reducing sugars, resins, saponins, steroids, tannins, terpenoids, xanthoproteins and essential oils (Paul et al. 2019).

The seven most commonly found essential fatty acids in *Brassica* vegetables are palmitic acid, stearic acid, oleic acid, linoleic acid, linolenic acid, eicosanoic acid and erucic acid. The presence of significant amounts of the long-chain monounsaturated fatty acids eicosenoic and erucic acids in *Brassica* vegetables, particularly in seeds, renders them unsuitable for human consumption but suitable for industrial use. For oils to be considered of good quality and suitable for human consumption, they must contain high concentrations of oleic and linoleic acids (Sharafi et al. 2015; Rai et al. 2018; Cartea et al. 2019; Jabeen 2020).

Brassica crops are rich in phytochemicals, including polyphenols, triterpenes and vitamins, as well as other health-promoting non-mineral compounds (Jahangir et al. 2009; Bhandari and Kwak 2015; Šamec et al. 2019; Salehi et al. 2021; Wang et al. 2025). *Brassica* vegetables are a valuable source of natural antioxidants, such

M. Ozturk et al., *Climate-Resilient* Brassicas,
https://doi.org/10.1007/978-3-032-12709-9_7

as glucosinolates, polyphenols, carotenoids and tocopherols (Hedges and Lister 2006).

The most common and diverse groups of polyphenols found in *Brassica* vegetables are flavonoids and hydroxycinnamic acids. Anthocyanins, which give cabbage and broccoli sprouts their red colouring, are also important compounds (Kapusta-Duch et al. 2012; Aires 2015). The flavonoids are categorised by their phytochemical structure, particularly the number and arrangement of carbon atoms and classified as flavonols, flavanones, isoflavones, flavones, flavan-3-ols and anthocyanidins, whereas the nonflavonoids are categorised as phenolic acids, hydroxycinnamates and stilbenes (Sanlier and Guler 2018).

The most common and diverse polyphenol groups found in *Brassica* species are flavonoids (mainly flavonols and anthocyanins) and hydroxycinnamic acids (Cartea et al. 2010). The characteristic taste and odour are due to glucosinolates and phenolic compounds. These bioactive substances exhibit various biological characteristics, including antioxidants, antimicrobials and anti-cancer features (Singh Chauhan et al. 2016; Salehi et al. 2021). In fact, various extracts from these plants are already commercially available (Li et al. 2023). *Brassica* crops have served as important sources of food; as such, they have been the subject of great scientific interest, particularly *B. oleracea* species, which have assumed an important role in human nutrition, as one of the important dietary sources of glucosinolates but also with high contents of phenolics and other antioxidant compounds (Tsunoda et al. 1980; Gómez-Campo 1999).

Glucosinolates are an important group of phytochemicals that are found in high concentrations in *Brassica* vegetables (Holst and Williamson 2004). These compounds have a basic chemical structure comprising a β-D-thioglucose group, a sulfonated oxime group and a variable side chain derived from amino acids, which comprises highly variable aliphatic, aromatic or heterocyclic groups. Even slight changes to the side chain result in glucosinolate groups with different chemical structures (Holst and Williamson 2004; Sanlier and Guler 2018). Approximately 130 different glucosinolate groups have been identified uptill now (Cartea et al. 2011); but the glucosinolates found in existing plant tissues are not biologically active due to their chemical stability. However, their hydrolysis products exhibit biological activity (Holst and Williamson 2004; Sanlier and Guler 2018). Depending on the chemical structure of the glucosinolate compounds in question, different hydrolytic products are formed, which include isothiocyanates, nitriles, sulfides, thiocyanates, epithiopyles, oxazolidine-2-thiones and indolyl compounds (Holst and Williamson 2004; Hanschen et al. 2015, 2017; Sanlier and Guler 2018). The main glucosinolates found in *Brassica* species include progoitrin, sinigrin, glucoraphanin and the thiocyanate precursors glucobrassican and neoglucobrassican (Bischoff 2016).

The unique organosulfur compounds found in the *Brassica* vegetables distinguish them from other plants and vegetables (Salehi et al. 2021). The two main organosulfur compounds among these are isothiocyanates, which give them their distinctive flavour and pungency (Nor et al. 2020), and S-methylcysteine sulfoxide, which is a precursor to the synthesis of sulphur-containing volatiles (Fusari et al. 2020). Latter is a precursor to the synthesis of these volatiles (Fusari et al. 2020).

The glucosinolates in commonly consumed *Brassica* vegetables become bioactive only through enzymatic hydrolysis during chewing or processing (Holst and Williamson 2004), and cooking these vegetables hydrolyses them due to the action of microflora in the human intestinal tract, thereby making glucosinolates bioactive (Dinkova-Kostova and Kostov 2012). As against this, some experimental studies have shown that cooking *Brassica* vegetables can significantly reduce the bioavailability of isothiocyanates by inactivating β-thioglucosidase (Song and Thornalley 2007; Sanlier and Guler 2018). Song and Thornalley (2007) have conducted a study demonstrating that processes such as microwaving, boiling and frying brassica vegetables generally do not significantly reduce glucosinolate levels; however, boiling can reduce glucosinolate levels by around 90% due to leaching into the cooking water.

The *Brassica*'s offer a promising option for food and nutritional security due to their high nutritional profiles, exceptional resilience to harsh environmental conditions, richness in macro and micronutrients, dietary fibres, essential amino acids, minerals, antioxidants, trace elements, proteins, lipids and carbohydrates (Gómez-Campo 1999; Lone et al. 2021; Wani et al. 2021; Ahmad et al. 2015; Jan et al. 2018). However, the vitamin, mineral and phytochemical content of *Brassica* vegetables depends heavily on genetic and environmental factors, as well as ecological conditions. The genetic variability among *Brassica* vegetable varieties, their hybrid status and exposure to abiotic stresses such as sunlight, photoperiod, drought, salinity and temperature, as well as biotic stresses resulting from environmental conditions leads to differences in soil chemical composition, fertilisation, harvesting, storage, transportation, preparation and consumption methods can all cause significant changes in their nutritional value (Fahey 2016; Salehi et al. 2021). These factors significantly affect the quantity and quality of phytochemicals and essential oils synthesised by *Brassica* vegetables (Salehi et al. 2021). The concentrations of phytochemicals and essential oils in *Brassica* vegetables also vary due to mechanical factors such as injury, rain, hail, wind and sandstorms. Various studies on this popular global crop report that the phytochemical content of *Brassica* vegetables can vary significantly during growth depending on agro-environmental factors such as light, temperature, water availability and soil fertility (Hedges and Lister 2006; Ongaro et al. 2019; Salehi et al. 2021).

In Europe, the most nutritious brassicas are considered broccoli, Brussels sprouts, cauliflower, cabbage and kale. As against this, commercial and nutritional interest in the East has focused on Chinese cabbage, bok choy, pak choi, mizuna, Chinese mustard greens and Chinese broccoli (Fahey 2016).

7.2 Pharmacological Activities and Human Health

Among the highest grown and consumed global crop types, the "*Brassica*" plant species have seen an increase in their output during the past few decades. These crops have a major positive impact on human health during consumption. The vegetables from this group are great plant-based providers of macronutrients, which are necessary for daily dietary requirements. The taste, fragrance and pungency are among the sensory qualities that are strongly associated with some of their metabolites. These crops offer a wide variety of macronutrients, such as proteins, soluble sugars, amino acids, organic acids and dietary fibres, but with the exception of those utilised to produce oil, others in general are not considered important sources of energy due to their small carbohydrate and fat concentrations. The latest investigations indicate that *Brassica* vegetables have special phytonutrients which are health-promoting; as such, their consumption may lower the risk of developing chronic illnesses, particularly some forms of cancer (Xiaomeng et al. 2025).

7.2.1 Anti-cancer Activity

Isothiocyanates, which are commonly found in *Brassica* vegetables, are glucosinolate derivatives and important phytochemicals, shown to exhibit antibacterial, antiviral and anticarcinogenic features (Wu et al. 2013; Bricker et al. 2014; Sanlier and Guler 2018). Sulforaphane, a natural isothiocyanate found in these vegetables, is one of the most important bioactive compounds said to have antioxidant and antitumour properties (Kalpana et al. 2013; Sanlier and Guler 2018). The data published by the US National Cancer Institute has revealed that sulforaphane is the most promising chemopreventive agent (Kelloff et al. 2000). Other studies emphasise that the concentration of sulforaphane in broccoli sprouts is approximately ten times higher than in mature broccoli. The clinical experimental studies on sulforaphane isolated from broccoli extracts have demonstrated its ability to inhibit tumour growth (Dinkova-Kostova et al. 2007; Sanlier and Guler 2018). Moreover, other important hydrolytic products of glucosinolates, such as glucoraphanin, gluconasturtiin and glucobrassicin, possess potential anticarcinogenic activity. In addition to these, indole-3-carbinol, a glucobrassicin metabolite, has been identified as having inhibitory effects on human breast and ovarian cancers (Sanlier and Guler 2018).

The scientific evidence suggests that consuming *Brassica* crops can reduce the risk of developing various types of cancers; such significant health benefits have been found to be related to the indoles, which are sulphur-containing compounds, generally glucosinolates and their hydrolysis products (Murillo and Mehta 2001; Wu and Lou 2007; Liu et al. 2013). As a result of these bioactive components, *Brassica* crops exhibit significant pharmacological features such as antioxidant, anti-inflammatory, anti-cancer and cardioprotective (Murillo and Mehta 2001).

Some clinical and risk assessment studies on different types of cancers have been conducted based on the consumption of *Brassica* vegetables. One of these has examined the comparative risk of rectal cancer associated with high versus low consumption of *Brassica* vegetables (Kim and Park 2009). It has also been reported that consuming 250 g of broccoli and Brussels sprouts daily significantly increases the excretion of carcinogenic substance 2-amino-1-methyl-6-phenylimidazo[4,5-b] pyridine (Walters et al. 2004). Moreover, the hepatoprotective effect of the antioxidant properties of volatile fatty acids obtained from white cabbage extract has also been emphasised (Morales-Lopez et al. 2017). A clinical experimental study by Wu et al. (2013) reports that people who consume large quantities of *Brassica* vegetables face a lower risk of stomach cancer. An experimental study carried out on the livers of rats has revealed that bioactive compounds such as the glucosinolates isolated from *Brassica* vegetables do exhibit anti-chemotherapy activity in the liver (Abdull Razis et al. 2011). Other researchers have hypothesised that isothiocyanate extract isolated from *Brassica* vegetables may prevent bladder cancer (Gupta 2011); similar findings have been reported by Zhang et al. (2006), Tang et al. (2006) and Zhao et al. (2007).

Another comparative study has examined the diets of women with breast cancer (aged 50–74) who consumed *Brassica* vegetables, and this has been compared with the diets of healthy women. The results obtained have revealed that eating one to two portions of these vegetables a day reduces the risk of breast cancer by 20–40 % (Terry et al. 2001). Other investigations have demonstrated that the edible portion of cauliflower contains bioactive substances that can significantly inhibit the growth of breast cancer cells (Gupta 2011). In the study using the urinary isothiocyanate biomarker, it has been found that Chinese women who consumed large quantities of *Brassica* vegetables had a significantly reduced risk of breast cancer (Fowke et al. 2003). A clinical case study of women living in the Caucasian region has revealed that living in a different region the broccoli consumption is negatively associated with breast cancer risk, particularly among premenopausal women (Ambrosone et al. 2004). Some studies have reported a reduced risk of prostate cancer in male patients under 65 years of age who consume high levels of *Brassica* vegetables (Cohen et al. 2000; Joseph et al. 2004).

7.2.2 Antioxidant Activity

The experimental antioxidant activity studies are typically conducted in vitro with the primary aim of determining free radical scavenging activity. Many of these investigations have demonstrated the high antioxidant potential of *Brassica* vegetables, particularly *B. oleracea*, *B. rapa* and *B. napus* (Tanongkankit et al. 2010; Georgieva et al. 2014; Salehi et al. 2021). In one of the recent studies, it has been found that organic extracts obtained from *B. oleracea* exhibit high antioxidant features (Salehi et al. 2021). Several in vitro activity reports by researchers like Šamec et al. (2011) and Pellegrini et al. (2010) have also emphasised the significant antioxidant capacity of aqueous extracts derived from *B. rapa*.

A clinical study carried out and involving experimental animals has revealed that aqueous extracts from Brussels sprouts have the potential to protect against oxidative DNA damage in rats (Deng et al. 1998); another study highlights the protective effect of extracts from the *Brassica rapa* plants against chromosomal damage and oxidative stress induced by gamma radiation, cyclophosphamide and urethane in mice (Tiku et al. 2008). In another experimental study, extracts from broccoli have been found to be highly effective at scavenging superoxide anion and hydrogen peroxide in *Drosophila melanogaster* (Li et al. 2008). Mollica et al. (2018) have also reported a significant reduction in oxidative/nitrosative stress and lipoperoxidation in an antioxidant activity study involving stimulated bladder, kidney and liver samples, which had been treated with lipopolysaccharides isolated from broccoli. Several clinical studies conducted on healthy individuals have demonstrated a protective effect of Brussels sprouts and fresh broccoli sprouts against DNA damage and reduced oxidative damage (Verhagen et al. 1997; Murashima et al. 2004).

7.2.3 *Anti-inflammatory Activity*

The bioactive compounds in *Brassica* vegetables, such as broccoli and cabbage, are reported to depict anti-inflammatory properties, which are achieved by activating detoxification enzymes, neutralising free radicals and boosting immune system function (Fimognari et al. 2012; Sanlier and Guler 2018). In addition to these reports, isothiocyanate compounds found in these vegetables are reported to slow down certain inflammatory processes. The cabbage has long been used in traditional medicine to alleviate the symptoms of gastrointestinal disorders as well as treat minor cuts, wounds and mastitis. It has also been reported to exhibit anti-inflammatory and antibacterial characteristics (Kapusta-Duch et al. 2012; Sanlier and Guler 2018).

7.2.4 *Neurodegenerative Activity*

Sulforaphane, a key bioactive compound, has been shown to significantly reduce cognitive deficits induced by fenciclidine when used as a pre-treatment. The consumption of a diet rich in broccoli sprouts containing sulforaphane during the juvenile and adolescent periods may prevent the onset of psychosis and alleviate cognitive impairment associated with schizophrenia (Shirai et al. 2015). Furthermore, broccoli sprout juice has been identified as a promising preventive and therapeutic approach for Alzheimer's disease (Masci et al. 2015).

7.2.5 *Anti-diabetic Activity*

The bioactive compound sulforaphane has been reported to possess the potential to prevent various diabetic complications, including cardiomyopathy, nephropathy, neuropathy and retinopathy (Xu et al. 2016; Sanlier and Guler 2018). A study undertaken on diabetic rats has revealed that red cabbage extract could be a natural therapeutic resource for treating diabetes, due to its antioxidant and antihyperglycaemic features (Kataya and Hamza 2008).

7.2.6 *Cholesterol-Lowering Activity*

An experimental study conducted on healthy individuals has revealed that consuming 100 g of fresh broccoli per day for 1 week significantly decreases total and LDL cholesterol levels, while increasing HDL cholesterol levels. Keeping these findings in view, eating broccoli can significantly improve cholesterol metabolism in humans (Murashima et al. 2004). In a clinical study conducted on a test group of around 80 men who consumed broccoli juice, it was observed that LDL cholesterol levels in the test group decreased by around 8.5% over 9 weeks (Gupta 2011). In another study it has been reported that administering cabbage extract to hypercholesterolemic hepatoma rats has resulted in a positive decrease in their serum cholesterol levels (Komatsu et al. 1998). A different clinical study reports that polyphenols isolated from red cabbage and Brussels sprouts reduce cholesterol concentration in the erythrocytes of hypercholesterolemic patients (Duchnowicz et al. 2012). An investigation undertaken using animals has revealed that broccoli sprout extract rich in bioactive compounds, such as glucoraphanin, when administered alongside a frozen broccoli extract containing simvastatin, reduced the hepatic cholesterol in all experimental animals after 7 weeks (Rodriguez-Cantu et al. 2011).

7.2.7 *Antimicrobial Activity*

The antimicrobial potential of plant extracts obtained from *Brassica* vegetables is a well-recognised fact (Sherman and Hodge 1936; Aires et al. 2012; Sharma et al. 2018; Salehi et al. 2021). The notable examples include the scientific demonstration of the antimicrobial activity of cabbage (*Brassica oleracea*) (Paul et al. 2012; Salehi et al. 2021). Another experimental study has reported that methanolic extracts obtained from *B. oleracea* are highly effective in controlling the growth of selected Gram-positive and Gram-negative bacteria (Ayshwarya and Sudha Rameshwari 2015). The studies covering investigations on the therapeutic potential of these extracts for use in medicine and pharmacology have confirmed the antibacterial bioactivity of fermented cabbage products (Bajpai et al. 2011, 2012). The extracts from broccoli and red cabbage sprouts have also been found to exhibit strong antibacterial activity (Vale et al. 2015).

These vegetables have been shown to exhibit antimicrobial activity due to the presence of volatile sulphur compounds originating from S-alk(en)yl-L-cysteine sulfoxides and glucosinolates (Kyung and Lee 2001). Moreover, the glucosinolate hydrolysis products, particularly allyl isothiocyanates, have been reported to effectively inhibit bacterial activity (Vig et al. 2009).

7.2.8 Gastrointestinal Activity

Carvalho et al. (2011) have studied the anti-ulcerogenic properties of cabbage extracts using an experimental animal model. It has been demonstrated that such an extract has the potential to significantly inhibit ulcer formation in animals, and it has since been confirmed by other studies conducted by other workers (Enye et al. 2013; Oguwike et al. 2014).

7.3 Side Effects and Toxicity

Despite the numerous health benefits of bioactive compounds derived from *Brassica* species, some studies suggest that consuming these plants may have minor toxic effects (Salehi et al. 2021). For example, a widespread outbreak in Southland-New Zealand, has resulted in deaths of over 400 pregnant and lactating cows following consumption of *B. napus* leaves (Matthews et al. 2020). Similar signs of liver toxicity have been observed in people who have consumed turnip greens (*Brassica rapa* ssp. *rapa* or *B. rapa* ssp. *rapifera*) (Collett et al. 2014). In an experimental study involving a rat in vivo model, a dose of 2–3 mmol/kg of progoitrin nitrile was reported to cause severe toxicity to the liver, pancreas and kidneys (Matthews et al. 2020).

A clinical trial involving around 300 women diagnosed with thyroid cancer in Malaysia has revealed a positive correlation between high consumption of *Brassica* vegetables and the disease. The presence of thiocyanates, produced when the body metabolises the thioglycosides found in these vegetables, has been linked to the inhibition of iodine transport and incorporation into thyroglobulin, which can lead to increased thyroid cell proliferation and thyroid carcinoma (Chandra 2010; Truong et al. 2010).

An irritant contact dermatitis following the application of *B. oleracea* cultivar leaves to the skin has also been reported (Milanesi and Gola 2016).

In fact, many significant risks to food safety and human health following bioaccumulation in edible plant parts pose a great threat (Uddin et al. 2021).

References

Abdull Razis AF, Bagatta M, De Nicola GR et al (2011) Up-regulation of cytochrome P450 and phase II enzyme systems in rat precision-cut rat lung slices by the intact glucosinolates, glucoraphanin and glucoerucin. Lung Cancer 71:298–305

Ahmad P, Hashem A, Abd-Allah EF et al (2015) Role of *Trichodermahar zianum* inmitigating NaCl stress in Indian mustard (*Brassica juncea* L) through antioxidative defense system. Front Plant Sci 6:868

Aires A (2015) Chapter 3 – Brassica composition and food processing A2 – preedy, victor, processing and impact on active components in food. Academic, San Diego, pp 17–25

Aires A, Carvalho R, Rosa E (2012) Glucosinolate composition of Brassica is affected by postharvest, food processing and myrosinase activity. J Food Process Preserv 36(3):214–224

Ambrosone CB, McCann SE, Freudenheim JL et al (2004) Breast cancer risk in premenopausal women is inversely associated with consumption of broccoli, a source of isothiocyanates, but is not modified by GST genotype. J Nutr 134:1134–1138

Ayshwarya M, Sudha Rameshwari K (2015) Antimicrobial activity of the plant extracts of Brassica oleracea var. capitata rubra. J Int Acad Res Multidiscip 3(10):149–156

Bajpai VK, Kang SC, Park E et al (2011) Diverse role of microbially bioconverted product of cabbage (Brassica oleracea) by Pseudomonas syringe pv. T1 on inhibiting Candida species. Food Chem Toxicol 49:403–407

Bajpai VK, Kang SC, Baek KH (2012) Microbial fermentation of cabbage by a bacterial strain of Pectobacterium atrosepticum for the production of bioactive material against Candida species. J Mycol Médicale 22:21–29

Bhandari SR, Kwak JH (2015) Chemical composition and antioxidant activity in different tissues of Brassica vegetables. Molecules 20(1):1228–1243

Bischoff KL (2016) Glucosinolates. In: Nutraceuticals. Elsevier, Amsterdam, pp 551–554

Bricker GV, Riedl KM, Ralston RA et al (2014) Isothiocyanate metabolism, distribution, and interconversion in mice following consumption of thermally processed broccoli sprouts or purified sulforaphane. Mol Nutr Food Res 58:1991–2000

Cartea ME, Francisco M, Soengas P, Velasco P (2010) Phenolic compounds in Brassica vegetables. Molecules 16(1):251–280

Cartea ME, Lema M, Francisco M et al (2011) Basic information on vegetable Brassica crops. In: Genetics, genomics and breeding of vegetable brassicas, pp 1–33

Cartea E, De Haro-Bailón A, Padilla G et al (2019) Seed oil quality of Brassica napus and Brassica rapa germplasm from Northwestern Spain. Foods 8(8):292

Carvalho CA, Fernandes KM, Matta SLP et al (2011) Evaluation of antiulcerogenic activity of aqueous extract of Brassica *oleracea* var. *capitata* (cabbage) on wistar rat gastric ulceration. Arq Gastroenterol 48:276–282

Chandra AK (2010) Chapter 42 – Goitrogen in food: cyanogenic and flavonoids containing plant foods in the development of goiter A2 – Watson, Ronald Ross. In: Bioactive foods in promoting health. Academic, San Diego, pp 691–716

Cohen JH, Kristal AR, Stanford JL (2000) Fruit and vegetable intakes and prostate cancer risk. J Natl Cancer Inst 92:61–68

Collett MG, Stegelmeier BL, Tapper BA (2014) Could nitrile derivatives of turnip (Brassica rapa) glucosinolates be hepato- or cholangiotoxic in cattle? J Agric Food Chem 62(30):7370–7375

Deng X-S, Tuo J, Poulsen HE, Loft S (1998) Prevention of oxidative DNA damage in rats by Brussels sprouts. Free Radic Res 28(3):323–333

Dinkova-Kostova AT, Kostov RV (2012) Glucosinolates and isothiocyanates in health and disease. Trends Mol Med 18:337–347

Dinkova-Kostova AT, Fahey JW, Wade KL et al (2007) Induction of the phase 2 response in mouse and human skin by sulforaphane-containing broccoli sprout extracts. Cancer Epidemiol Biomarkers Prev 16:847–851

Duchnowicz P, Bors M, Podsedek A et al (2012) Effect of polyphenols extracts from Brassica vegetables on erythrocyte membranes (in vitro study). Environ Toxicol Pharmacol 34:783–790

Enye JC, Chineke HN, Onubeze DPM, Nweke I (2013) Evaluation of the healing effects of aqueous extracts of Musa Paradisiaca (unripe plantain) and Brassica oleracea (cabbage) on peptic ulcer. IOSR J Dental Med Sci 8:40–46

Fahey JW (2016) Brassica: characteristics and properties. In: Encyclopedia of food and health. Elsevier Inc., pp 469–477

Fimognari C, Turrini E, Ferruzzi L et al (2012) Natural isothiocyanates: genotoxic potential versus chemoprevention. Mutat Res 750:107–131

Fowke JH, Chung FL, Jin F et al (2003) Urinary isothiocyanate levels, brassica, and human breast cancer. Cancer Res 63(14):3980–3986

Fusari CM, Nazareno MA, Locatelli DA et al (2020) Phytochemical profile and functionality of Brassicaceae species. Food Biosci 36:100606

Georgieva S, ChristovaBagdassarian V, Atanassova M (2014) Comparative evaluation of the polyphenol composition and antioxidant capacity of propolis and Echinacea purpurea. J Exp Integr Med 4(1):51

Gómez-Campo C (1999) Biology of Brassica coenospecies. Elsevier, Amsterdam

Gupta US (2011) Brassica vegetables, what's new about crop plants. CRC Press, pp 378–402

Hanschen FS, Herz C, Schlotz N et al (2015) The Brassica epithionitrile 1-cyano-2,3-epithiopropane triggers cell death in human liver cancer cells in vitro. Mol Nutr Food Res 59:2178–2189

Hanschen FS, Klopsch R, Oliviero T et al (2017) Optimizing isothiocyanate formation during enzymatic glucosinolate breakdown by adjusting pH value, temperature and dilution in Brassica vegetables and Arabidopsis thaliana. Sci Rep 7(1):40807

Hedges LJ, Lister C (2006) Nutritional attributes of Brassica vegetables. Crop & Food Research Confidential report

Holst B, Williamson G (2004) A critical review of the bioavailability of glucosinolates and related compounds. Nat Prod Rep 21:425–447

Jabeen N (2020) Agricultural, economic and societal importance of Brassicaceae plants. In: The plant family Brassicaceae: biology and physiological responses to environmental stresses. Springer, pp 45–128

Jahangir M, Kim HK, Choi YH, Verpoorte R (2009) Health-affecting compounds in Brassicaceae. Compr Rev Food Sci F 8:31–43

Jan N, Qazi HA, Ramzan S, Riffat John R (2018) Developing stress-tolerant plants through ın vitro tissue culture: family Brassicaceae. In: Biotechnologies of crop improvement, vol 1. Springer, pp 327–372

Joseph MA, Moysich KB, Freudenheim JL et al (2004) Cruciferous vegetables, genetic polymorphisms in glutathione S-transferases M1 and T1, and prostate cancer risk. Nutr Cancer 50:206–213

Kalpana D, Gayathri R, Gunassekaran GR et al (2013) Apoptotic role of natural isothiocyanate from broccoli (Brassica oleracea italica) in experimental chemical lung carcinogenesis. Pharm Biol 51:621–628

Kapusta-Duch J, Kopec A, Piatkowska E et al (2012) The beneficial effects of Brassica vegetables on human health. Rocz Panstw Zakl Hig 63:389–395

Kataya HA, Hamza AA (2008) Red cabbage (Brassica oleracea) ameliorates diabetic nephropathy in rats. Evid Based Complement Alternat Med 5:281–287

Kelloff GJ, Crowell JA, Steele VE et al (2000) Progress in cancer chemoprevention: development of diet-derived chemopreventive agents. J Nutr 130:467–471

Kim MK, Park JHY (2009) Cruciferous vegetable intake and the risk of human cancer: epidemiological evidence. Proc Nutr Soc 68:103–110

Komatsu W, Miura Y, Yagasaki K (1998) Suppression of hypercholesterolemia in hepatoma-bearing rats by cabbage extract and its component, S-methyl-l-cysteine sulfoxide. Lipids 33:499–503

Kyung KH, Lee YC (2001) Antimicrobial activities of sulfur compounds derived from S-alk (en) yl-L-cysteine sulfoxides in Allium and Brassica. Food Rev Int 17(2):183–198

Li YM, Chan HYE, Huang Y, Chen ZY (2008) Broccoli (Brassica oleracea var. botrytis L.) improves the survival and up-regulates endogenous antioxidant enzymes in Drosophila melanogaster challenged with reactive oxygen species. J Sci Food Agric 88(3):499–506

Li Y, Zhou B, Teng Z et al (2023) Improved metabolomic approach for evaluation of phytochemicals in mustard, kale, and broccoli microgreens under different controlled environment agriculture conditions. J Agric Food Res 14:100719

Liu X, Song Q, Tang Y et al (2013) Human health risk assessment of heavy metals in soil-vegetable system: a multi-medium analysis. Sci Total Environ 463:530–540

Lone WA, Majeed N, Umer Yaqoob U, Riffat John R (2021) Exogenous brassinosteroid and jasmonic acid improve drought tolerance in *Brassica rapa* L. genotypes by modulating osmolytes, antioxidants and photosynthetic system. Plant Cell Rep. https://doi.org/10.1007/s00299-021-02763-9

Masci A, Mattioli R, Costantino P et al (2015) Neuroprotective effect of Brassica oleracea sprouts crude juice in a cellular model of Alzheimer's disease. Oxidative Med Cell Longev 2015:781938

Matthews ZM, Parton KH, Collett MG (2020) Investigating the cause of Brassica-associated liver disease (BALD) in cattle: progoitrinderived nitrile toxicosis in rats. Toxicon 5:100021

Milanesi N, Gola M (2016) Irritant contact dermatitis caused by Savoy cabbage. Contact Derm 74(1):60–61

Mollica A, Stefanucci A, Zengin G et al (2018) Polyphenolic composition, enzyme inhibitory effects ex-vivo and in-vivo studies on two Brassicaceae of north-central Italy. Biomed Pharmacother 107:129–138

Morales-Lopez J, Centeno-Alvarez M, Nieto-Camacho A et al (2017) Evaluation of antioxidant and hepatoprotective effects of white cabbage essential oil. Pharm Biol 55:233–241

Murashima M, Watanabe S, Zhuo XG et al (2004) Phase 1 study of multiple biomarkers for metabolism and oxidative stress after one-week intake of broccoli sprouts. Biofactors 22:271–275

Murillo G, Mehta RG (2001) Cruciferous vegetables and cancer prevention. Nutr Cancer 41(1–2):17–28

Nor NDM, Lignou S, Bell L et al (2020) The relationship between glucosinolates and the sensory characteristics of steamed-pureed turnip (Brassica Rapa subsp. Rapa L). Foods 9(11):1719

Oguwike FN, Offor CC, Nwadighoha AN, Ebede SO (2014) Evaluation of efficacy of cabbage juice (Brassica oleracea linne) as potential antiulcer agent and its effect on the haemostatic mechanism of male albino wistar rats. IOSR J Dental Med Sci 13:92–97

Ongaro A, Povolo C, Zagotto G, Ribaudo G (2019) HPLC and NMR quantification of bioactive compounds in flowers and leaves of Brassica rapa: the influence of aging. Nat Prod Res 34(9):1288–1291

Paul S, Bin Emran T, Saha D, Zahid Hosen S (2012) Phytochemical and antimicrobial activity of the plant extracts of Brassica oleracea against selected microbes. Asian J Pharm Med Sci 2(2):30–34

Paul S, Geng CA, Yang TH et al (2019) Phytochemical and health-beneficial progress of turnip (Brassica rapa). J Food Sci 84(1):19–30

Pellegrini N, Chiavaro E, Gardana C et al (2010) Effect of different cooking methods on color, phytochemical concentration, and antioxidant capacity of raw and frozen Brassica vegetables. J Agric Food Chem 58(7):4310–4321

Rai GK, Bagati S, Rai PK et al (2018) Fatty acid profiling in rapeseed mustard (Brassica species). Int J Curr Microbiol App Sci 7(05):148–157

Rodriguez-Cantu LN, Gutierrez-Uribe JA, Arriola-Vucovich J et al (2011) Broccoli (Brassica oleracea var. italica) sprouts and extracts rich in glucosinolates and isothiocyanates affect cholesterol metabolism and genes involved in lipid homeostasis in hamsters. J Agric Food Chem 59:1095–1103

Salehi B, Quispe C, Butnariu M et al (2021) Phytotherapy and food applications from Brassica genus. Phytother Res 35(7):3590–3609

Šamec D, Piljac-Žegarac J, Bogovic M et al (2011) Antioxidant potency of white (Brassica oleracea L. var. capitata) and Chinese (Brassica rapa L. var. pekinensis [Lour.]) cabbage: the influence of development stage, cultivar choice and seed selection. Sci Hortic 128(2):78–83

Šamec D, Kruk V, Ivanišević P (2019) Influence of seed origin on morphological characteristics and phytochemicals levels in *Brassica oleracea* var. *acephala*. Agronomy 9(9):502

Sanlier N, Guler SM (2018) The benefits of Brassica vegetables on human health. J Hum Health Res 1(1):1–13

Sharafi Y, Majidi MM, Goli SAH, Rashidi F (2015) Oil content and fatty acids composition in Brassica species. Int J Food Prop 18(10):2145–2154

Sharma A, Rai PK, Prasad S (2018) GC-MS detection and determination of major volatile compounds in Brassica juncea L. leaves and seeds. Microchem J 138:488–493

Sherman JM, Hodge HM (1936) The bactericidal properties of certain plant juices. J Bacteriol 31:96

Shirai Y, Fujita Y, Hashimoto R et al (2015) Dietary intake of sulforaphane-rich broccoli sprout extracts during juvenile and adolescence can prevent phencyclidine-induced cognitive deficits at adulthood. PLoS One 10:e0127244

Singh Chauhan E, Tiwari A, Singh A, Anamika Tiwari C (2016) Phytochemical screening of red cabbage (Brassica oleracea) powder and juice – a comparative study. J Med Plants Stud 4(5):196–199

Song L, Thornalley PJ (2007) Effect of storage, processing and cooking on glucosinolate content of Brassica vegetables. Food Chem Toxicol 45:216–224

Tang L, Zhang Y, Jobson HE et al (2006) Potent activation of mitochondria-mediated apoptosis and arrest in S and M phases of cancer cells by a broccoli sprout extract. Mol Cancer Ther 5:935–944

Tanongkankit Y, Chiewchan N, Devahastin S (2010) Effect of processing on antioxidants and their activity in dietary fiber powder from cabbage outer leaves. Dry Technol 28(9):1063–1071

Terry P, Wolk A, Persson I, Magnusson C (2001) Brassica vegetables and breast cancer risk. JAMA 285:2975

Tiku AB, Abraham SK, Kale RK (2008) Protective effect of the cruciferous vegetable mustard leaf (Brassica campestris) against in vivo chromosomal damage and oxidative stress induced by γ-radiation and genotoxic chemicals. Environ Mol Mutagen 49(5):335–342

Truong T, Baron-Dubourdieu D, Rougier Y, Guenel P (2010) Role of dietary iodine and cruciferous vegetables in thyroid cancer: a countrywide case-control study in New Caledonia. Cancer Causes Control 21:1183–1192

Tsunoda S, Hinata K, Gómez-Campo C (1980) *Brassica* crops and Wild Allies: biology and breeding. Japan Scientific Press, p 354

Uddin MM, Zakeel MCM, Zavahir JS et al (2021) Heavy metal accumulation in rice and aquatic plants used as human food: a general review. Toxics 9(12):360

Vale APP, Santos J, Melia N et al (2015) Phytochemical composition and antimicrobial properties of four varieties of Brassica oleracea sprouts. Food Control 55:248–256

Verhagen H, de Vries A, Nijhoff WA et al (1997) Effect of Brussels sprouts on oxidative DNA-damage in man. Cancer Lett 114(1-2):127–130

Vig AP, Rampal G, Thind TS, Arora S (2009) Bio-protective effects of glucosinolates – a review. LWT Food Sci Technol 42(10):1561–1572

Walters DG, Young PJ, Agus C et al (2004) Cruciferous vegetable consumption alters the metabolism of the dietary carcinogen 2-amino-1-methyl-6-phenylimidazo[4,5-b]pyridine (PhIP) in humans. Carcinogenesis 25:1659–1669

Wang C, Kuang L, Tian Z et al (2025) Genome-wide association study reveals the genetic basis of vitamin C content in rapeseed (Brassica napus L.) seedlings. Front Plant Sci 16:1649023

Wani UM, Majeed ST, Raja V et al (2021) Ectopic expression of a novel cold-resistance protein 1from *Brassica oleracea* promotes tolerance to chilling stress in transgenic tomato. Sci Rep 11:16574

Wu YD, Lou YJ (2007) Brassinolide, a plant sterol from pollen of *Brassica napus* L., induces apoptosis in human prostate cancer PC-3 cells. Die Pharmazie. Int J Pharm Sci 62(5):392–395

Wu QJ, Yang Y, Wang J (2013) Cruciferous vegetable consumption and gastric cancer risk: a meta-analysis of epidemiological studies. Cancer Sci 104:1067–1073

Xiaomeng Z, Qiong J, Xin J et al (2025) *Brassica* vegetables – an undervalued nutritional goldmine. Hortic Res 12(2):uhae302

Xu Z, Wang S, Ji H et al (2016) Broccoli sprout extract prevents diabetic cardiomyopathy via Nrf2 activation in db/db T2DM mice. Sci Rep 6:30252

Zhang Y, Munday R, Jobson HE et al (2006) Induction of GST and NQO1 in cultured bladder cells and in the urinary bladders of rats by an extract of broccoli (Brassica oleracea italica) sprouts. J Agric Food Chem 54:9370–9376

Zhao H, Lin J, Grossman H et al (2007) Dietary isothiocyanates, GSTM1, GSTT1, NAT2 polymorphisms and bladder cancer risk. Int J Cancer 120:2208–2213

Chapter 8
Economic Importance

8.1 Traditional Uses

*Brassica*s can be divided into three main categories according to the part of the plant used: oilseeds, vegetables and condiments. The economically important members of the *Brassica* group are widely used as cooked vegetables and are widely accepted as functional foods and dietary supplements (Faust et al. 2017; Li et al. 2023). Many *Brassica* crops continue to be consumed following traditional gastronomic practices such as microwave heating, ice cream production, fermentation, steaming, boiling and pan-frying (Murillo and Mehta 2001). The crops from this group generally include vegetables and forage plants, which play an important role in the human diet, being consumed globally and standing among the most important food crops in China, Japan, the Indian Subcontinent and Europe.

The following list categorises popular *Brassica* cultivars together with their edible parts and vernacular names (Cartea et al. 2011; Abdel-Razzak 2021; Petruzzello 2023):

Leaves:

Brassica oleracea var. *acephala*—Kale
Brassica oleracea var. *alboglabra*—Chinese kale
Brassica rapa var. *chinensis*—Pak choi, bok choy, Chinese mustard
Brassica rapa var. *narinosa*—Chinese flat cabbage, taitsai
Brassica rapa var. *nipposinica*—Mibuna, mizuna
Brassica rapa var. *pekinensis*—Chinese cabbage
Brassica rapa var. *perviridis*—Komatsuna, tendergreen
Brassica rapa var. *parachinensis*—Choy sum
Brassica rapa var. *rapa*—Turnip
Brassica napus var. *pabularia*—Leaf rape

Terminal leaf buds (heads):

M. Ozturk et al., *Climate-Resilient* Brassicas,
https://doi.org/10.1007/978-3-032-12709-9_8

Brassica oleracea var. *capitata*—Cabbage
Brassica oleracea var. *capitata sabauda*—Savoy cabbage

Loose heads:

Brassica oleracea var. *costata*—Tronchuda

Vegetative buds:

Brassica oleracea var. *gemmifera*—Brussels sprouts

Inflorescences:

Brassica oleracea var. *botrytis*—Cauliflower
Brassica oleracea var. *italica*—Broccoli

Stem:

Brassica oleracea var. *gongylodes*—Kohlrabi

Seeds:

Brassica rapa var. *dichotoma*—Brown sarson, toria
Brassica rapa var. *oleifera*—Turnip rape, rapeseed
Brassica rapa var. *trilochularis*—Yellow sarson

Roots:

Brassica rapa var. *rapa*—Turnip
Brassica napus var. *napobrassica*—Swede, rutabaga

Shoots:

Brassica rapa var. *rapa*—Turnip
Brassica rapa var. *ruvo*—Broccoleto

The species belonging to the *Brassica* are at the same time used in traditional medicine and have also been used to treat farm animals (Kumar and Bharati 2013).

In Asia, a phytosterol-enriched *Brassica rapa* pollen product produced in the form of Qianlie Kang Pule'an tablets is reported to prove beneficial for patients with chronic prostatitis, urinary incontinence and swollen joints (Wang et al. 2015). According to national traditional medicinal data, various *Brassica* species, including the wild variety of *B. oleracea*, are reported to be effective against neuralgia and rheumatic diseases in Lebanon, in Bangladesh, *B. oleracea* has been used to treat heart disease and in Europe, in particular Italy; cabbage has traditionally been used to treat bruises, rheumatic pain, wounds and scurvy (Marc et al. 2008; Passalacqua et al. 2007; Emran et al. 2015). Another Italian ethnobotanical study has highlighted the use of heated *B. rupestris* and *B. oleracea* leaves to treat rheumatic pain (Tuttolomondo et al. 2014). A study from Romania reports the use of the roots and leaves of *B. napus* to treat abscesses, boils, acne and eczema (Gilca et al. 2018). In Australia, an ointment made from the leaves and seeds of *B. rapa* has been used to relieve sore muscles and disinfect wounds (Rahman et al. 2018), whereas in Brazil, it has been demonstrated that juice made from *B. oleracea* leaves alleviates symptoms of stomach ulcers (Gonçalves et al. 2012).

The traditional uses of some of the most commonly cultivated *Brassica* vegetables are summarised below:

8.1.1 Brassica oleracea *Group Vegetables*

Cabbage Historically, cabbage juice has been recognised as an effective laxative and treatment for constipation, as it is reported to have been commonly used during ancient civilisations, namely the Greeks, Romans and Egyptians (Hatfield 2004; Fahey 2016). Works published by figures such as Pythagoras, Diogenes and Cato emphasise cabbage's potential to prevent disease and enhance quality of life when incorporated into daily consumption routines. During the Middle Ages, cabbage was also used for medicinal purposes, such as making cough syrup and dressing wounds. In the 1700s, sailors consumed cabbage for its vitamin C content as well as to treat gangrene (Fahey 2016).

Keeping all these facts in view, the traditional use of cabbage in our society continues to this day. In particular, the raw cabbage is still considered one of the most effective traditional remedies for headaches, which has also been used in traditional folk medicine to prevent sunstroke, reduce fever and alleviate foot pain (Hatfield 2004; Fahey 2016). In traditional European medicine, we come across many reports where the cabbage has been evaluated to treat various ailments such as sore throats, hoarseness, rheumatism, colic and melancholy (Hatfield 2004; Fahey 2016). The ethnobotanical data also suggest that cabbage leaf compresses effectively the reduction in breast engorgement during breastfeeding (Ayers 2000).

Cabbage is a popular staple of the global human diet, which is largely due to local production methods, consumer preferences, affordability and availability in local markets. However, there are variations between countries and regions (Fig. 8.1). This vegetable plays an important role in the fresh produce as well as processed food sectors in many parts of the world. In North America and Europe, it is one of the ten most commonly sold and voluminous vegetables (Fahey 2016).

White cabbage is often used in salads or boiled and also consumed as sauerkraut, a fermented product (Fig. 8.2) (Nilnakara et al. 2009; Samec et al. 2017; Fahey 2016). Another popular traditional dish is pickled white cabbage, prepared by slicing raw cabbage into thin slices and storing these in brine water (Wouters et al. 2013; Samec et al. 2017).

This type of consumption is common in many cultures around the world. For example, indigenous people in the southern Appalachian region of the United States have traditionally used boiled or fried cabbage, cabbage juice and sauerkraut to treat stomach upsets and cleanse the intestines (Cavender 2006). The clinical studies have demonstrated that drinking cabbage juice can significantly aid the healing of

Fig. 8.1 Some of the *Brassica* vegetables sold in local markets in Turkiye

peptic ulcers, and even today, many diet and nutritional therapists recommend cabbage for gastrointestinal health (Cheney 1949; Meletis and Zabriskie 2008).

In the light of growing consumer demand for probiotic products and beverages in recent years, new methods have been developed to produce healthier, safer and tastier products. One such approach is to use functional yeast cultures with probiotic characteristics in sauerkraut and sauerkraut juice in order to prepare and sell added-value food products (Suskovic et al. 2010). Many studies have also been conducted

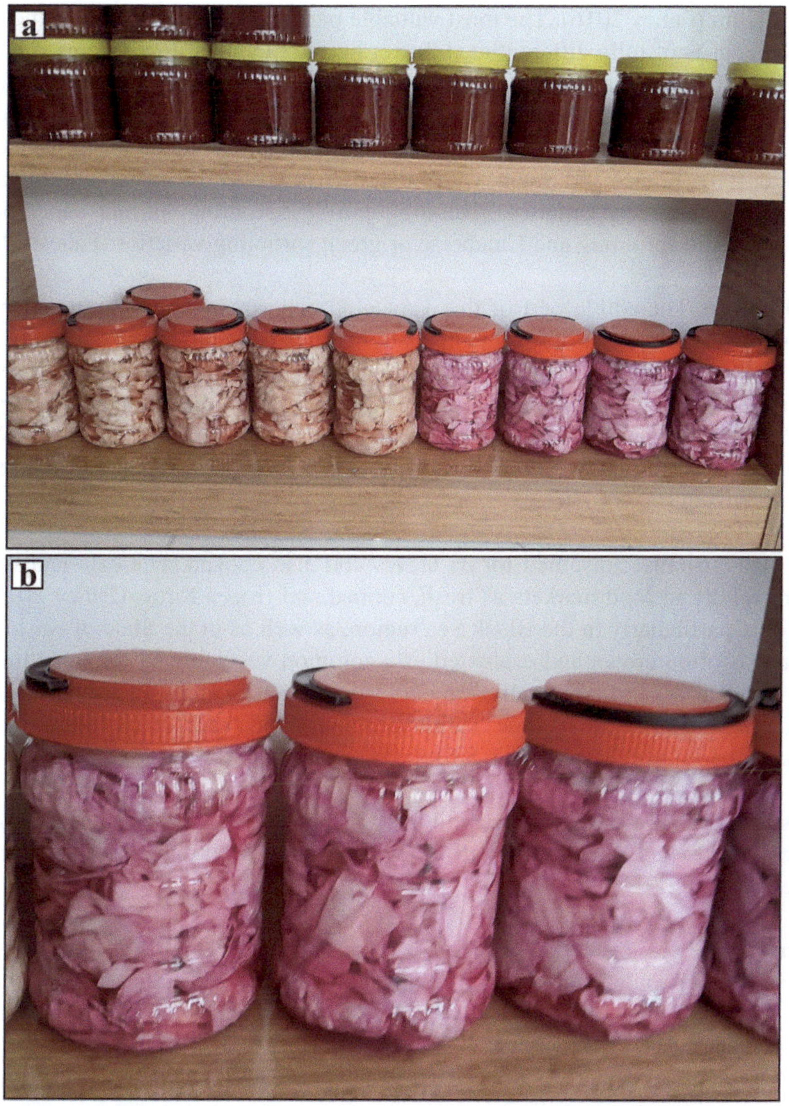

Fig. 8.2 Pickled white cabbage, prepared and sold by the locals in Turkiye. (**a**) General view of the pickled cabbage from a shop, (**b**) A closer view of the plastic bottles with pickles

to evaluate the suitability of white cabbage as a raw material for producing probiotic cabbage juice. The production process uses different strains of *Lactobacillus plantarum*, *Lactobacillus casei* and *Lactobacillus delbrueckii* to create healthier products (Yoon et al. 2006).

Broccoli Over the past three decades, broccoli has become a popular vegetable because of its health benefits. These include anticarcinogenic features, high fibre content, and an abundance of bioactive compounds such as vitamins C and E, folate and calcium (Fahey 2016). The most valuable part of the broccoli plant is the head, which consists of fully differentiated flower buds and tender upper stems (Fig. 8.1) (Fahey 2016). Commercially, the vegetable is usually sold as fresh or frozen florets, but it is often consumed in the form of salads and also commonly cooked and served on its own. The main regions where broccoli is cultivated are North America, Mexico, Italy, France and Spain. Currently, there are over 100 commercially available hybrid broccoli varieties, including purple sprouting, purple cape, purple Sicilian, white sprouting and Calabrese or green sprouting varieties (Fahey 2016).

Cauliflower The edible part of this vegetable consists of undifferentiated, white (chlorophyll-free) shoot tips, growing as tight clusters on top of the fleshy stem tissue (Fig. 8.1). These tips are commonly known as "curds" (Fahey 2016). The vegetable is widely cultivated commercially in France, Italy, the United Kingdom and North America, but the purple, green and orange varieties are also available. While similar to broccoli, this vegetable is also pickled. It is sold in food markets, either fresh or frozen (Fahey 2016).

Kale It is usually consumed for its leaves and also cooked. The kale is also sold commercially in food markets as fresh, canned and frozen forms (Fahey 2016). In Anatolia, particularly in the Black Sea region, as well as in the State of Kashmir in India, it is often grown and evaluated as a common vegetable by the local inhabitants. The leaves of this vegetable are harvested in season and sold at local markets. The popular local dishes include "karalahana soup" and "karalahana dolma" (stuffed black cabbage) and "Haak" (Fig. 8.3).

Brussels sprouts Also known as miniature cabbages, these vegetables are usually eaten after being steamed or boiled. They are sold in food markets, either fresh or frozen (Fahey 2016).

Chinese kale This vegetable is usually eaten for its flower buds, stalks and young leaves. It is a popular ingredient in salads and stir-fries in Southeast Asia, particularly in China and Taiwan. It has lately become popular in Japanese, Western European and American cuisines (Fahey 2016).

8.1.2 Brassica rapa *Group Vegetables*

Turnip This starchy vegetable is usually consumed in the cooked form. It can be pickled or added to mixed vegetables, and the greens can also be eaten fresh when in season (Sun 2015; Fahey 2016).

Broccoli raab This vegetable has thinner stems and smaller florets than broccoli and tastes bitter. It is considered a popular delicacy in Italy and is often lightly sautéed with garlic (Fahey 2016).

Fig. 8.3 Locally produced kale and its derivatives. (**a**) General view of kale; (**b**) kale being cooked, (**c**) kale soup, (**d**) stuffed kale being cooked

Chinese cabbage It has been cultivated and consumed in northern China for 1600 years, and is also an important crop in Korea, Taiwan and Japan (Sun 2015; Fahey 2016).

Chinese mustards—pak choi Throughout history, pak choi has been one of the most important vegetables cultivated in Asia, particularly in China. Its leaves are usually eaten fresh. However, during periods of drought when fresh vegetables are scarce, they can be boiled and dried for later use (Fahey 2016).

Tendergreens The large, dark green and mildly aromatic leaves of this vegetable can be consumed fresh or pickled. It is popular in Korea, Taiwan and Japan (Fahey 2016).

Turnip rape The oil obtained from the seeds of this vegetable is used for cooking. Its leaves can also be used as a garnish (Sun 2015; Fahey 2016).

8.1.3 **Brassica napus** ***Group Vegetables***

Swede It is commonly cultivated for use as animal feed, but the leaves are also cooked and consumed by humans, also used for pickling (Fahey 2016).

Colza It is a popular vegetable widely consumed in India and China. The leaves and sprouted seeds can be eaten in salads, and the florets can be prepared and eaten like broccoli (Fahey 2016).

8.2 Industrial Uses

Biomass, bioethanol and biodiesel are attracting attention as vital components of a sustainable, renewable energy system. Currently, bioethanol and biodiesel are emerging as key alternatives to fossil fuels in the liquid fuel sector. Biodiesel is popular in Europe and Asia, while bioethanol dominates in the United States (Milazzo et al. 2013; Ansari et al. 2023; Ozturk et al. 2015a, b, 2020, 2023a, b, c, 2024a, b; Ozturk and Hakeem 2018, 2019a, b). Despite concerns about potential ecosystem and land degradation, as well as disruption to food supplies, biodiesel is gaining global popularity due to its potential to enhance future energy security and reduce greenhouse gas emissions (Searchinger et al. 2008; Milazzo et al. 2013; Surriya et al. 2014). Its popularity is primarily driven by political, economic, environmental, health and security considerations, particularly with regard to ensuring national energy security, balancing national and regional trade and promoting rural development. Its popularity is also driven by its non-ecotoxicity and lower greenhouse gas emissions (Milazzo et al. 2013). Some of the *Brassica* species are also evaluated in phytoremediation (Ozturk et al. 2012).

In addition to its role in sustainable agriculture, some of the species of *Brassica* are evaluated in biodiesel production. The reason for this is the presence of high oil content, i.e., 31–48% in dry seeds, and a favourable fatty acid profile, characterised by a low polyunsaturated to high monounsaturated fatty acid ratio. The rapeseed-based biodiesel is reported to possess the ability to reduce CO_2 emissions by 65–95% compared to traditional diesel (Verma et al. 2019). The presence of oxygen atoms in rapeseed oil contributes to lower emissions of carbon monoxide (CO), hydrocarbons (HC) and particulate matter (PM). However, this same oxygen content raises combustion temperatures, which can enhance combustion efficiency but also lead to increased nitrogen oxide (NOx) emissions (Ge et al. 2017). In addition to this, the rapeseed straw can also be employed in the generation of succinic acid and cellulosic bioethanol following dilute-acid pre-treatment (Kuglarz et al. 2018).

The rapeseed oil is one of the most important vegetable oils used in food production within the EU (Milazzo et al. 2013). Although its market is highly volatile, the EU has traditionally sourced most of its biodiesel from domestically grown rapeseed, a relatively expensive feedstock. However, the EU now imports rapeseed primarily from Canada, India and Ukraine (Rosillo-Campo et al. 2009; Lamers et al. 2011; Milazzo et al. 2013). The main European countries cultivating large areas of rapeseed in north-central Europe are France, Germany, Poland and the United Kingdom (Stephenson et al. 2008; Milazzo et al. 2013).

"Canola" is a trademarked term referring to a product resulting from traditional plant breeding techniques, being suitable for human and animal consumption, as it contains less than 2% erucic acid and less than 30 mmol of glucosinolates (Milazzo et al. 2013). Since 1965, around 90% of the canola produced in Canada has been genetically modified to be herbicide-resistant for more sustainable use (Milazzo et al. 2013).

In addition to the facts cited above, the rapeseed oil holds substantial industrial potential beyond biodiesel production. It serves as a renewable raw material in the manufacture of various products such as paints, detergents, inks and laundry agents (Mohamed et al. 2025). Additionally, it is employed in the formulation of hair conditioners, photographic materials, and biodegradable plastics. The high content of long-chain fatty acids and triglycerides in rapeseed oil makes it particularly suitable for these applications. Experimental studies have demonstrated its effectiveness in producing biopolymer precursors and surfactant agents, highlighting its role as a sustainable alternative to petroleum-derived inputs in the industrial sector (Mohamed et al. 2025).

China is one of the major producers of rapeseed; in Chile, it is one of the main oilseed crops cultivated for potential biodiesel production (Milazzo et al. 2013). Currently, Chilean rapeseed is primarily used as feed for the national salmon industry, which is intended for export (Milazzo et al. 2013). Moreover, one of the most valuable by-products of the biodiesel production process is rapeseed meal, which can be used as animal feed, for direct combustion or for biogas production (Gartner and Reinhardt 2003). The crop pressings have also been reported to have potential value as biopesticides (Milazzo et al. 2013).

When it comes to producing protein-rich animal feed, canola meal, derived from rapeseed, is the second most important ingredient after soybeans. The rapeseed contains up to 40% oil and 20% protein; the heat-treated rapeseed meal has the potential to significantly increase the nutritional value of feed for milk production (Milazzo et al. 2013; Friedt et al. 2018; Hagos et al. 2020).

The remaining straw from rapeseed cultivation can occasionally be harvested for use as fuel. In Denmark, such straw has been used as a biofuel (Schmidt 2010; Milazzo et al. 2013). However, it is more commonly used for its natural ability to improve soil fertility and increase humus content following its use during recropping. This environmentally friendly practice has led to the widespread incorporation of rapeseed straw into soil in countries such as Germany, Sweden, the United Kingdom and Australia (Schmidt 2010; Milazzo et al. 2013; Friedt et al. 2018; Hagos et al. 2020).

8.3 Other Uses

The waste produced when processing white cabbage leaves has been reported to have the potential to be used as fertiliser and animal feed, due to its high dietary fibre and phytochemical contents (Nilnakara et al. 2009; Samec et al. 2017). Similarly, the high crude protein levels in turnips—a variety of *Brassica*—offer considerable economic benefits in terms of animal nutrition (Neilsen et al. 2008; Tiryakioglu and Turk 2012). Another study reports that turnips increase the nutrient intake and growth performance of fish in aquaculture systems, thus highlighting their potential as a natural source of fish feed (Haliloğlu et al. 2012; Abdel-Razzak 2021). *Brassica napus* has also been reported as a potential natural source of nutrition for cattle, poultry, and fish due to its high-quality edible oil and high-protein flour (Jabeen 2020).

Brassica vegetables contain important pigments, the anthocyanins, which have been highlighted in the reports as extracts worth being used as natural dyes (Favela-González et al. 2020). Haddar et al. (2018) have conducted a study on dyeing wool and silk fibres, demonstrating the effectiveness of using anthocyanin-rich plant extracts from red cabbage for dyeing fabrics. When evaluated in the washing, rubbing and exposure to light, these extracts are found to be an excellent source of natural dyes.

The oil obtained from *Brassica napus* is commonly used for cooking, but has also been reported to have various potential industrial applications in sectors such as food production, pharmaceuticals, cosmetics and textiles. In the food industry, it is used in sandwich spreads, coffee whiteners and creams. In the pharmaceutical and cosmetics industries, it is used for eliminating dandruff, as a hair tonic and skin ointment, and as a laxative. Furthermore, oil from *B. napus* has reportedly been used in printing inks, plastic packaging and as insecticides (Jabeen 2020).

In addition to the multitude of applications cited above, other uses of *Brassica* species include their use as ornamental plants, being frequently employed in urban landscape design. However, their evaluation in the urban areas, where anthropogenic influences are intense, can vary depending on the area's socioeconomic structure, geographical location, urban history, land use, the attitudes of local government and personal preferences (Altay 2012). The use of *Brassica oleracea* var. *acephala* (ornamental kale) as a seasonal ornamental plant has recently been observed in various urban areas (Altay et al. 2010; Osma et al. 2010; Altay 2012; Eskin et al. 2012).

References

Abdel-Razzak HS (2021) Turnip (Brassica rapa var. rapa L.) breeding. In: Advances in plant breeding strategies: vegetable crops: volume 8: bulbs, roots and tubers. Springer International Publishing, Cham, pp 345–405

Altay V (2012) Mustafa Kemal Üniversitesi Tayfur Ata Sökmen Kampüsü (Hatay)'nün süs bitkileri. Karadeniz Fen Bilim Derg 3(1):11–26

Altay V, Ozyiğit İİ, Yarci C (2010) Urban flora and ecological characteristics of the Kartal District (Istanbul): a contribution to urban ecology in Turkiye. Sci Res Essays 5(2):183–200

Ansari MKA, Unal BT, Ozturk M, Owens P (2023) Plants as medicine and aromatics: pharmacognosy, ecology and conservation. CRC Press, Taylor & Francis, Boca Raton, pp 1–330

Ayers JF (2000) The use of alternative therapies in the support of breastfeeding. J Hum Lact 16:52–56

Cartea ME, Francisco M, Soengas P, Velasco P (2011) Phenolic compounds in Brassica vegetables. Molecules 16:251–280

Cavender A (2006) Folk medical uses of plant foods in southern Appalachia, United States. J Ethnopharmacol 108:74–84

Cheney G (1949) Rapid healing of peptic ulcers in patients receiving fresh cabbage juice. Calif Med 70:10–15

Emran TB, Rahman MA, Uddin MMN et al (2015) Effects of organic extracts and their different fractions of five Bangladeshi plants on in vitro thrombolysis. BMC Complement Altern Med 15(1):128

Eskin B, Altay V, Özyiğit İİ, Serin M (2012) Urban vascular flora and ecologic characteristics of the Pendik District (Istanbul-Turkey). Afr J Agric Res 7(4):629–646

Fahey JW (2016) Brassica: characteristics and properties. In: Encyclopedia of food and health. Elsevier Inc., pp 469–477. https://doi.org/10.1016/B978-0-12-384947-2.00083-0

Faust D, Nikolova T, Wätjen W et al (2017) The Brassica-derived phytochemical indolo [3, 2-b] carbazole protects against oxidative DNA damage by aryl hydrocarbon receptor activation. Arch Toxicol 91(2):967–982

Favela-González KM, Hernández-Almanza AY, De la Fuente-Salcido NM (2020) The value of bioactive compounds of cruciferous vegetables (Brassica) as antimicrobials and antioxidants: a review. J Food Biochem 44(10):e13414

Friedt W, Tu J, Fu T (2018) Academic and economic importance of *Brassica napus* rapeseed. In: The Brassica napus genome. Springer, Cham, pp 1–20

Gartner SO, Reinhardt GA (2003) Life cycle assessment of biodiesel: update and new aspects. IFEU, Heidelberg; Proj No 530/025

Ge JC, Yoon SK, Choi NJ (2017) Using canola oil biodiesel as an alternative fuel in diesel engines: a review. Appl Sci 7(9):881

Gilca M, Tiplica GS, Salavastru CM (2018) Traditional and ethnobotanical dermatology practices in Romania and other Eastern European countries. Clin Dermatol 36(3):338–352

Gonçalves ALM, Lemos M, Niero R et al (2012) Evaluation of the genotoxic and antigenotoxic potential of Brassica oleracea L. var. acephala D.C. in different cells of mice. J Ethnopharmacol 143(2):740–745

Haddar W, Ben Ticha M, Meksi N, Guesmi A (2018) Application of anthocyanins as natural dye extracted from Brassica oleracea L. var. capitata f. rubra: dyeing studies of wool and silk fibres. Nat Prod Res 32(2):141–148

Hagos R, Shaibu AS, Zhang L et al (2020) Ethiopian mustard (*Brassica carinata* A. Braun) as an alternative energy source and sustainable crop. Sustainability 12(18):7492

Haliloğlu Hİ, Arslan M, Lee B-J, Dabrowski K (2012) The effects of dietary turnip (Brassica rapa) and biofuel algae on growth and chemical composition in rainbow trout (Oncorhynchus mykiss) juveniles. Turk J Fish Aquat Sci 12:323–329

Hatfield G (2004) Encyclopedia of folk medicine: old world and new world traditions. ABC-CLIO, Santa Barbara, pp 59–60

Jabeen N (2020) Agricultural, economic and societal importance of Brassicaceae plants. In: The plant family Brassicaceae: biology and physiological responses to environmental stresses. Springer Nature Singapore, Singapore, pp 45–128

Kuglarz M, Alvarado-Morales M, Dąbkowska K, Angelidaki I (2018) Integrated production of cellulosic bioethanol and succinic acid from rapeseed straw after dilute-acid pretreatment. Bioresour Technol 265:191–199

Kumar R, Bharati KA (2013) New claims in folk veterinary medicines from Uttar Pradesh, India. J Ethnopharmacol 146(2):581–593

Lamers P, Hamelinck C, Junginger M, Faaij A (2011) International bioenergy trade-a review of past developments in the liquid biofuels market. Renew Sustain Energy Rev 15:2655–2676

Li Y, Zhou B, Teng Z et al (2023) Improved metabolomic approach for evaluation of phytochemicals in mustard, kale, and broccoli microgreens under different controlled environment agriculture conditions. J Agric Food Res 14:100719

Marc EB, Nelly A, Annick DD, Frederic D (2008) Plants used as remedies antirheumatic and antineuralgic in the traditional medicine of Lebanon. J Ethnopharmacol 120:315–334

Meletis CD, Zabriskie N (2008) Supporting gastrointestinal health with nutritional therapy. Altern Complement Ther 14:132–139

Milazzo MF, Spina F, Vinci A et al (2013) Brassica biodiesels: past, present and future. Renew Sust Energ Rev 18:350–389

Mohamed IA, Shalby N, El-Badri AM et al (2025) Multipurpose uses of rapeseed (*Brassica napus* L.) crop (food, feed, industrial, medicinal, and environmental conservation uses) and improvement strategies in China. J Agric Food Res 2025:101794

Murillo G, Mehta RG (2001) Cruciferous vegetables and cancer prevention. Nutr Cancer 41(1–2):17–28

Neilsen JE, Rowe BA, Lane PA (2008) Vegetative growth and development of irrigated forage turnip (Brassica rapa var. rapa). Grass Forage Sci 63:438–446

Nilnakara S, Chiewchan N, Devahastin S (2009) Production of antioxidant dietary fiber powder from cabbage outer leaves. Food Bioprod Process 87:301–307

Osma E, Özyiğit İİ, Altay V, Serin M (2010) Urban vascular flora and ecological characteristics of Kadıköy district, Istanbul, Turkey. Maejo Int J Sci Technol 4(1):64–87

Ozturk M, Hakeem KR (2018) Plant and human health-Vol. 1-Ethnobotay and physiology. Springer Science, New York, pp 1–805

Ozturk M, Hakeem KR (2019a) Plant and human health-Vol. 2-Phytochemistry and molecular aspects. Springer Science, New York, 697 pp

Ozturk M, Hakeem KR (2019b) Plant and human health-Vol. 3-Phytochemistry and therapeutic uses. Springer, New York, 385 pp

Ozturk M, Memon AR, Gucel S, Sakcali MS (2012) Brassicas in Turkey and their possible role in the phytoremediation of degraded habitats. In: The plant family Brassicaceae: contribution towards phytoremediation. Environmental pollution book series 21. Springer Verlag, New York, pp 265–288

Ozturk M, Ashraf M, Aksoy A, Ahmad MSA (2015a) Phytoremediation for green energy. Springer Science+Business Media, New York, pp 1–191

Ozturk M, Ashraf M, Aksoy A, Ahmad MSA (2015b) Plants, pollutants & remediation. Springer Science+Business Media, New York, pp 1–407

Ozturk M, Egamberdieva D, Milica P (2020) Biodiversity and biomedicine-our future. Academic Press-Elsevier Inc., UK-USA, p 561

Ozturk M, Bhat RA, Ashraf M, Tonelli FMP, Unal BT, Dar GH (2023a) Phytohormones and stress responsive secondary metabolites. Academic Press-Elsevier, UK, pp 1–287

Ozturk M, Roy A, Bhat RA, Vardar-Sukan F, Tonelli FMP (2023b) Synthesis of bionanomaterials for biomedical applications. Elsevier, Netherlands, 558 pp

Ozturk M, Altay V, Nazish M et al (2023c) Medicinal halophytes of Asia. Springer Cham, Switzerland, p 213

Ozturk M, Sridhar KR, Sarwat M et al (2024a) Ethnic knowledge and perspectives of medicinal plants. Vol. I. Curative properties and treatment strategies. AAP-Taylor & Francis, FL, USA, pp 1–664

Ozturk M, Sridhar KR, Sarwat M et al (2024b) Ethnic knowledge and perspectives of medicinal plants. Vol. II. Nutritional & dietary benefits. AAP-Taylor & Francis, FL, USA, pp 1–668

Passalacqua NG, Guarrera PM, De Fine G (2007) Contribution to the knowledge of the folk plant medicine in Calabria region (Southern Italy). Fitoterapia 78:52–68

Petruzzello M (2023) List of economically important members of the family Brassicaceae. Encyclopedia Britannica, ttapscenter.com

Rahman M, Khatun A, Liu L, Barkla BJ (2018) Brassicaceae mustards: traditional and agronomic uses in Australia and New Zealand. Molecules 23(1):231

Rosillo-Campo F, Pelkmans L, Walter A (2009) A global overview of vegetable oils with reference to biodiesel. Report to IEA Bioenergy (Task40). London

Samec D, Pavlović I, Salopek-Sondi B (2017) White cabbage (Brassica oleracea var. capitata f. alba): botanical, phytochemical and pharmacological overview. Phytochem Rev 16(1):117–135

Schmidt JH (2010) Comparative life cycle assessment of rapeseed oil and palmoil. Int J Life Cycle Assess 15:183–197

Searchinger T, Heimlich R, Houghton RA et al (2008) Use of US croplands for biofuels increases greenhouse gases through emissions fromland-use change. Science 319:1238–1240

Stephenson AL, Dennis JS, Scott SA (2008) Improving the sustainability of petroleum from oil-seed rape in the UK. Process Saf Environ Prot 86:427–440

Sun R (2015) Economic/academic importance of Brassica rapa. In: The Brassica rapa genome. Springer Berlin Heidelberg, Berlin, Heidelberg, pp 1–15

Surriya O, Waqar K, Sarah SS et al (2014) Bio-fuels: a blessing in disguise. In: Phytoremediation and green energy. Springer Verlag, USA, pp 1–54

Suskovic J, Kos B, Beganović J et al (2010) Antimicrobial activity-the most important property of probiotic and starter lactic acid bacteria. Food Technol Biotech 48:296–307

Tiryakioglu H, Turk M (2012) Effects of different sowing and harvesting times on yield and quality of forage turnip (Brassica rapa L.) grown as a second crop. Turk J Field Crops 17(2):166–170

Tuttolomondo T, Licata M, Leto C et al (2014) Popular uses of wild plant species for medicinal purposes in the Nebrodi Regional Park (North-Eastern Sicily, Italy). J Ethnopharmacol 157:21–37

Verma P, Stevanovic S, Zare A et al (2019) An overview of the influence of biodiesel, alcohols, and various oxygenated additives on the particulate matter emissions from diesel engines. Energies 12(10):1987

Wang R, Kobayashi Y, Lin Y et al (2015) A phytosterol enriched refined extract of Brassica campestris L. pollen significantly improves benign prostatic hyperplasia (BPH) in a rat model as compared to the classical TCM pollen preparation Qianlie Kang Pule'an Tablets. Phytomedicine 22(1):145–152

Wouters D, Grosu-Tudor S, Zamfir M, DeVuyst L (2013) Bacterial community dynamics, lactic acid bacteria species diversity and metabolite kinetics of traditional Romanian vegetable fermentations. J Sci Food Agric 93:749–760

Yoon KY, Woodams EE, Hang YD (2006) Production of probiotic cabbage juice by lactic acid bacteria. Bioresour Technol 97:1427–1430

Chapter 9
Global Perspectives and Future Approaches

9.1 Global Perspective of *Brassica*

The Brassicaceae family comprises 4636 accepted taxa, classified into 340 genera and 52 tribes, with the genus *Brassica* standing out due to its wide range of economically significant crops, including oilseeds, condiments and cruciferous vegetables. Over the past five decades, the global yield of *Brassica* crops has shown a consistent upward trend, with *Brassica napus* production reaching approximately 72 Gt in 2020, highlighting its substantial economic value (Zandberg et al. 2022). These crops are now cultivated in more than 124 countries worldwide. According to FAO estimates, the combined gross production value of key *Brassica* crops—such as broccoli, cabbage, cauliflower, mustard and rapeseed—was approximately 67.5 billion USD in 2020 (Subramanian et al. 2023). Beyond their economic significance, *Brassica* crops offer a wide array of benefits, ranging from human health enhancement to promoting sustainable agricultural practices. This section explores the global potential of *Brassica*, emphasising its diverse roles and multifaceted contributions across agricultural, environmental and nutritional domains.

9.1.1 Role in Meeting Global Food and Health Needs

Brassicas are primarily classified into three categories based on the parts consumed: oilseeds, vegetables and condiments. Vegetable category primarily includes *B. juncea* (mustard greens), *B. napus* (rutabaga) and *B. rapa* (Chinese cabbage, turnip, mizuna, tatsoi, winter rape and pak choi), as well as *B. oleracea* (broccoli, kohlrabi, kale, cauliflower, Brussels sprouts, cabbage and cabbage mustard). Oilseeds predominantly comprise *B. napus* (rapeseed), and condiments typically utilise seeds from *B. juncea* and *B. nigra* (Zhang et al. 2025). Condiments consist of *B. oleracea*,

M. Ozturk et al., *Climate-Resilient* Brassicas,
https://doi.org/10.1007/978-3-032-12709-9_9

B. juncea and *B. carinata* (Zhang et al. 2025). In 2022, the production of major *Brassica* crops—including cabbage, cauliflower, broccoli, mustard seeds and rape or colza seeds—totalled over 186 million tons, covering ~44 million hectares, with a combined value exceeding 92 billion US dollars (Zhang et al. 2025). Basically, *Brassica* is an excellent source of macronutrients required for our daily needs, while also having metabolites that enhance taste and aroma. They can improve the growth of consumers. In addition to this, biofortification can enhance its nutritional value and be cost-effective for all (Jakubikova et al. 2005).

Vitamin C is also essential for good health as it has the property of regulating ROS production in the body and also enhancing anti-ageing properties. So, the consumption of *Brassica* can fulfil 50% of the daily demand of vitamin C in the human diet (Pennington and Fisher 2010). In a report, it was found that 100 mg FW of *Brassica* contained 12.9 to 176 mg of vitamin C (Kaulmann et al. 2014). Other than this phylloquinone (vitamin K1) also present in high amount of about 435 µg and 459 µg per 100 g FW in the genotypes of mustard (Farnham et al. 2012). *B. oleracea* has 308.7 to 576.9 µg per 100 g FW in younger leaves and from 369.2 to 616.5 µg per 100 g FW in older leaves. Phylloquinone has an important role in blood coagulation as well, and it can also strengthen the quality of bone (Farnham et al. 2012). Moreover, lutein and zeaxanthin have the property of preventing neurodegenerative disease (Mrowicka et al. 2022). *Brassica* also has carotenoids and tocopherols, which have biotherapeutic properties that can reduce the chances of night blindness. Instead of this, *B. oleracea* has a high level of phenolic compounds, i.e., 27.0 mg per 100 g FW, which can counteract chronic diseases such as cardiovascular diseases and type 2 diabetes (Kaulmann et al. 2014). Among *Brassicas*, Kale, one of the nutrient-dense *Brassica* vegetables, is an excellent source of key minerals like calcium, magnesium, and potassium, all of which contribute significantly to blood pressure regulation and overall cardiovascular well-being (Zhang et al. 2025).

Brassica species have high nutritional value due to their rich content of unsaturated fatty acids, particularly linolenic and linoleic acids. They also contain low levels of saturated fats, total cholesterol, and low-density lipoprotein (LDL), making them an important component of a nutritious diet (Mohamed et al. 2025). Additionally, due to the higher content of sulphur compounds in canola oil than soyabean oil, which have antioxidative properties as well as being more tolerant against oxidation (Fadda et al. 2022). Rapeseed oil contains different bioactives such as ferulic acid, phospholipids, glucoraphanin, indole-3-carbinol, squalene, carotenoids, flavonoids, vitamin E and sterols. All these constitute having various health-promoting properties related to anti-cancer, neuroprotective, anti-microbial, anti-inflammatory, anti-obesity, anti-diabetic and cardioprotective activities (Shen et al. 2023). Furthermore, rapeseed oil comprises numerous bioactive components that have demonstrated potential medicinal applications. Notably, it exhibits anticarcinogenic effects (Savva and Kafatos 2016; Shahtoori et al. 2015), possesses antioxidant properties (Amarowicz et al. 2000), and contributes to the reduction of blood cholesterol levels and cardiovascular disease risk (Yang et al. 2019). Moreover, certain studies have indicated that canola oil contains two chemotherapeutic agents, i.e. tamoxifen and cerulenin, which exhibit inhibitory activity against breast cancer and are capable of suppressing tumour growth (Cho et al. 2010; Mabasa et al. 2013). Moreover, *Brassica* contains S and N-containing secondary metabolites, which show anticancerous properties. Some reports suggest that the risk of cancers like colorectal, prostate, gastric and breast cancers can be reduced. In addition to this, allyl isothiocyanate, a product of sinigrin hydrolysis, induces cell cycle arrest and apoptosis in drug-resistant cancer cells (Jakubikova et al. 2005). A recent report by Castelão-Baptista et al. (2021) found that broccoli, a species of *Brassica*, contains bioactive compounds such as glucosinolates, chlorogenic and sinapic acids, and flavonoids, which exhibit antioxidant activity by scavenging free radicals and inhibiting LDL oxidation. These compounds also show antiangiogenic effects that may reduce cancer risk, while sulforaphane present in broccoli is associated with lower obesity rates, especially in developed countries. Building on this, *Brassica* crops have demonstrated notable therapeutic potential against multiple diseases, with several bioactive effects already known and others yet to be uncovered.

9.1.2 Utilising **Brassica** *Crops to Promote Sustainable Farming*

Brassica species can serve as effective green manure, significantly enhancing soil fertility by improving microbial activity, physical and chemical properties of soil, thereby supporting sustainable agricultural practices. Among them, rapeseed offers multiple agronomic benefits, such as high biomass production, excellent nutrient scavenging capacity of which contributes to improved soil health and sustainability (Wang et al. 2018). In addition to this, soil properties can be improved by increasing nitrogen availability, which enhances cation exchange capacity. Organic acids such as malate and citrate secreted by rapeseed roots aid in the solubilization of insoluble nutrients by stimulating enzymatic activity in the rhizosphere (Wang et al. 2015). *Brassica* crops, especially mustard, contribute significantly to sustainable agriculture through practices like intercropping and crop rotation with wheat and chickpea, enhancing soil fertility and land efficiency. Their high secondary metabolite content allows use as natural biofumigants for managing soil-borne diseases and weeds (Hagos et al. 2020). In a report, it was found that rice productivity can be increased by 6–10% after the cultivation of rapeseed (Mohamed et al. 2025). Moreover, the weed production can be reduced by the *B. napus* as it has the property of competition due to the presence of allelochemicals such as isothiocyanate allyl and isothiocyanate benzyl, causing inhibition of germination and growth of weeds (Zhou et al. 2015). Likewise, phytoremediation is a remarkable trait of *Brassica* species, where heavy metals such as lead and cadmium are removed from the soil through accumulation in harvestable plant parts as well, and heavy metals also lead to disruption in physiological and biochemical functions, resulting in heavy loss in yield and productivity (Dąbrowska et al. 2017; Candan et al. 2023). In *Brassica napus*, it was reported that the application of plant growth-promoting rhizobacteria can significantly enhance the phytoextraction of zinc and cadmium (Dąbrowska et al. 2017), also enhancing soil properties and reducing the heavy metal availability to the crop plants (Ok et al. 2011).

Next to it, sustainable agriculture is closely related to biodiesel production. Due to the presence of high oil content, i.e. 31–48% in dry seeds, and a favourable fatty acid profile, characterised by a low polyunsaturated to high monounsaturated fatty acid ratio. Rapeseed-based biodiesel has been reported to reduce CO_2 emissions by 65–95% compared to traditional diesel (Verma et al. 2019). The presence of oxygen atoms in rapeseed oil contributes to lower emissions of carbon monoxide (CO), hydrocarbons (HC) and particulate matter (PM). However, this same oxygen content raises combustion temperatures, which can enhance combustion efficiency but also lead to increased nitrogen oxide (NOx) emissions (Ge et al. 2017). In addition to utilising rapeseed oil for biodiesel production, rapeseed straw can also be employed in the generation of succinic acid and cellulosic bioethanol following dilute-acid pre-treatment (Kuglarz et al. 2018). Similarly, rapeseed oil holds substantial industrial potential beyond biodiesel production. It serves as a renewable raw material in the manufacture of various products such as paints, detergents, inks and laundry agents (Mohamed et al. 2025). Additionally, it is employed in the formulation of hair conditioners, photographic materials and biodegradable plastics. The high content of long-chain fatty acids and triglycerides in rapeseed oil makes it particularly suitable for these applications. Experimental studies have demonstrated its effectiveness in producing biopolymer precursors and surfactant agents, highlighting its role as a sustainable alternative to petroleum-derived inputs in the industrial sector (Mohamed et al. 2025). Mustard straw serves as a renewable biofuel for energy in rural areas and is being explored for jet fuel in developed countries. The protein-rich residue after oil extraction is highly valuable for animal feed, especially ruminants and organic fertiliser (Friedt et al. 2018; Hagos et al. 2020). These diverse uses make *Brassica* a resilient, economically valuable crop for sustainable farming under changing climates (Hagos et al. 2020).

9.2 Future Approaches for *Brassica*

Future *Brassica* research and cultivation approaches will likely focus on enhancing crop resilience, improving nutritional content and developing sustainable farming practices. Genetic engineering and advanced breeding techniques may create varieties with increased resistance to pests, diseases and environmental stresses such as drought and salinity (Hamdan and Tan 2024). Precision agriculture technologies, including remote sensing and data analytics, could optimise resource use and improve yield predictions. Research into the *Brassica* microbiome may also lead to novel strategies for enhancing plant growth and health (Akhter and Sofi 2022). Efforts to develop *Brassica* varieties with enhanced nutritional profiles, particularly beneficial compounds like glucosinolates, may contribute to improved human health outcomes. Sustainable cultivation methods, such as intercropping and conservation tillage, will likely gain prominence in *Brassica* production systems to address environmental concerns and promote long-term agricultural sustainability.

9.2.1 Enhancing the Capacity for Phytoremediation or Resilience to Heavy-Metal Toxicity

Many agricultural soils worldwide are contaminated with heavy metals like arsenic (As), lead (Pb), mercury (Hg), cadmium (Cd) and cobalt (Co), primarily due to industrial activities and human interventions. These pollutants hinder microbial functions, making the soil unsuitable for crop cultivation. Phytoremediation, a plant-based approach to soil rehabilitation, is a cost-effective, visually appealing and eco-friendly solution for restoring contaminated lands (Rafique et al. 2025; Gul et al. 2025; Ozturk et al. 2025).

The genetic alteration of the GSHI gene, which overexpresses gamma-ECS in plastids, has been suggested as a practical approach to augment the phytoremediation capabilities of certain crop plants in combating various heavy metals that contribute to soil toxicity (Thakur et al. 2020). Reisinger et al. (2008) augmented the expression of γ-glutamylcysteine (γ-ECS) and glutathione synthase (GS) genes in Indian mustard to boost its tolerance to various metalloids. The adjustments were evaluated separately in agar medium and hydroponic systems (Reisinger et al. 2008).

9.2.2 Techniques for the Enhancement of Brassica by Sequencing

Next-generation sequencing (NGS) technology advancements have intensified competition for *Brassica* species in commercial breeding and consumer markets. Breeders are progressively pursuing novel varieties that need little input, particularly designed for specific locations, with a direct emphasis on enhancing agronomic performance and nutritional advantages (Collard and Mackill 2008). In 2003, the genomes of *Brassica* were sequenced for the first time on a smaller scale, using the less complex diploid progenitor genomes of canola (Snowdon and Iniguez Luy 2012). Cutting-edge next-generation DNA sequencing techniques revolutionise genetic research, enabling large-scale genome analysis with single-base precision. Modern advancements include full-length cDNA sequencing and emerging non-coding RNA technologies, further refining genetic exploration. Next-generation sequencing supports essential biological research, including complete canola genome sequencing and addressing future breeding challenges. It also aids in discovering novel genes from ancient DNA while expanding environmental metagenomic analysis (Margulies et al. 2005; Zhang and Parkin 2025; Wang et al. 2025a, b).

9.2.3 Marker-Assisted Selection (Molecular Markers) Techniques for Brassica

Breeders and academics generally agree that genetic variety is present in spring canola for breeding programs. Within the canola gene pool, genetic diversity, ascertained using molecular marker methods used for distinctly different canola species, has been conserved for future application, and these gene pools may facilitate advancements in canola genomics (Cowling 2007). Molecular markers, present in all crops, play a vital role in genome sequencing and genetic characterisation. The research aims to deepen the understanding of genetic diversity and its potential in crop development (Ramesh et al. 2020). Validated alleles and nucleotide sequences offer key insights into diversity sourced from wild ancestors and elite varieties, whose preserved germplasm in gene banks ensures future agricultural advancements (Buckler et al. 2006).

The use of genome-wide association studies (GWAS) has resulted in the methodology known as genomic selection (GS), a potential instrument for formulating innovative breeding strategies and creating new marker-based models for genetic assessment (Rustagi et al. 2022). The critical factor for the successful and effective implementation in crop species is the availability of genome-wide high-throughput, cost-effective and flexible molecular markers that exhibit low ascertainment bias and are suitable for large population sizes, applicable to both model and non-model crop species, regardless of the presence of a reference genome sequence (Bhat et al. 2016; Janni et al. 2020).

9.2.4 Enhancement of Indian Mustard Using Genome Editing

Genome editing techniques provide methods for functionally characterising genes and are effectively used for the genetic enhancement of various agricultural crops. Recently, clustered regularly interspersed short palindromic repeats (CRISPR) and CRISPR-associated protein 9 (Cas9) have emerged as a promising tool for genome editing in many plants (Gaj et al. 2016). *B. juncea*, being an amphidiploid species, may possess two or more copies of the same gene inside its genome. The primary problem regarding the implementation of technology in this crop is that there have been no published reports on the use of genome editing to enhance *B. juncea* (Rustagi et al. 2022).

9.2.5 *Brassica and Climate Resilience*

Modern breeding aims to combine conventional breeding methods with cutting-edge molecular, genomic and phenomics technologies to produce climate-ready plants with high production potential (Panjabi et al. 2019; Abdel-Razzak 2021; Quezada-Martinez et al. 2021). Modern breeding methods, such as DH (doubled haploid), MABC (marker-aided backcrossing), MARS (marker-assisted recurrent selection) and GS (genomic selection), are currently used to accelerate the process of breeding and enhance testing for specific abiotic and biotic challenges. Breeding techniques powered by the field of biotechnology, including marker-assisted breeding and genetic modifications, can contribute to and accelerate genetic advancements. Early maturing cultivars, which allow plants to finish important developmental stages before stress starts, are becoming more and more necessary as a result of climate change. Genome editing makes it possible to domesticate novel crops derived from Crop Wild Relatives (CWR) with critical traits that allow them to withstand hard environments more quickly. For the purpose of discovering possible genes relevant to agronomic, physiological and phenological features related to changes in the climate, functional genetics and editing of genomes are being utilised to assist crops in more effectively adjusting to changing climatic conditions. Genetic modification combined with traditional plant breeding techniques makes it easier to create and implement climate-smart crops that are suited to producers' requirements.

The capacity of a plant to withstand and bounce out from the consequences of climate change is known as climatic resilience. Throughout the history of mankind, plant breeding has been essential for feeding the world's expanding population and revolutionising agriculture. The main goal of plant breeding is to create a superior genetic genotype that can be cultivated under certain circumstances, both generally and specifically, to increase yield. With molecular and biotechnological advances, a vast amount of biological information pertaining to genes and QTLs (Quantitative Trait Loci) was produced in recognition of the value of genomic resources in accelerating breeding efforts. The advancements in accurate phenotyping and genotyping have enormous potential for creating varieties of crops that adapt to changing climate circumstances. This enhances the breeding efforts to create climate-resilient genotypes. In order to guarantee the availability of food in unfavourable environments, climate-resilient cultivars are being developed using smart breeding techniques. Sustainable development urgently needs QTL mapping, identification of candidate genes by next-generation sequencing, the establishment of gene-based markers or significant potential gene pyramiding of stress-responsive genes in prominent cultivars for environmental sensitivity. GS to boost genetic gain despite variations in climate is a crucial factor; likewise, GWAS to identify QTL linked to adaptive alterations, because natural crossing over and linkage predominate in the population. The secret to resistance involves genotyping arrays, genotyping by sequencing (GBS) and pan-genomics, which take into account a wide range of diversity and utilise genomes with fewer breeding cycles (Roy et al. 2021).

Modern developments in breeding programmes have revolutionised the discipline by providing a potent substitute for conventional breeding techniques. In order to speed up breeding programs and identify desired characteristics and genetic variants which improve the productivity of crops and resistance to stress, genome analysis of plants has become an essential technique. Plant genomics, a fundamental area of omics research, is the scientific investigation of complete plant genomes to determine their genetic makeup and functionality. The organisation of DNA sequences provides information about molecular evolutionary relationships as well as genomic evolution. Additionally, genomics aids in clarifying the functions of genes as well as how they interplay with one another to regulate growth and survival over a variety of stressful conditions. Comprehensive knowledge regarding the genetic basis of crop characteristics is now possible primarily because of the establishment of fast-forward genotyping and phenotyping tools, which have made it possible for scientists to perform multi-trait correlation analyses using genome-wide association studies (GWAS). The identification of agronomically significant features was additionally made easier by quantitative trait loci (QTL) analysis, which demonstrated the intricate connections among genotype and phenotype. Additionally, new opportunities for agricultural enhancement are made possible by developments in haplotype-based breeding, allele mining for resistance, gene cloning and characterisation and the use of naturally occurring variants. Extensive genomics records are anticipated to yield a multitude of knowledge on plant genomes, and genomics-assisted breeding (GAB) is positioned to emerge as a key instrument for expediting crop improvement and design. Efficient determination of genes by next-generation sequencing (NGS)-based trait mapping has been made possible by an extensive knowledge of plant genomes. The modern age of genomics has been inaugurated through current developments in genome sequencing technology, which have transformed plant breeding by making large plant genomes accessible. Scientists can now investigate the broad range of genetic variants which govern complicated traits because of sophisticated NGS techniques. Single-nucleotide polymorphisms (SNPs) are one of several genomic markers that are essential for identifying particular genes in the genome. Due to their abundance, affordability and accuracy, SNPs have been utilised extensively to evaluate a huge number of crop cultivars, providing important insights into genetic diversity and characteristics such as resistance to stress. Despite its strength, SNP screening alone may not be able to fully capture extensive genomic variants. Methods like as optical mapping, which produce long-read maps, are being utilised to deal with issues and reveal a wider range of genetic variation. Genotyping by sequencing (GBS) in conjunction with NGS is currently extensively utilised for multiple sample sequencing of crops containing larger genomes. Providing a less data-intensive and more accurate substitute for NGS-based methods, static SNP sequencing arrays may have been employed. Plant breeding chips and other genotyping tools are currently being created for use in genomics-assisted breeding (GAB) (George and Darshana 2024). Crop development tactics stand to benefit greatly from novel methodologies which employ information gathered from pan genomics, or multi-omics layers, including transcriptomics, proteomics, metabolomics, and genomes. Genetic characterisation, evaluation of gene expression, including high-throughput capabilities, and sequencing of genomes have all increased due to current developments in molecular biology techniques. In order to decode the genome utilising highly efficient NGS technology, genomic DNA must be isolated, then it must be multiplied using the polymerase chain reaction (PCR), sequenced, and its quality evaluated. Studies of the activity of genes are made possible through the genetic sequencing and assembly of DNA, which is accompanied by the structural and functional characterisation of the gene. This process also clarifies the cellular and organismal relationships between gene products (Roychowdhury et al. 2023).

References

Abdel-Razzak HS (2021) Turnip (Brassica rapa var. rapa L.) breeding. In: Advances in plant breeding strategies: vegetable crops: vol. 8: bulbs, roots and tubers. Springer, Cham, pp 345–405

Akhter R, Sofi SA (2022) Precision agriculture using IoT data analytics and machine learning. J King Saud Univ Comput Inf Sci 34(8):5602–5618

Amarowicz R, Naczk M, Shahidi F (2000) Antioxidant activity of crude tannins of canola and rapeseed hulls. J Am Oil Chem Soc 77(9):957

Bhat JA, Ali S, Salgotra RK et al (2016) Genomic selection in the era of next generation sequencing for complex traits in plant breeding. Front Genet 7:221

Buckler ES, Gaut BS, McMullen MD (2006) Molecular and functional diversity of maize. Curr Opin Plant Biol 9(2):172–176

Candan F, Ozturk M, Altay V, Yalcin IE (2023) An experimental study on the effects copper and lead on the seedlings of some economically important vegetable species. Notulae Botanicae Horti Agrobotanici Cluj-Napoca 51(4):13158–13158

Castelão-Baptista JP, Barros A, Martins T et al (2021) Three in one: the potential of Brassica by-products against economic waste, environmental hazard, and metabolic disruption in obesity. Nutrients 13(12):4194

Cho K, Mabasa L, Fowler AW et al (2010) Canola oil inhibits breast cancer cell growth in cultures and in vivo and acts synergistically with chemotherapeutic drugs. Lipids 45:777–784

Collard BC, Mackill DJ (2008) Marker-assisted selection: an approach for precision plant breeding in the twenty-first century. Philos Trans R Soc B Biol Sci 363(1491):557–572

Cowling WA (2007) Genetic diversity in Australian canola and implications for crop breeding for changing future environments. Field Crop Res 104(1–3):103–111

Dąbrowska G, Hrynkiewicz K, Trejgell A, Baum C (2017) The effect of plant growth-promoting rhizobacteria on the phytoextraction of Cd and Zn by *Brassica napus* L. Int J Phytoremediat 19(7):597–604

Fadda A, Sanna D, Sakar EH et al (2022) Innovative and sustainable technologies to enhance the oxidative stability of vegetable oils. Sustainability 14(2):849

Farnham MW, Lester GE, Hassell R (2012) Collard, mustard and turnip greens: effects of genotypes and leaf position on concentrations of ascorbic acid, folate, βresearches-carotene, lutein and phylloquinone. J Food Compos Anal 27(1):1–7

Friedt W, Tu J, Fu T (2018) Academic and economic importance of *Brassica napus* rapeseed. The Brassica napus Genome:1–20

Gaj T, Sirk SJ, Shui SL, Liu J (2016) Genome-editing technologies: principles and applications. Cold Spring Harb Perspect Biol 8(12):a023754

Ge JC, Yoon SK, Choi NJ (2017) Using canola oil biodiesel as an alternative fuel in diesel engines: a review. Appl Sci 7(9):881

George RM, Darshana AS (2024) The role of plant breeding in developing climate-resilient crops. Compr Insights Into Environ Sci 2024:65
Gul A, Ul Huda N, Batool TS et al (2025) Nanobiosensors to detect environmental pollution. In: Agricultural biotechnology. Apple Academic Press, pp 231–264
Hagos R, Shaibu AS, Zhang L et al (2020) Ethiopian mustard (*Brassica carinata* A. Braun) as an alternative energy source and sustainable crop. Sustainability 12(18):7492
Hamdan MF, Tan BC (2024) Genetic modification techniques in plant breeding: a comparative review of CRISPR/Cas and GM technologies. Hortic Plant J 2024:2024
Jakubikova J, Bao Y, Sedlák J (2005) Isothiocyanates induce cell cycle arrest, apoptosis and mitochondrial potential depolarization in HL-60 and multidrug-resistant cell lines. Anticancer Res 25(5):3375–3386
Janni M, Gullì M, Maestri E et al (2020) Molecular and genetic bases of heat stress responses in crop plants and breeding for increased resilience and productivity. J Exp Bot 71(13):3780–3802
Kaulmann A, Jonville MC, Schneider YJ et al (2014) Carotenoids, polyphenols and micronutrient profiles of *Brassica oleraceae* and plum varieties and their contribution to measures of total antioxidant capacity. Food Chem 155:240–250
Kuglarz M, Alvarado-Morales M, Dąbkowska K, Angelidaki I (2018) Integrated production of cellulosic bioethanol and succinic acid from rapeseed straw after dilute-acid pretreatment. Bioresour Technol 265:191–199
Mabasa L, Cho K, Walters MW et al (2013) Maternal dietary canola oil suppresses growth of mammary carcinogenesis in female rat offspring. Nutr Cancer 65(5):695–701
Margulies M, Egholm M, Altman WE et al (2005) Genome sequencing in microfabricated high-density picolitre reactors. Nature 437(7057):376–380
Mohamed IA, Shalby N, El-Badri AM et al (2025) Multipurpose uses of rapeseed (*Brassica napus* L.) crop (food, feed, industrial, medicinal, and environmental conservation uses) and improvement strategies in China. J Agric Food Res 2025:101794
Mrowicka M, Mrowicki J, Kucharska E, Majsterek I (2022) Lutein and zeaxanthin and their roles in age-related macular degeneration-neurodegenerative disease. Nutrients 14(4):827
Ok YS, Usman AR, Lee SS et al (2011) Effects of rapeseed residue on lead and cadmium availability and uptake by rice plants in heavy metal contaminated paddy soil. Chemosphere 85(4):677–682
Ozturk M, Fatima S, Maryam S et al (2025) Recent advances in the use of bionanotechnology for bioremediation. In: Agricultural biotechnology. Apple Academic Press, pp 177–215
Panjabi P, Yadava SK, Kumar N et al (2019) Breeding Brassica juncea and B. rapa for sustainable oilseed production in the changing climate: progress and prospects. In: Genomic designing of climate-smart oilseed crops. Springer, Cham, pp 275–369
Pennington JA, Fisher RA (2010) Food component profiles for fruit and vegetable subgroups. J Food Compos Anal 23(5):411–418
Quezada-Martinez D, Addo Nyarko CP, Schiessl SV, Mason AS (2021) Using wild relatives and related species to build climate resilience in Brassica crops. Theor Appl Genet 134(6):1711–1728
Rafique K, Gul A, Altay V et al (2025) Environmental pollution and technological applications. In: Agricultural biotechnology. Apple Academic Press, pp 279–300
Ramesh P, Mallikarjuna G, Sameena S et al (2020) Advancements in molecular marker technologies and their applications in diversity studies. J Biosci 45:1–15
Reisinger S, Schiavon M, Terry N, Pilon-Smits EA (2008) Heavy metal tolerance and accumulation in Indian mustard (*Brassica juncea* L.) expressing bacterial γ-glutamylcysteine synthetase or glutathione synthetase. Int J Phytoremediat 10(5):440–454
Roy A, Purkaystha S, Bhattacharyya S (2021) Advancement in molecular and fast breeding programs for climate-resilient agriculture practices. In: Harsh environment and plant resilience. Springer, Cham. https://doi.org/10.1007/978-3-030-65912-7_4
Roychowdhury R, Das SP, Gupta A et al (2023) Multi-omics pipeline and omics-integration approach to decipher plant's abiotic stress tolerance responses. Gene 14(6):1281

Rustagi A, Rajwanshi R, Kumar D et al (2022) Genetic modification of *Brassica juncea*: current scenario and future prospects. In: The *Brassica juncea* genome. Springer, Cham, pp 171–196

Savva SC, Kafatos A (2016) Vegetable oils: dietary importance. Elsevier

Shahtoori PM, Emtyazjoo M, Bonakdar S, Rabani M (2015) The anticancer effects of hexane extract of rainbow trout (*Oncorhynchus mykiss*) muscle enriched with vegetable oil *Brassica* sp. Cancer Res J 3:63–67

Shen J, Liu Y, Wang X et al (2023) A comprehensive review of health-benefiting components in rapeseed oil. Nutrients 15(4):999

Snowdon RJ, Iniguez Luy FL (2012) Potential to improve oilseed rape and canola breeding in the genomics era. Plant Breed 131(3):351–360

Subramanian P, Kim SH, Hahn BS (2023) Brassica biodiversity conservation: prevailing constraints and future avenues for sustainable distribution of plant genetic resources. Front Plant Sci 14:1220134

Thakur AK, Parmar N, Singh KH, Nanjundan J (2020) Current achievements and future prospects of genetic engineering in Indian mustard (*Brassica juncea* L. Czern & Coss.). Planta 252:1–20

Verma P, Stevanovic S, Zare A et al (2019) An overview of the influence of biodiesel, alcohols, and various oxygenated additives on the particulate matter emissions from diesel engines. Energies 12(10):1987

Wang YL, Almvik M, Clarke N et al (2015) Contrasting responses of root morphology and root-exuded organic acids to low phosphorus availability in three important food crops with divergent root traits. AoB Plants 7:plv097

Wang B, Song L, Wang Z et al (2018) Production and feeding technology of fodder-rapeseed in China [J]. Chin J Oil Crop Sci 40(5):695–701

Wang R, Han M, Han T et al (2025a) Unraveling yield heterosis in Chinese cabbage hybrid by comparative transcriptomic analysis and LHCB1 gene function analysis. Front Plant Sci 16:1627259

Wang C, Zhang Q, Liu Y et al (2025b) Genome-wide identification of MYBL2 in Brassicaceae, with a focus on the expression pattern of regulating anthocyanin synthesis in Brassica crops. Front Plant Sci 16:1629560

Yang R, Xue L, Zhang L et al (2019) Phytosterol contents of edible oils and their contributions to estimated phytosterol intake in the Chinese diet. Foods 8(8):334

Zandberg JD, Fernandez CT, Danilevicz MF et al (2022) The global assessment of oilseed brassica crop species yield, yield stability and the underlying genetics. Plants 11(20):2740

Zhang L, Parkin IA (2025) Unveiling unique metabolomic and transcriptomic profiles in three Brassicaceae crops. Front Plant Sci 16:1597905

Zhang X, Jia Q, Jia X et al (2025) Brassica vegetables-an undervalued nutritional goldmine. Hortic Res 12(2):uhae302

Zhou Y, Wu W, Xu Y et al (2015) Effects of returned rapeseed straw on soil fertility and yield of subsequent rice. J Yangzhou Univ 36:53–58

Index

M. Ozturk et al., *Climate-Resilient* Brassicas,
https://doi.org/10.1007/978-3-032-12709-9

MIX
Papier aus verantwortungsvollen Quellen
Paper from responsible sources
FSC® C105338

If you have any concerns about our products,
you can contact us on
ProductSafety@springernature.com

In case Publisher is established outside the EU,
the EU authorized representative is:
Springer Nature Customer Service Center GmbH
Europaplatz 3, 69115 Heidelberg, Germany

Printed by Libri Plureos GmbH
in Hamburg, Germany